Introduction to

MATHEMATICAL PHYSICS

UNIVERSITY PHYSICS SERIES

A Group of Text Books for Intermediate and Advanced Courses

WALTER C. MICHELS, PH.D.

CONSULTING EDITOR

Introduction to Optics. FOURTH EDITION. *By* John K. Robertson, F.R.S.C., *Professor of Physics, Queen's University.*

Principles of Electricity. THIRD EDITION. *By* Leigh Page, Ph.D., *Late Professor of Mathematical Physics, Yale University;* and Norman Ilsley Adams, Jr., Ph.D., *Professor of Physics, Yale University.*

Physical Mechanics. SECOND EDITION. *By* R. Bruce Lindsay, Ph.D., *Hazard Professor of Physics, Brown University.*

Introduction to Mathematical Physics. *By* William Band, *Professor of Physics, Washington State University.*

Introduction to
MATHEMATICAL PHYSICS

BY

WILLIAM BAND
Professor of Physics
Washington State University

D. VAN NOSTRAND COMPANY, INC.
PRINCETON, NEW JERSEY
TORONTO LONDON
NEW YORK

D. VAN NOSTRAND COMPANY, INC.
120 Alexander St., Princeton, New Jersey (*Principal office*)
257 Fourth Avenue, New York 10, New York

D. Van Nostrand Company, Ltd.
358, Kensington High Street, London, W.14, England

D. Van Nostrand Company (Canada), Ltd.
25 Hollinger Road, Toronto 16, Canada

Published simultaneously in Canada by
D. Van Nostrand Company (Canada), Ltd.

Library of Congress Catalogue Card No. 59–11564

PRINTED IN THE UNITED STATES OF AMERICA

To my wife

Claire M. Band

PREFACE

This text is designed for use in courses on theoretical physics for college senior and graduate students.

Most existing texts on theoretical physics for undergraduates have been worked out from the point of view of nineteenth century physics. Such courses fail to prepare the student for performing the functions expected of physicists, either in modern academic life or in industry and engineering. Both mathematics and physics have undergone revolutionary changes in content and point of view in the past thirty years. The new point of view is no more difficult for a young person to understand today than were the classical ideas of the last century. There seems to be no good reason against presenting as much of twentieth century mathematical physics to the undergraduate as he is capable of digesting, and leaving to his later interest, as circumstance or need arises, the self-study of the more historic aspects of the subject. The purpose of the present course is twofold: (a) to introduce to the student the most powerful mathematical tools available today and (b) to survey theoretical physics from a modern unified point of view.

Mathematical physics is never an easy course; in fact, it is often regarded as about the toughest course in the college program. To master it, a student must have an unusually good aptitude in mathematics and a sound feeling for physical principles. No textbook can ever gloss over these difficulties; to attempt to do so is to deceive the student. During the past few decades too many American educators have assumed that the youth of this country does not have the capacity to face any really tough intellectual challenge, and the education of the more gifted student has been neglected. The disastrous results of this neglect are now too manifest in the current shortage of skilled scientists. It is recognized that we need more scientists; but more urgent still, we need more mature minds. These do not grow in an atmosphere of animal fun and frolic, but require serious challenge on an intellectual level. Our responsibility is to present a continually expanding and intensifying challenge to the rising generation.

There are many hopeful signs that this re-appraisal of our task will be accomplished, and perhaps ten years from now we can expect considerable improvement. Meanwhile, those of us who feel the need should do our part to accelerate the change. This text is therefore dedicated to the proposition that the normal college student has a natural zest for intellectual achievement, which can be satisfied by presenting him with sufficiently challenging problems, culminating in rewarding glimpses of still higher levels of understanding.

It is impossible to thank all of the writer's former students who have been of assistance in the preparation of this text. But special thanks are due to Dr. Arthur A. Evett for checking over the manuscript, and to D. R. Curran and D. G. Doran for their help in correcting the proofs.

The writer is most deeply indebted to Professor Henry Margenau for his encouragement and his frank and thorough evaluation of the entire manuscript. Finally he wishes to express his gratitude to Professor Walter C. Michels, Editor of the University Series in Physics, for his enthusiastic help in preparing the work for publication.

William Band

Pullman, Washington
April 1959

CONTENTS

CHAPTER I

Introduction: The Language of Mathematical Physics

CHAPTER II

The Continuum Theory of Matter

CHAPTER III

The Molecular Theory of Matter

Chapter I

INTRODUCTION: THE LANGUAGE OF MATHEMATICAL PHYSICS

Chapter I

INTRODUCTION: THE LANGUAGE OF MATHEMATICAL PHYSICS

I.1 Philosophical Remarks. It has not been customary, in college texts on mathematical or theoretical physics, to attempt either a definition of the objectives of the course or any discussion of a unifying philosophical point of view in regard to the course in its relation with other disciplines. Although too lengthy a discussion of such topics would obviously be out of place, it does seem that a teacher owes more to his students than the bare outline provided by the chapter headings of the text. With this in mind we give in this introductory section a brief résumé of the modern physicist's philosophy of his science and try to place in perspective the role of mathematical physics in this general setting.

Empirical science may be approximately defined as the systematic investigation of those aspects of things experienced in perception that can be regarded as possessing space and time orderliness. Theoretical science is the study of this space-time orderliness and the associated correlations among the observations of empirical science. Those parts of science that are classified as physics, or as chemistry, etc., are determined at the present time more by arbitrary convention or historical accident than by any logical principle of differentiation. In earlier times such differentiation was a simple matter: scientific observations fell naturally into groups among which only the vaguest correlations were apparent. With the increased knowledge available today, all such gaps in the relatedness of things have disappeared, and the so-called "borderline fields" are rich in material and in general even more challenging than the older ones.

A useful approximation to a logical classification may be stated in terms of the atomic theory and the concept of different levels of organization. Thus, nuclear physics is the study of the space-time relations between what we tentatively regard as the fundamental particles of the physical world—neutrons, protons, electrons, etc.—and those interactions that lead to stable structures called atomic nuclei. Pure physics is then the study of the relations between atoms, as well as those interactions leading to stable molecules. Chemistry includes all interactions between molecules—even those leading to the formation of the living cell. Biology studies the interactions of living cells, between living cells and inanimate matter, up to those interactions leading to the growth of organisms like the human body. Social sciences study the relations among living organisms, and so on up the scale of complexity in organization. This scheme of classification does not systematize all actual boundaries between the conventional fields of study: *e.g.* geology and psychology cut across different levels, and many boundaries exist within each single level. The scheme, however, serves to clarify the places of physics among the sciences.

On this scheme of classification, the lower the level of organization the more "basic" the subject is felt to be. The implication of this is that the phenomena observed at the higher levels are to be explained in terms of the phenomena at the lower levels. Nuclear physics may claim to be more basic than pure physics, the objects studied being at a lower level of organization. In the materialistic atmosphere of nineteenth century science, it was natural to hope that the deeper that man probed into the microscopic structure of matter, the simpler would be the properties of the objects discovered, and thus a satisfying explanation of the physical world could be achieved. The early nuclear physicists generally believed that they were indeed discovering the ultimate building blocks of the physical world. Today this belief is generally regarded as a naïve oversimplification of truth. The deeper the nuclear physicist probes, the more difficult it becomes to interpret the subnuclear objects as the building blocks of a materialistic explanation of the physical world: it is probably safe to assert that all active workers in this field have been forced into a much more sophisticated interpretation of their discoveries. It is hoped that the student of this text will begin to appreciate these modern sophistications before the end of the course.

The major objective of classical theoretical physics may be stated to be an explanation of the complicated physical world in terms of the simple properties and interactions of elementary particles. To serve this objective the emphasis of introductory courses in theoretical physics has naturally been on dynamics and differential equations. With the failure of the building-block scheme to stand up under the findings of modern research, a new emphasis has become inevitable. Instead of attempting to set up an explanation of events in terms of a mechanical model, modern theoretical physics has turned to less intuitive models as the basis of explanation. This amounts to the acceptance of a less restrictive language in terms of which to describe the physical world.

The function of physical theory may be outlined briefly as follows: events are observed, classified, measured, correlated, and predicted. The scheme of classification depends strongly on hypothetical factors and qualitative judgments. The system of measurement and the units chosen in turn depend on the scheme of classification. The discovery of correlations among the quantitative measurements requires a highly developed mathematical technique. When predictions based on these correlations are found to be successful, the physicist accepts the results as verification of the qualitative judgments upon which the whole theory rests. The final objective is the discovery of one grand harmonious descriptive system in the form of a universal law embracing the whole of physical reality. What this "means" in any wider sense is a question for metaphysics and philosophy; the physicist as such is motivated solely by the faith that such a universal law does exist and is comprehensible to the human mind.

This general outline is not new; it applies equally well to the materialistic building-block era of theoretical physics. The point here is that the building-block concept has failed as the basis of classification. The character of theoretical physics has consequently undergone a major change during the past few decades, with its emphasis shifting away from mechanical models toward abstract mathematical considerations. It is for this reason that we have elected to entitle our course an introduction to mathematical physics, rather than theoretical physics, although the subject area remains the same. The proper scope of the present text is the study of the correlations that have been discovered among the empirical results of experimental physics. The material with which we work therefore consists of the results of measurements; every such result is a number or a set of numbers, and the theory of the relations between sets of numbers is a part of mathematics.

There is a vast literature concerning the definitions of the various kinds of number known to modern mathematics. They form the "alphabet" of mathematical physics and are usually accepted by the physicist as given; although it is a matter of recorded history that the demands of mathematical physics have stimulated the discovery or invention of new kinds of number. In this chapter we review briefly the kinds of number now in everyday use by mathematical physicists, without inquiring into their detailed definitions; we then develop the "grammar" of mathematical physics—*i.e.* the rules of the mathematical language required to describe the correlations between the results of measurements in the physical laboratory.

Simple measurements of the basic fundamental quantities like length, area, volume, mass, charge, time, etc., in terms of any arbitrary system of units, may be recorded as integral numbers, 1, 2, 3, . . . Rates of change, length in time, mass per volume, etc., may be recorded in terms of the rational numbers, $\frac{1}{2}$, $\frac{1}{3}$, $\frac{3}{4}$, etc. The relations between length and area necessitate the use of irrational numbers $2^{1/2}$, pi (π), etc. The discussion of periodic processes, alternating currents, and simple harmonic motion is facilitated by the use of the complex numbers or ordered number pairs, represented by points each having two coordinates in an Argand diagram. The discussion of directed vectors, forces, displacements, velocities, etc., requires the use of number triplets, the three numbers of a triplet being the three components manipulated according to convenient algebraic rules set up to match the observed properties of such entities. The discussion of still more complicated phenomena, *e.g.* the behavior of systems having many degrees of freedom, is facilitated by the use of ordered sets of N numbers, called N-vectors, where N is arbitrarily determined for convenience. In mathematical physics it is customary to assume that the student is already familiar with this "alphabet" up to and including the complex number. Some detailed discussion of the three-dimensional vector and of all higher-order concepts therefore forms the first part of this course.

Exercise I.1

1.1. Distinguish between inductive reasoning and deductive reasoning; give an example of each.

1.2. Distinguish between the number zero and the concept of "nothing."

1.3. Distinguish among the meanings of "five" in the following three instances: (*a*) the abstract number five; (*b*) the number of fingers on one hand; (*c*) the number of cents in a nickel.

1.4. Prove that the square root of two is irrational.

1.5. Discuss how the rules of addition and subtraction set up in modern algebra have been invented for convenience in everyday situations. Give examples of entities for which these rules would be meaningless. Give examples of entities for which the rules would have meaning but be false.

1.6. Define the measure of angle.

1.7. Give an analytical definition of the function $\sin\theta$ and also the geometrical definition of the same function. On what basis do we believe these two definitions to be equivalent?

1.8. The complex number $x+iy$ may also be written in the form (x, y). Prove that the usual rules for addition and subtraction and multiplication of complex numbers in the "i-notation" may be expressed instead in the form

$$(x, y)+(u, v) = (x+u, y+v)$$

$$(x, y)\cdot(u, v) = (xu-yv, xv+yu).$$

1.9. On the Argand diagram $x+iy$ becomes $r\cdot e^{i\theta}$ and can also be written in the form (r, θ). Prove that the usual rules for addition and multiplication can also be written in the forms:

$$(r, \theta)+(s, \phi) = ((R, \cos^{-1}[(r \cos \theta+s \cos \phi)/R)]; \text{ (find } R)$$
$$(r, \theta)\cdot(s, \phi) = (rs, \theta+\phi).$$

1.10. From the results of Ex. 1.9, prove that

$$[(r, \theta)+(s, \phi)]\cdot(t, \psi) = (r, \theta)\cdot(t, \psi)+(s, \phi)\cdot(t, \psi).$$

I.2 Vector Algebra in Three Dimensions. We are familiar in elementary physics with directed vectors like displacement, velocity, force, etc. They are usually defined as directed magnitudes. In practice, however, we know that it is convenient to measure three coordinates to locate a point, three components to determine a force or velocity, and so on. Therefore in this book we prefer to set up a more explicit definition of such vectors. The *vector* is here defined as the set of three numbers, considered as a whole and obeying a certain algebra which consists of mutually consistent rules for adding, multiplying, differentiating such entities. The concept of a directed magnitude can be derived from this definition and is not treated as a primitive concept. This point of view results in an immediate economy of symbolism and, eventually, in an economy of thought. This economy of thought is perhaps the most important objective of all theoretical physics, because it permits the mind to proceed to more abstract concepts with greater ease.

From the physical point of view the rules of vector algebra are invented primarily to handle actual physical processes. Let the vector **A** be the ordered set of numbers (A_1, A_2, A_3), and the vector **B** be the ordered set (B_1, B_2, B_3); then the sum or difference of these two vectors is defined as the ordered set:

$$\mathbf{A}\pm\mathbf{B} = (A_1\pm B_1, A_2\pm B_2, A_3\pm B_3). \tag{2.1}$$

This agrees generally with the results of experiments on the resultant of two physical vectors, as we know from elementary physics. The rule satisfies the usual distributive law familiar in the algebra of ordinary numbers.

Two vectors are defined as equal to each other if and only if their respective components are equal to each other separately. A vector equation in three dimensions is then equivalent to three ordinary equations.

The *scalar product* of two vectors, written $\mathbf{A}\cdot\mathbf{B}$, is defined by:

$$\mathbf{A}\cdot\mathbf{B} = A_1B_1+A_2B_2+A_3B_3 = \mathbf{B}\cdot\mathbf{A}. \tag{2.2}$$

This quantity is a single ordinary number, not a vector. It satisfies

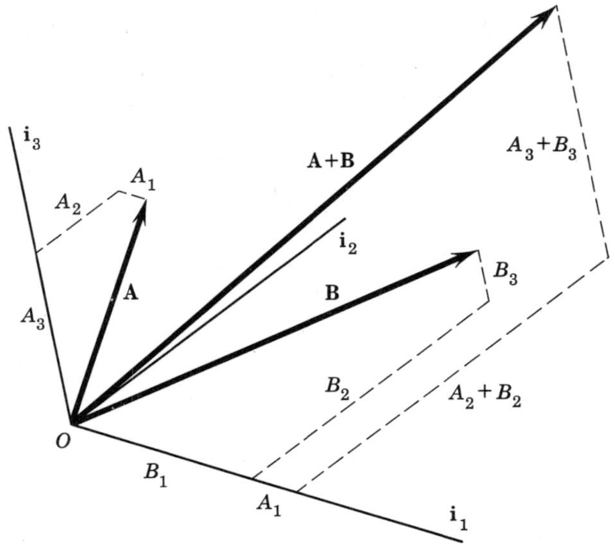

FIG. 2.1. Pictorial representation of the vector sum of two concurrent vectors **A** and **B**

the commutation rule indicated in the second part of the equation. The physical usefulness of this product lies in the fact that, if **A** represents a force and **B** a displacement, the scalar product equals the observed work done by the force if its point of application moves through the displacement.

The *vector product*, written $\mathbf{A} \times \mathbf{B}$, is defined by:

$$\mathbf{A} \times \mathbf{B} = (A_2B_3 - A_3B_2, A_3B_1 - A_1B_3, A_1B_2 - A_2B_1) = -\mathbf{B} \times \mathbf{A}. \tag{2.3}$$

This is another vector. The vector product anticommutes. Although it is not at once obvious from this definition the vector product is useful in the following way: if **A** is a displacement from a fixed point to a movable point, and **B** is a force applied at the movable point, then $\mathbf{A} \times \mathbf{B}$ is the torque exerted by the force about the fixed point.

The *length* of a vector **A** is defined as*

$$A = (\mathbf{A} \cdot \mathbf{A})^{1/2} = (A_1^2 + A_2^2 + A_3^2)^{1/2}. \tag{2.4}$$

* The term "length", as used here, is consistent with the common usage in mathematics texts—it is a pure, dimensionless number. When it is used in the future, it will be clear from the context whether the word is used in this sense or as an indication of the physical dimension having the same name. Whenever a vector, taken as a set of three numbers, is multiplied by a physical unit (*e.g.* meter, meter per second, newton), the result is a directed physical magnitude. The vector (6, 6, 7) has a "length" 11; (6, 6, 7) meter is a directed displacement having a "length" of 11 meters.

The angle θ between two vectors **A** and **B** is defined by

$$\cos\theta = (\mathbf{A}\cdot\mathbf{B})/AB, \tag{2.5}$$

where A and B are the lengths of the vectors **A** and **B**. Two non-zero vectors are said to be perpendicular or *orthogonal* if their scalar product vanishes.

The three vectors (1, 0, 0), (0, 1, 0) and (0, 0, 1) are called *unit vectors*, and a convenient notation for these vectors is:

$$\mathbf{i}_1 = (1, 0, 0); \quad \mathbf{i}_2 = (0, 1, 0); \quad \mathbf{i}_3 = (0, 0, 1). \tag{2.6}$$

From these definitions it is easy to prove the following relations:

$$\mathbf{i}_m\cdot\mathbf{i}_n = \delta_{mn} \quad (m, n = 1, 2, 3), \tag{2.7}$$

where δ_{mn} is the *Kronecker delta* symbol, standing for unity when $m = n$ and for zero when $m \neq n$. Also

$$\mathbf{i}_m\times\mathbf{i}_n = \sum_p e_{mnp}\mathbf{i}_p \quad (m, n = 1, 2, 3), \tag{2.8}$$

where p is to be summed over the three values $p = 1, 2, 3$. The e-symbol has the value unity if mnp is cyclic with 1, 2, 3; minus unity

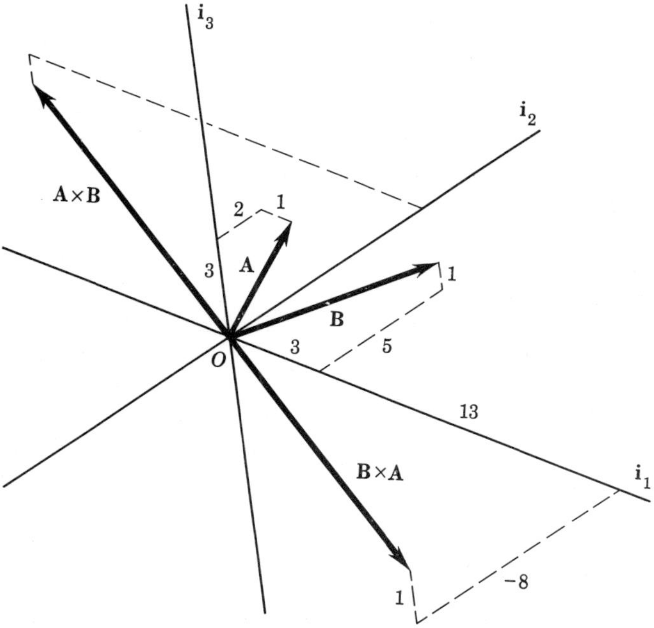

FIG. 2.2. Pictorial representation of the vector product of two vectors **A** and **B**. The numbers show the components of the vectors along the axes $\mathbf{i}_1$, $\mathbf{i}_2$, $\mathbf{i}_3$

if *mnp* is anticyclic with 1, 2, 3; and zero if two or more of the subscripts are equal:

$$\left.\begin{aligned} e_{123} &= e_{231} = e_{312} = 1. \\ e_{213} &= e_{132} = e_{321} = -1. \\ e_{112} &= e_{131} = e_{223} = \text{etc.}, = 0. \end{aligned}\right\} \tag{2.9}$$

Making use of the definitions in Eqs. 2.1 and 2.6, we can prove easily that any vector can be expressed as a linear combination of the three unit vectors, thus:

$$\mathbf{A} = \sum_m A_m \mathbf{i}_m. \tag{2.10}$$

Moreover, by using Eq. 2.7 on this relation, we show that any component of the vector **A** can be written in the form:

$$A_n = \mathbf{A} \cdot \mathbf{i}_n. \tag{2.11}$$

Using Eq. 2.3 and 2.10, we can prove that the vector product can be written in the convenient form of a determinant:

$$\mathbf{A} \times \mathbf{B} = \begin{vmatrix} i_1 & i_2 & i_3 \\ A_1 & A_2 & A_3 \\ B_1 & B_2 & B_3 \end{vmatrix}. \tag{2.12}$$

The *triple product* $\mathbf{A} \cdot (\mathbf{B} \times \mathbf{C})$ can be worked out by using Eq. 2.3 for the vector product $\mathbf{B} \times \mathbf{C}$ and then Eq. 2.2 for the scalar product of **A** and $\mathbf{B} \times \mathbf{C}$; the result can be expressed as the following determinant:

$$\mathbf{A} \cdot (\mathbf{B} \times \mathbf{C}) = \begin{vmatrix} A_1 & A_2 & A_3 \\ B_1 & B_2 & B_3 \\ C_1 & C_2 & C_3 \end{vmatrix}. \tag{2.13}$$

From the definition of angle and length of vectors, Eqs. 2.5 and 2.4, we can prove that the triple product is equal in magnitude to the volume of the parallelepiped enclosed between the three vectors as concurrent sides. It is invariant to any cyclic permutation of the three vectors, but changes sign for any anticyclic (or odd) permutation. If the determinant of Eq. 2.13 vanishes, the three vectors must be co-planar.

Given any three non-co-planar vectors **A**, **B**, and **C**, we can express any other vector **V** as a linear combination of **A**, **B**, and **C** thus:

$$\mathbf{V} = V_a \mathbf{A} + V_b \mathbf{B} + V_c \mathbf{C}, \tag{2.14}$$

where V_a, V_b, and V_c are the components of V with respect to the set of vectors **A**, **B**, and **C**. Let us write **V** in the form (V_1, V_2, V_3) and the vectors **A** and **B** in their component notations: (A_1, A_2, A_3) and (B_1, B_2, B_3), and use these forms in Eq. 2.14 to write it as follows:

$$V_m = V_a A_m + V_b B_m + V_c C_m \quad (m = 1, 2, 3). \tag{2.15}$$

These are a set of three simultaneous equations for the three unknowns V_a, V_b, and V_c, and they can be solved uniquely if and only if the determinant of their coefficients does not vanish. Comparison with Eq. 2.13 then shows the truth of the assertion of Eq. 2.14.

Given a set of three non-co-planar vectors **A**, **B**, and **C**, the set of three vectors **a**, **b**, and **c** defined by:

$$\mathbf{a} = \mathbf{B}\times\mathbf{C}/(\mathbf{A}\cdot\mathbf{B}\times\mathbf{C}); \quad \mathbf{b} = \mathbf{C}\times\mathbf{A}/(\mathbf{A}\cdot\mathbf{B}\times\mathbf{C}); \quad \mathbf{c} = \mathbf{A}\times\mathbf{B}/(\mathbf{A}\cdot\mathbf{B}\times\mathbf{C}); \tag{2.16}$$

is called the *reciprocal set* to **A**, **B**, **C**, and has the following properties:

$$\mathbf{a}\cdot\mathbf{B} = \mathbf{a}\cdot\mathbf{C} = 0; \quad \mathbf{a}\cdot\mathbf{A} = 1, \tag{2.17}$$

with analogous properties for **b** and **c**. In words, **a** is orthogonal to both **B** and **C** and has a component along **A** equal to the reciprocal of A. The set of points defined by the position vectors

$$\mathbf{k} = k_1\mathbf{A} + k_2\mathbf{B} + k_3\mathbf{C}, \tag{2.18}$$

where k_1, k_2, and k_3 are any three integers, is called a *lattice*. The set of points defined by the vectors

$$\mathbf{g} = g_1\mathbf{a} + g_2\mathbf{b} + g_3\mathbf{c}, \tag{2.19}$$

where g_1, g_2, and g_3 are any integers, is called the *reciprocal lattice*. These concepts are used in the study of crystal symmetries.

In more elementary introductions to vector algebra, the vector was supposed to be defined as a directed magnitude, like a physical displacement, and the relations like Eqs. 2.1 and 2.2 were deduced from observation of physical entities. In this text we take the opposite view and have defined the vector purely analytically through its algebraic properties. We then confirm experimentally that many physical entities like velocity, etc., do seem to behave, within experimental error, like these algebraic vectors. The physical entity is said to be represented by the algebraic vector. If eventually we discover that certain physical entities, at first thought to behave like algebraic vectors, do in fact behave somewhat differently, we do not therefore abandon the rules; we simply admit that these entities cannot be represented exactly by algebraic vectors, and we try to invent some other algebraic entity to describe them. A particularly important example of this is the discovery that displacement and velocity, when the latter is very large, do not behave like ordinary vectors in three dimensions. In the theory of relativity a four-dimensional vector algebra is invented for a more exact description of these entities.

Exercise I.2

2.1. Multiplication of a vector **A** by a scalar number c is usually defined by the equation:

$$c\mathbf{A} = (cA_1, cA_2, cA_3).$$

Show that this equation can be derived from Eq. 2.1 even when the number c is not an integer.

2.2. Prove that Eq. 2.5 agrees with the usual definition of angle between two vectors in Euclidean geometry. Hint: draw the two vectors from a common origin and take the plane containing the two vectors as the $x-y$ plane, letting $A_3 = B_3 = 0$.

2.3. Prove analytically, *i.e.* without using any other information than is implied in Eqs. 2.3, 2.4, and 2.5, and the well-known analytical properties of the circular functions, that the length of the vector $\mathbf{A} \times \mathbf{B}$ is $AB \sin \theta$. Hint: again take the $x-y$ plane containing the two vectors.

2.4. The direction cosines of a vector are the cosines of the angles between the direction of the vector and the three coordinate axes, respectively. Prove that the angle θ between two vectors whose direction cosines are (l_1, l_2, l_3) and (m_1, m_2, m_3) is given by

$$\cos \theta = l_1 m_1 + l_2 m_2 + l_3 m_3.$$

2.5. A force **F** whose components are F_1, F_2, and F_3 is applied to a point P whose coordinates are x, y, z. Prove that it exerts a torque about the origin given by $\mathbf{r} \times \mathbf{F}$, where **r** is the vector from the origin to the point P, and the vector product is defined by Eq. 2.3.

2.6. Two vectors, **A** and **B**, are drawn from a common point. Prove that the area of the parallelogram based on the two vectors as edges is equal to the magnitude of $\mathbf{A} \times \mathbf{B}$ as defined in Eq. 2.3. Hint: use Ex. 2.2 above.

2.7. Prove each example of Eqs. 2.7 and 2.8 using the definitions Eqs. 2.2 and 2.3 for scalar and vector products.

2.8. Give the proof in detail of Eq. 2.10 using Eqs. 2.1 and 2.6.

2.9. Derive Eq. 2.11.

2.10. Multiply out the determinant in Eq. 2.12 and compare the result with Eq. 2.3.

2.11. Derive Eq. 2.13 as outlined in the text.

2.12. Use the results of Ex. 2.2 and Ex. 2.6 to prove that the triple scalar product of Eq. 2.13 does in fact equal the volume of the parallelepiped described in the text.

2.13. Prove that the triple scalar product changes sign when the order of the vectors in the product is given an odd permutation.

2.14. Given three vectors (0, 1, 1), (1, 0, 1), and (1, 1, 0), find the set reciprocal to them. Draw a diagram to illustrate.

2.15. Consider the set of three numbers given as the angles between any given direction and the three coordinate axes: $\boldsymbol{\theta} = (\theta_1, \theta_2, \theta_3)$. Discuss the possibility of giving a meaning to the equations

$$\boldsymbol{\theta}+\boldsymbol{\phi} = (\theta_1+\phi_1, \theta_2+\phi_2, \theta_3+\phi_3) \quad \text{and} \quad \boldsymbol{\theta}\cdot\boldsymbol{\phi} = \theta_1\phi_1+\theta_2\phi_2+\theta_3\phi_3.$$

Does this set of numbers behave like a vector in the analytic sense?

2.16. Prove that $\mathbf{a}\cdot\mathbf{b}\times\mathbf{c} = 1/(\mathbf{A}\cdot\mathbf{B}\times\mathbf{C})$ where **a**, **b**, and **c** are defined by Eq. 2.16.

I.3 Function Space. Consider an ordered set of N numbers as an N-dimensional vector:

$$\mathbf{F} = (F_1, F_2, F_3, \cdots F_N) \tag{3.1}$$

or, more briefly,

$$F = \{F_n\} \quad (n = 1, 2, 3, \cdots N). \tag{3.2}$$

The totality of such number-sets, *i.e.* all the vectors obtained by giving to the components F_n all the possible numerical values, is called an *N-dimensional space.* The word "space" implies that we are about to set up algebraic rules for operating on the N-vectors. In fact, we shall define the vector sum in exact analogy with Eq. 2.1 in the new space:

$$\mathbf{F}+\mathbf{G} = (F_1+G_1, F_2+G_2, \cdots, F_N+G_N), \tag{3.3}$$

and the scalar product of two N-vectors in exact analogy with Eq. 2.2:

$$\mathbf{F}\cdot\mathbf{G} = \sum_n F_nG_n, \tag{3.4}$$

Now let N increase indefinitely. We may, for instance, represent the values of the subscript in Eq. 3.2 by points on a line of given length and increase N by plotting the points closer and closer together. When we write down a vector like that given in Eq. 3.2, we assign a definite number, F_n to the nth point on the line. If we make the points suffi-

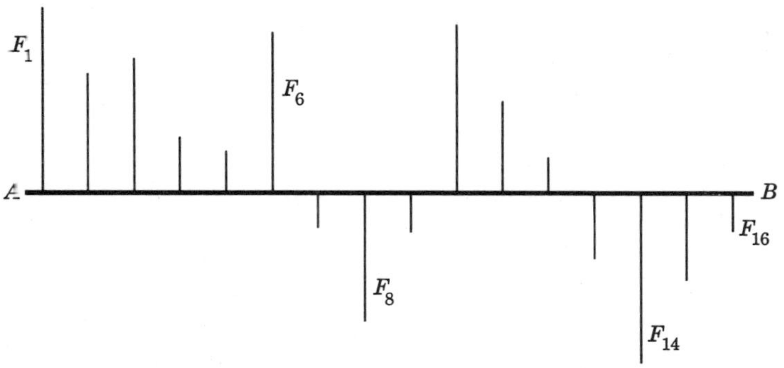

FIG. 3.1. Graphical representation of a sixteen-dimensional vector **F**

ciently close together, we may say that the vector $\{F_n\}$ actually assigns a number to each point on the line—in other words, it defines a function of position on the line.

Reversing the foregoing discussion, let us consider some definite function of position in the usual notation $F(x)$; this assigns a definite number to each value of x within the range of the definition of the function, *i.e.* to each point on the x-axis in its *domain*. We can regard the function $F(x)$ as a vector whose component is indicated by x instead of by n, and whose dimensionality is equal to the number of x-values in the domain—formally infinite. The totality of functions of x forms a space in the same sense as the totality of N-dimensional vectors, and we call it the *function space*. The following rules at once suggest themselves by analogy with ordinary vector algebra.

The vector sum of two functions $F(x)$ and $G(x)$ is exactly the usual sum of the two functions, *viz.*: $F(x)+G(x)$; this is evident if we use the generalization of Eq. 3.3. The scalar product of two functions is defined in terms of an integral:

$$\mathbf{F}\cdot\mathbf{G} = \int F(x)G(x)dx, \tag{3.5}$$

where the integration extends over the domain of the functions; in this definition we regard the differential dx as a measure of the number of points in the interval, and this definition is then a natural generalization of Eq. 3.4. The length of a function considered as a vector in function space is F where

$$F^2 = \mathbf{F}\cdot\mathbf{F} = \int [F(x)]^2 dx, \tag{3.6}$$

which is clearly analogous to Eq. 2.4.

Two functions are *orthogonal* if their scalar product vanishes:

$$F(x)\perp G(x) \quad \text{if} \quad \int F(x)G(x)dx = 0. \tag{3.7}$$

The question arises: Can we discover a set of functions that will serve as unit vectors in function space, as do the unit vectors $\mathbf{i}_1$, $\mathbf{i}_2$, and $\mathbf{i}_3$ of Eq. 2.6 in three dimensions? In N-dimensional space the generalization is fairly obvious; the unit vectors are

$$(1, 0, 0, \cdots 0); \quad (0, 1, 0, \cdots 0); \quad \cdots \quad (0, 0, 0, \cdots 1). \tag{3.8}$$

In function space the improper *Dirac delta functions* are the direct analogs of these unit vectors:

$$\delta(x-x') \quad \text{(all values of } x'\text{)}. \tag{3.9}$$

These functions are defined to be zero everywhere except at the one point $x = x'$, and to be so great at this point that the area under the curve is unity. There is a delta function associated with every point

x', so we can formally set up an infinity of such functions for the whole of function space. Handling such improper functions requires special techniques beyond the scope of the present course, and only brief discussion will be attempted later in this chapter. Fortunately, it has been discovered that there are many sets of proper functions that can serve the same purpose as unit vectors in function space. It is only necessary that such a set of functions shall all be mutually orthogonal, have unit length in the sense of Eq. 3.6, and be sufficiently numerous to span the infinity of dimensions implied by function space.

The importance of this discovery lies in the analog of Eq. 2.10. According to this, any function of the space can be expressed as a linear combination of the unit functions of the space. Put the other way round, if we can select an infinity of unit functions that have the required orthogonality and are sufficiently numerous (technically they must form a closed set), we can proceed to form all possible linear combinations of these unit functions to set up the entire function space.

A particularly simple set of unit functions is the following:

$$\mathbf{i}_n(x) = (1/L)^{1/2} \sin n\pi x/L, \quad (-L \leqslant x \leqslant L), \tag{3.10}$$

where n is any positive integer. These functions have the property:

$$\int_{-L}^{L} \mathbf{i}_n(x)\mathbf{i}_m(x)dx = \delta_{nm}, \tag{3.11}$$

meaning that different functions, $m \neq n$, are orthogonal, while each one itself has unit length as defined in Eq. 3.6. The class of odd functions of x, such that $F(x) = -F(-x)$, can be built up from linear combinations of these unit functions:

$$F(x) = \sum_n A_n \mathbf{i}_n(x) = \sum_n A_n (1/L)^{1/2} \sin n\pi x/L, \tag{3.12}$$

where A_n are the components of $F(x)$ with respect to this set of unit functions. To find these coefficients for any given function $F(x)$, we may proceed with the well-known methods used in the Fourier theorem, or we may appeal to the space analogy with Eq. 2.11 and write down directly the result:

$$\begin{aligned} A_n = \mathbf{F} \cdot \mathbf{i}_n &= \int F(x)\mathbf{i}_n(x)dx \\ &= (1/L)^{1/2} \int_{-L}^{L} F(x) \sin (n\pi x/L)dx. \end{aligned} \tag{3.13}$$

To obtain the class of even functions, $G(x) = G(-x)$, we use instead the set of unit functions

$$\mathbf{j}_n(x) = (1/L)^{1/2} \cos n\pi x/L \quad (-L \leqslant x \leqslant L), \tag{3.14}$$

where n is any positive integer. These functions have the same properties as the $\mathbf{i}(x)$ functions:

$$\int_{-L}^{L} \mathbf{j}_n(x)\mathbf{j}_m(x)dx = \delta_{nm}; \tag{3.15}$$

and moreover we also have every $\mathbf{j}(x)$ orthogonal to every $\mathbf{i}(x)$:

$$\int_{-L}^{L} \mathbf{i}_n(x)\mathbf{j}_m(x)dx = 0. \tag{3.16}$$

The general even function is expressible in the form:

$$G(x) = \sum_n B_n\mathbf{j}_n(x). \tag{3.17}$$

Any function $f(x)$ can be written as the sum of an even function and an odd function in a unique way:

$$f(x) = \tfrac{1}{2}[f(x)+f(-x)]+\tfrac{1}{2}[f(x)-f(-x)]. \tag{3.18}$$

Therefore, any function $f(x)$ can be expressed as the sum of two such series as given in Eqs. 3.12 and 3.17:

$$f(x) = \sum_n A_n\mathbf{i}_n(x)+\sum_n B_n\mathbf{j}_n(x), \tag{3.19}$$

where, again in analogy with Eq. 2.11, we have

$$A_n = \int f(x)\mathbf{i}_n(x)dx \quad \text{and} \quad B_n = \int f(x)\mathbf{j}_n(x)dx. \tag{3.20}$$

As this course proceeds we shall find a number of other sets of orthogonal functions that can be used instead of the sine and cosine functions as unit functions in function space. They lead to the same classes of functions, *i.e.* essentially to the same function space, and can be thought of as analogous with the various alternative coordinate systems that are available in ordinary geometry.

We shall introduce one important generalization here. This occurs if we allow our vectors—or our functions—to have complex values. Thus, in N-dimensional space, the N components may all be complex numbers. The scalar product of two such complex vectors is then conventionally defined in such a way that the length of any vector is real:

$$\mathbf{A}\cdot\mathbf{B} = \sum_n A_nB_n^*, \tag{3.21}$$

where the asterisk indicates the complex conjugate. The length of any vector $\mathbf{A}$ in this algebra is then the square root of the sum of the squares of the moduli of the components and is, therefore, a real quantity. This may be generalized to function space, when we write

$$\mathbf{F}\cdot\mathbf{G} = \int F(x)G^*(x)dx. \tag{3.22}$$

A suitable set of orthogonal unit functions for this kind of space is

$$I_n(x) = (1/2L)^{1/2} \exp(in\pi x/L), \tag{3.23}$$

and the space derived by all linear combinations of these unit functions is called a *Hilbert space.*

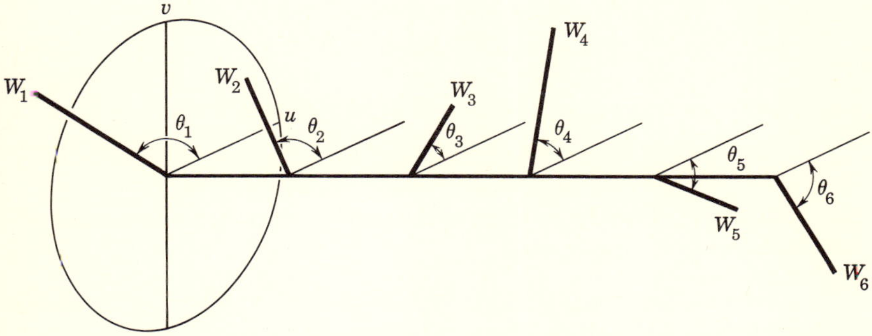

FIG. 3.2. Graphical representation of a six-dimensional vector **W** with complex components. An Argand diagram is set up at each of the six points representing the six dimensions; **u** is the real axis and **v** is the imaginary axis

In ordinary vector space, the algebraic functions of position assign a number (real or complex) to each point in the space; any algebraic function is in fact a rule by which we can assign a number to each point or vector in the vector space. The obvious analog to this concept, appropriate for function space, is some rule to assign a number to each vector in that space—*viz.* each function in function space. Just as there are many kinds of rules for assigning a number to each point in

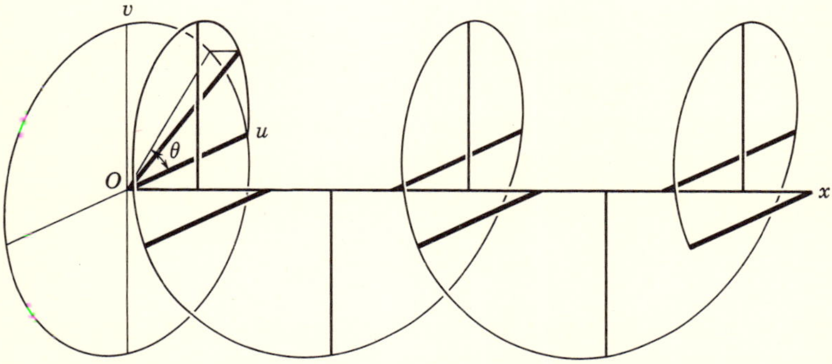

FIG. 3.3. Graph of the function e^{ikx}. The Argand diagram is drawn at every point on the x-axis, in a plane perpendicular to the x-axis. The spiral lies everywhere on the cylinder of unit radius and the angle $\theta = kx$. The period on the x-axis is $2\pi/k$.

ordinary space, so there are many kinds of rules for assigning a number to each function in function space; these rules are called *functionals.* A suitable symbol to represent a functional would naturally be of the form $F(\Phi)$, where F stands for the rule, and Φ for the function to which the rule assigns the number $F(\Phi)$. A very simple rule would be:

$$F(\Phi) = \int_a^b f(x)\Phi(x)dx, \tag{3.24}$$

where $f(x)$ is some other function simply used to define the rule or the functional. This kind of rule is capable of assigning a number to every function $\Phi(x)$ that is continuous everywhere, differentiable any number of times, and vanishes together with all its derivatives at both ends, a and b, of its domain. This kind of functional is usually called a *distribution.* It is linear in the following sense:

$$\begin{aligned} F(\Phi+\Psi) &= \int_a^b f(x)[\Phi(x)+\Psi(x)]dx \\ &= \int_a^b f(x)\Phi(x)dx + \int_a^b f(x)\Psi(x)dx = F(\Phi)+F(\Psi). \end{aligned} \tag{3.25}$$

There is a linear functional distribution for every one of a large class of functions $f(x)$, and there are also linear functionals or distributions for which no well-defined function $f(x)$ can be set up. Thus the simple rule

$$D(\Phi) = \Phi(x_0) \tag{3.26}$$

assigns to the function $\Phi(x)$ its value at the fixed point x_0. This is a linear functional rule in the above sense, but the function fitting it to the form of Eq. 3.24 would have to be the Dirac delta function:

$$D(\Phi) = \int_a^b \delta(x-x_0)\Phi(x)dx = \Phi(x_0). \tag{3.27}$$

The significance of this is that, although the Dirac delta function itself is not well-defined, the *Dirac delta functional* of Eq. 3.26 is well-defined. A great many other singular functions, or improper functions, do in fact correspond to well-defined functionals. Actually, many of the functions used in mathematical physics have singular points or infinities at which they are strictly not properly defined, and they can usually be replaced by well-defined functionals. Thus it becomes possible to carry out rigorously, in terms of functional analysis, many of our mathematical operations that were formerly nonrigorous and depended mainly on physical intuition for their acceptability. We do not propose to undertake this program here, but the possibility is mentioned so that the student will realize that, although some of the steps in our arguments may seem insecure, they can generally be made rigorous by a suitable generalization.

Exercise I.3

3.1. Write down the scalar products of the following pairs of 4-vectors, using both the rules of Eqs. 3.4 and 3.21:

(a) $(x, y, z, t)\cdot(x, y, z, t)$,
(b) $(x, y, z, it)\cdot(x, y, z, -it)$,
(c) $(x+iy, x-iy, z, t)\cdot(x+iy, -x+iy, z, t)$.

3.2. Find the lengths of the following functions as defined in Eq. 3.6:

(a) $f(x) = mx$, $0 \leqslant x \leqslant L$, m a constant;
(b) $f(x) = q/x$, $0 \leqslant x \leqslant L$, q a constant;
(c) $f(x) = e^{-ax}$, $-b < x < \infty$, b a positive constant;
(d) $f(x) = e^{-a\sqrt{x}}$, $-\infty < x < \infty$, a a constant.

Note: make use of Eq. 3.22 in discussing part (d).

3.3. Prove Eq. 3.11 from Eq. 3.10.

3.4. Prove Eqs. 3.15 and 3.16 from Eq. 3.14.

3.5. Find the analogs of Eqs. 3.10 and 3.13 when the functions are defined over the interval $-\frac{1}{2}L \leqslant x \leqslant \frac{1}{2}L$.

3.6. Prove that the unit functions of Eq. 3.23 are mutually orthogonal.

3.7. If the function $f(x)$ of Eq. 3.24 is $(2/L)^{1/2}\sin(k\pi x/L)$, where k is some integer, find the numbers assigned by the functional $F(\Phi)$ to the functions $\Phi_m(x) = (2/L)^{1/2}\sin(m\pi x/L)$ defined over the interval $0 \leqslant x \leqslant L$.

3.8. Compare Eq. 3.13 with Eq. 3.24 and discuss Eq. 3.13 in terms of the functional concept.

3.9. What are the Fourier coefficients of the Dirac delta function?

I.4 Transformation Theory. Let the components (F_1, F_2, F_3) of a given physical vector $\mathbf{F}$, *e.g.* some force, be measured relative to some arbitrary Cartesian coordinate system and its unit vectors $\mathbf{i}_1$, $\mathbf{i}_2$, and $\mathbf{i}_3$. Let the same force be measured also in some other coordinate system, rotated relative to the first one so that its unit vectors are $\mathbf{i}_1'$, $\mathbf{i}_2'$, $\mathbf{i}_3'$. These new unit vectors are also mutually orthogonal:

$$\mathbf{i}_m'\cdot\mathbf{i}_n' = \delta_{mn}.$$

We wish to find the relations between the new components F_1', F_2', F_3' and the original ones. Using the notation of Eq. 2.10, we have two different expressions for the same vector, so that

$$\mathbf{F} = \sum_m F_m\mathbf{i}_m = \sum_m F_m'\mathbf{i}_m'. \tag{4.1}$$

Take the scalar product of $\mathbf{i}_n'$ with this equation, and we find the desired relations:

$$\sum_m F_m\mathbf{i}_m\cdot\mathbf{i}_n' = F_n'. \tag{4.2}$$

The quantities $\mathbf{i}_m \cdot \mathbf{i}_n'$ are evidently the direction cosines of the angles between the mth old axis and the nth new axis of coordinates. Another notation for Eq. 4.2 is often used, *viz.*:

$$l_{mn} = \mathbf{i}_m \cdot \mathbf{i}_n'; \quad F_n' = \sum_m l_{mn} F_m, \tag{4.3}$$

where l_{mn} is the cosine of the angle between the mth old and the nth new axis of coordinates.

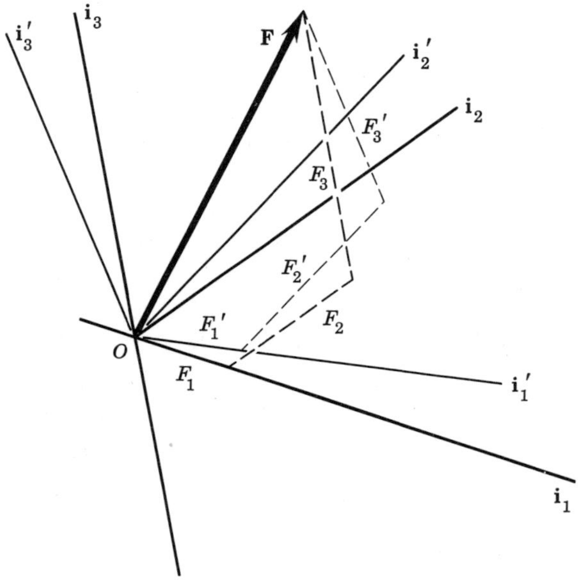

FIG. 4.1. Graphical representation of the transformation of the components of a vector **F** due to rotation of the coordinate system

Put Eq. 4.3 into Eq. 4.1 and obtain

$$\sum_m F_m \mathbf{i}_m = \sum_m \sum_q l_{qm} F_q \mathbf{i}_m'. \tag{4.4}$$

The components of **F** in the original coordinate system are involved here on both sides of the equation. The equation must be true for any vector we care to choose, for example, $\mathbf{F} = \mathbf{i}_k$, whatever the index k may be, $k = 1, 2$, or 3. Such a vector has components $F_m = \delta_{mk}$, and Eq. 4.4 now reads

$$\mathbf{i}_k = \sum_m l_{km} \mathbf{i}_m'. \tag{4.5}$$

The coefficients l_{mn} define a linear transformation, Eq. 4.3, from the components F_m to the components F_n'; at the same time they define a linear transformation, Eq. 4.5, from the unit vectors $\mathbf{i}_m'$ to the vectors $\mathbf{i}_n$.

From Eq. 4.5 we may derive identities satisfied by the transformation coefficients; taking the scalar product of any pair of original unit vectors we write:

$$\delta_{kq} = \mathbf{i}_k \cdot \mathbf{i}_q = \sum_m \sum_n l_{kn}\mathbf{i}_n' \cdot l_{qm}\mathbf{i}_m' = \sum_m \sum_n l_{kn} l_{qm} \mathbf{i}_n' \cdot \mathbf{i}_m',$$

and because the $\mathbf{i}'$ vectors are still orthogonal, this results in the identities:

$$\sum_n l_{kn} l_{qn} = \delta_{kq}. \tag{4.6}$$

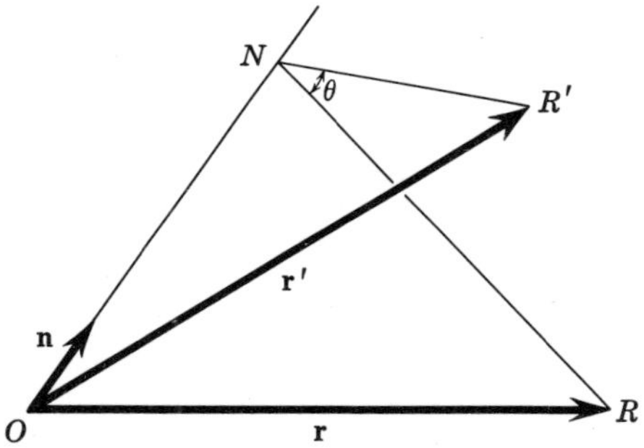

FIG. 4.2. Diagram to illustrate Exercise 4.6. The vectors **r** and **r**′ are equal in length, and the lines *RN* and *R′N* are perpendicular to the axis *ON*

In elementary work, the pure rotations are often the only transformations studied. It is important, however, to consider more general transformations. For example, the following simple linear transformations are not rotations:

Inversion:

$$\mathbf{i}_m' = -\mathbf{i}_m \quad (m = 1, 2, 3). \tag{4.7}$$

Reflection in the plane of $\mathbf{i}_2$ *and* $\mathbf{i}_3$*:*

$$\mathbf{i}_1' = -\mathbf{i}_1, \quad \mathbf{i}_2' = \mathbf{i}_2, \quad \mathbf{i}_3' = \mathbf{i}_3. \tag{4.8}$$

From Eq. 4.5 the linear transformation coefficients for the inversion are easily seen to be

$$l_{mn} = -\delta_{mn} \tag{4.9}$$

and the components of any vector simply reverse sign under inversion:

$$A_m' = -A_m. \tag{4.10}$$

In the vector product, as defined by Eq. 2.3, the components A_m and B_m both change sign on inversion, so the products do not, and the components of the vector product are unchanged by inversion. Therefore, as soon as we include more general transformations than simple rotations, it becomes advisable to distinguish between two types of vector: the true vector that reverses sign under an inversion of the coordinate system, and the *pseudo-vector* or axial vector (like the vector product) that does not change under an inversion. Under reflection, a true vector changes only its component normal to the plane of reflection, whereas the reflection of a pseudo-vector reverses the sign of both components lying in the reflecting plane.

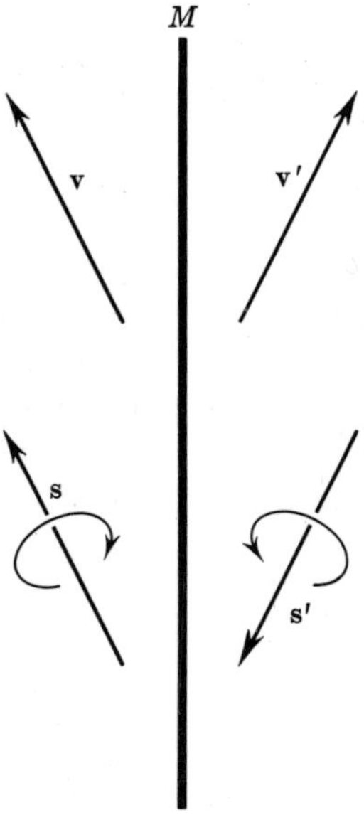

FIG. 4.3. The effects of reflection on two kinds of vector. The true vector **v** represents a translational velocity and **v**′ is its mirror image. Only the component normal to the mirror is reversed. The vector **s** represents a spin, which may be represented as a vector product, and is therefore a pseudo-vector, in the sense indicated. Its mirror image is **s**′, with spinning sense reversed. The use of the right-hand rule for associating an axis with the spin reverses the sign of the components parallel to the mirror and leaves unchanged the component perpendicular to the mirror

A scalar is defined as a number that remains unchanged by coordinate transformations; *e.g.* the length of a vector is a scalar with respect to rotations. True scalars do not change sign either under inversion reflection or rotation, by definition. But the triple product defined in Eq. 2.13 reverses sign both under inversion and under reflection; such quantities are called *pseudo-scalars*.

Any function of position (scalar, vector, etc.) that remains unchanged by inversion of the coordinate system is said to be an even function, whereas one that reverses sign under inversion is said to be an odd function. The general term for evenness or oddness is *parity*: an even function is said to have positive parity, and an odd function has negative parity.

Consider now a very general transformation of coordinates defined by three functions:

$$x_m' = x_m'(x_1, x_2, x_3) \quad (m = 1, 2, 3). \tag{4.11}$$

The three functions are to be identified with the new coordinates of a point that was at the position (x_1, x_2, x_3) in the original coordinate system. This is, in general, a nonlinear transformation, because the functions need not be linear functions. It can, however, be expressed as a result of linear transformations in every small region, different small regions of space undergoing different linear transformations. Thus we can differentiate Eq. 4.11 and write:

$$dx_m' = \sum_n (\partial x_m'/\partial x_n)dx_n. \tag{4.12}$$

Then if we regard the displacement $d\mathbf{x}$ as an example of the vector $\mathbf{F}$ in a more general linear transformation than Eq. 4.3, we can write Eq. 4.12 in the same form as Eq. 4.3 with linear coefficients:

$$l_{nm} = \partial x_m'/\partial x_n \tag{4.13}$$

which are functions of position determined by Eq. 4.11. We can set up new base vectors for the transformed coordinate system, $\mathbf{e}_m'$, where

$$d\mathbf{x} = \sum_m dx_m \mathbf{i}_m = \sum_n dx_n' \mathbf{e}_n'. \tag{4.14}$$

These base vectors are related to the original ones by an equation analogous with Eq. 4.5, *viz.*:

$$\mathbf{i}_m = \sum_n (\partial x_n'/\partial x_m)\mathbf{e}_n'. \tag{4.15}$$

The base vectors $\mathbf{e}_n'$ are not, in general, either of unit length or orthogonal, because the transformation coefficients $\partial x_n'/\partial x_m$ do not necessarily satisfy the identities, Eq. 4.6. This formalism will be taken up again in the section on curvilinear coordinates.

Exercise I.4

4.1. A coordinate system is rotated through 30° in the $x-y$ plane. Write out all the quantities l_{mn} for this rotation and verify the identities of Eq. 4.6 in detail.

4.2. A transformation is set up by means of the equations:

$$x_1' = (x_1^2+x_2^2)^{1/2},\ x_2' = \tan^{-1}(x_2/x_1),\ x_3' = x_3.$$

Find the coefficients l_{mn} of Eq. 4.13 for this transformation, and verify that they do not satisfy the identities of Eq. 4.6. What kind of transformation is this?

4.3. Substitute Eq. 4.9 into Eq. 4.6 and discuss the result. Prove that, if the determinant of the coefficients l_{mn} is required to be positive, we shall exclude both the inversion and the reflections of Eqs. 4.7 and 4.8. Discuss the transformation:

$$x_1' = -x_1,\ x_2' = -x_2,\ x_3' = x_3.$$

4.4. Prove that the triple scalar product reverses sign both under inversion and under reflection, but not under rotation.

4.5. Let a true vector **V** be equal to a pseudo-vector **P** in some coordinate system, so that $V_m = P_m$ ($m=1, 2, 3$); show that following a reflection $V_m' \neq P_m'$, so that the equation between **V** and **P** is not invariant for such transformations.

4.6. A vector **r** is rotated into another vector **r**′ of equal magnitude, the rotation being through an angle θ in a plane perpendicular to a unit vector **n** (see Fig. 4.2). Show that

$$\mathbf{r}' - \mathbf{r} = \mathbf{n} \times (\mathbf{r}' + \mathbf{r}) \tan \tfrac{1}{2}\theta.$$

4.7. Use the result of Ex. 4.6 to derive the transformation matrix for a rotation through an angle θ about an axis of arbitrary direction in space; show that the matrix can be expressed in the form:

$$\begin{matrix} \cos\theta + n_1^2(1-\cos\theta) & n_1n_2(1-\cos\theta)+n_3\sin\theta & n_1n_3(1-\cos\theta)-n_2\sin\theta \\ n_1n_2(1-\cos\theta)-n_3\sin\theta & \cos\theta+n_2^2(1-\cos\theta) & n_2n_3(1-\cos\theta)+n_1\sin\theta \\ n_3n_1(1-\cos\theta)+n_2\sin\theta & n_3n_2(1-\cos\theta)-n_1\sin\theta & \cos\theta+n_3^2(1-\cos\theta) \end{matrix}$$

4.8. Verify that the matrix of Ex. 4.7 satisfies the identities of Eq. 4.6.

I.5 Matrices, Tensors, and Dyadics. A *matrix* is an array of numbers written out in rectangular form with rows and columns. It can be expressed conveniently in subscript notation, such as A_{mn}, where m and n are integers; n indicates the column and m the row in which the number A_{mn} is located in the matrix. We shall here consider only square matrices with the number of rows equal to the number of columns. Many important physical entities can be expressed as matrices. We have already discussed one such entity, *viz.* the set of direction cosines l_{mn} appearing in Eq. 4.3. Another example which will form the subject of later sections is the set of stresses at any point in an elastic material; here the mn^{th} component would represent the force per unit area in the nth direction acting across a surface in the material whose normal is in the mth direction. The algebra of matrices, like the algebra of vectors, is invented for its usefulness in describing the observed properties of these entities.

Consider two matrices A and B; C is defined as the *matrix sum* of A and B if the components of C are each equal to the sum of the corresponding components of A and B:

$$C = A+B \quad \text{if} \quad C_{mn} = A_{mn}+B_{mn} \tag{5.1}$$

for all m, n values. The matrix C is defined as the *matrix product* of A and B if its components satisfy the following relations:

$$C = AB \quad \text{if} \quad C_{mn} = \sum_q A_{mq}B_{qn} \tag{5.2}$$

for all m, n values. Note that the sum can exist only if A and B both have the same dimensions, *i.e.* the same number of rows and the same number of columns. The matrix product can exist only if the first factor matrix has the same number of columns as the number of rows in the second factor matrix. In general, the matrix product is non-commutative:

$$BA \neq AB, \quad \text{because} \quad \sum_q B_{mq}A_{qn} \neq \sum_q A_{mq}B_{qn}. \tag{5.3}$$

The matrix A is called *symmetrical* if its components satisfy

$$A_{mn} = A_{nm} \tag{5.4}$$

for all values of the subscripts; the matrix is *antisymmetrical* if

$$A_{mn} = -A_{nm} \tag{5.5}$$

for all values of the subscripts. The matrix whose components are A_{nm} is called the *transpose* of the matrix whose components are A_{mn}. If we write $\tilde{A}$ for the transpose of A, we have

$$\tilde{A}_{nm} = A_{mn}. \tag{5.6}$$

Any symmetrical matrix is equal to its transpose, and any antisymmetrical matrix is the negative of its transpose.

Any matrix can be expressed as the sum of a symmetrical and an antisymmetrical matrix; thus we can write down the identity

$$A = \tfrac{1}{2}(A+\tilde{A})+\tfrac{1}{2}(A-\tilde{A}) \tag{5.7}$$

and prove that $A+\tilde{A}$ is a symmetrical matrix, and $A-\tilde{A}$ an antisymmetrical matrix.

A *diagonal matrix* is one that has non-zero elements only on the diagonal; it is necessarily a symmetrical matrix. The product of two diagonal matrices commutes.

The *identity matrix*, E, is a diagonal matrix with each diagonal element equal to unity; it has the components δ_{mn}. From the rule of matrix multiplication, we have

$$AE = EA = A, \tag{5.8}$$

where A is any arbitrary matrix with the same dimensions as E.

The *reciprocal* of a matrix A must be such that, when multiplied into A, it yields the identity E. Thus, if we write A^{-1} for the reciprocal we want it to satisfy,

$$A^{-1}A = AA^{-1} = E. \tag{5.9}$$

This will be the case if the components of A^{-1} are chosen so that

$$A_{mn}^{-1} = (\text{the cofactor of } A_{nm} \text{ in } |A|)/|A|, \tag{5.10*}$$

where $|A|$ is written for the determinant of the components of A. If this determinant happens to be zero, as often happens in matrices of physical interest, then the reciprocal does not exist, and the matrix is said to be *singular*.

We introduce now a useful notation called the *dyadic* notation, analogous to the unit vector notation in vector algebra. We define the *dyad product* of two unit vectors $\mathbf{i}_m$ and $\mathbf{i}_n$ as the matrix having all zero elements except a unity at the mn^{th} position, and write this matrix as $\mathbf{i}_m\mathbf{i}_n$. Any matrix A can then be written out explicitly in the form

$$A = \sum_{m,n}\sum A_{mn}\mathbf{i}_m\mathbf{i}_n, \tag{5.11†}$$

and its transpose is

$$\tilde{A} = \sum_{m,n}\sum A_{mn}\mathbf{i}_n\mathbf{i}_m. \tag{5.12}$$

If we take the scalar product of two matrices in this notation:

$$A\cdot B = \sum_{m,n}\sum A_{mn}\mathbf{i}_m\mathbf{i}_n\cdot\sum_{p,q}\sum B_{pq}\mathbf{i}_p\mathbf{i}_q = \sum_{m,n,p,q}\sum\sum\sum A_{mn}B_{pq}\mathbf{i}_m\mathbf{i}_n\cdot\mathbf{i}_p\mathbf{i}_q,$$

where we have transposed scalar factors and summation operations, but kept all the unit vectors in proper relation with the scalar product operator. Now applying Eq. 2.7 to the dot product and taking out the scalar factor δ_{np}, we find

$$A\cdot B = \sum_{m,n,q}\sum\sum A_{mn}B_{nq}\mathbf{i}_m\mathbf{i}_q = \sum_{m,q}\sum C_{mq}\mathbf{i}_m\mathbf{i}_q = C, \tag{5.13}$$

where

$$C_{mq} = \sum_n A_{mn}B_{nq}.$$

Thus, comparing this with Eq. 5.2 we see that the scalar product is equivalent to the matrix product.

* Multiplying this equation by A_{pm} and summing over the index m, we find

$$\sum_m A_{pm}A_{mn}^{-1} = (\sum_m A_{pm} \times \text{cofactor of } A_{nm} \text{ in } |\mathbf{A}|)/|\mathbf{A}|.$$

The numerator on the right is just the determinant, by definition of the cofactor, for any choice of either the index p or n, provided $p = n$: it would be obtained by using the pth row as the basis of calculating the determinant. On the other hand, if we choose $p \neq n$, the result is the same as if we substituted the pth row for the nth row before computing the determinant, and this is then a determinant with two identical rows, necessarily zero. Thus we prove that

$$\sum_m A_{pm}A_{mn}^{-1} = \delta_{pn},$$

which are the components of the identity matrix E, as in Eq. 5.9.

† Written in this form, A is often called a *dyadic*. The term "matrix" is then reserved for the simple array of components A_{mn} alone.

Any symmetric 3×3 matrix can be represented geometrically by a quadratic surface. Thus consider any symmetrical matrix G with components G_{mn}, and look for all points (x_1, x_2, x_3) such that

$$\sum_{m,n}\sum G_{mn}x_m x_n = 1. \tag{5.14}$$

This equation defines a quadratic surface that is determined uniquely by the components of the matrix G.

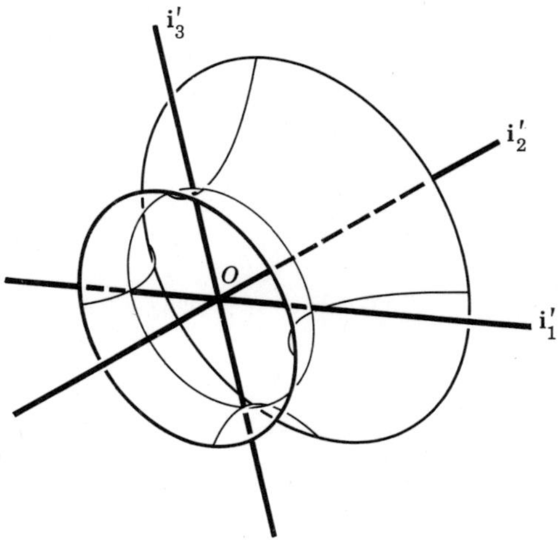

FIG. 5.1. A quadratic surface with its principal axes. The surface represents a matrix whose principal components g_{11} and g_{33} are positive while g_{22} is negative, in the notation of Eqs. 5.15 and 5.16.

This geometrical representation suggests that we may rotate the coordinate system into the principal axes of the quadratic surface, in which case it will take on the form of an equation containing no cross products:

$$g_{11}x_1^2 + g_{22}x_2^2 + g_{33}x_3^2 = 1,$$

where g_{11}, etc., are three new coefficients. This matrix of new coefficients is evidently diagonal in form; it is called the *normal form* of the matrix G. The important question is: Given any nondiagonal symmetrical matrix, can we find its normal form?

Let the coordinates into which we have to rotate the system, in order to get the normal form, have unit vectors $\mathbf{i}_1'$, $\mathbf{i}_2'$, $\mathbf{i}_3'$. Let us use the dyadic notation and write the matrix G in the forms

$$G = \sum_{m,n}\sum G_{mn}\mathbf{i}_m\mathbf{i}_n = \sum_m g_{mm}\mathbf{i}_m'\mathbf{i}_m'. \tag{5.15}$$

Take the scalar product of this with any one of the new unit vectors:

$$G\cdot\mathbf{i}_n' = g_{nn}\mathbf{i}_n' \quad (n = 1, 2, \text{or } 3). \tag{5.16}$$

This kind of equation is an *eigenvalue equation*;* it states that the product of the matrix and the unit vector $\mathbf{i}_n'$ is another vector parallel to $\mathbf{i}_n'$. Only three characteristic directions, $\mathbf{i}_1'$, $\mathbf{i}_2'$, and $\mathbf{i}_3'$ are such that this equation is true; they are the *principal directions* of the matrix. The equation gives us a clue for finding the principal directions for any matrix that has been given in nondiagonal form. Thus, given G we seek vectors $\mathbf{a}$ such that

$$G\cdot\mathbf{a} = g\mathbf{a}, \tag{5.17}$$

where g is the *eigenvalue* sought. Comparing Eq. 5.17 with Eq. 5.16, we note that there should be three solutions of Eq. 5.17, one corresponding to each value of n in Eq. 5.16. If we write the unknown vector $\mathbf{a}$ in the form

$$\mathbf{a} = \sum_n a_n\mathbf{i}_n,$$

we see that Eq. 5.17 is equivalent to the three equations:

$$\begin{aligned} &\mathbf{i}_1[G_{11}a_1+G_{12}a_2+G_{13}a_3 = ga_1] \\ &\mathbf{i}_2[G_{21}a_1+G_{22}a_2+G_{23}a_3 = ga_2] \\ &\mathbf{i}_3[G_{31}a_1+G_{32}a_2+G_{33}a_3 = ga_3] \end{aligned} \tag{5.18}$$

These three equations can be solved simultaneously for the three unknowns a_1, a_2, and a_3 if and only if the determinant of the coefficients vanishes:

$$\begin{vmatrix} G_{11}-g & G_{12} & G_{13} \\ G_{21} & G_{22}-g & G_{23} \\ G_{31} & G_{32} & G_{33}-g \end{vmatrix} = 0. \tag{5.19}$$

This is a cubic equation in g and can be solved for the three values of g corresponding to the eigenvalues, yielding the diagonal elements

* An eigenvalue equation is defined to have the general form:

$$G\cdot\Phi = g\Phi,$$

where G is some operator, Φ some function, and g a number, which is called an eigenvalue of the operator. Eq. 5.16 is of this form, and it is interesting to generalize it to function space. The analogs of the unit vectors are the unit functions $\mathbf{i}_n(x)$, and the analog of a matrix is a function of two variables, $G(x, x')$. The analog of Eq. 5.16 is then

$$\int G(x, x')\mathbf{i}_n(x')dx' = g\mathbf{i}_n(x).$$

This is a typical integral equation, with the function $G(x, x')$ called the *kernel*. If the kernel is given, it is possible to solve the integral equation for the functions $\mathbf{i}_n(x)$ and the eigenvalues g, using methods analogous with the method used in this section in the case of ordinary vector space.

of the normal form of G. If G is not a symmetrical matrix, Eq. 5.19 leads to complex values of g; only if G is symmetrical does it have a real normal form. For matrices having only two rows and two columns, the determinant corresponding to Eq. 5.19 is a 2 by 2 form and is then simply a quadratic equation.

The dyadic notation and the geometrical representation of a matrix have led us to a useful rule for the transformation of matrices; thus Eq. 5.15 is essentially a transformation rule associating the matrix G in the original coordinate system with a different matrix—the diagonal form—in another coordinate system. We can easily generalize this rule. Let any matrix G be represented on two different coordinate systems whose unit vectors are, respectively, $\mathbf{i}_m$ and $\mathbf{i}_m'$ ($m = 1, 2, 3$), and let the components of G on the two systems be related through the equation

$$G = \sum_{m,n}\sum G_{mn}\mathbf{i}_m\mathbf{i}_n = \sum_{m,n}\sum G_{mn}'\mathbf{i}_m'\mathbf{i}_n'. \tag{5.20}$$

Multiply this equation from the left by scalar product with $\mathbf{i}_p'$ and from the right by scalar product with $\mathbf{i}_q'$, and so find

$$G_{pq}' = \sum_{m,n}\sum G_{mn}\mathbf{i}_p'\cdot\mathbf{i}_m\mathbf{i}_n\cdot\mathbf{i}_q' = \sum_{m,n}\sum l_{mp}G_{mn}l_{nq}, \tag{5.21}$$

where l_{mn}, etc., are the direction cosines between the two systems of coordinates. This transformation rule is called the *tensor transformation rule*, and any physical entity that is represented by a matrix of components that obeys this rule when the coordinate system is rotated is called a *tensor*. In fact, the stress system in an elastic material is a tensor of this kind. The deduction of Eq. 5.21 from Eq. 5.20 should be compared with the deduction of Eq. 4.3 from Eq. 4.1 in the transformation rule for vectors.

If we write L for the matrix whose components are l_{mn} in Eq. 5.21, that equation becomes, in pure matrix notation:

$$G' = \tilde{L}GL. \tag{5.22}$$

Note the order of the subscripts in the last expression in Eq. 5.21.

Restricting our transformations to be pure rotations, the components of L must obey the identities of Eq. 4.6, which in pure matrix notation become

$$L\tilde{L} = E \quad \text{or} \quad \tilde{L} = L^{-1}. \tag{5.23}$$

A matrix satisfying this relation is called an *orthogonal matrix*. When this is the case, we can prove a theorem that will be useful later: The determinant of the components of any tensor is invariant under rotations of coordinates. The proof follows from a lemma that the determinant of the matrix product of two matrices equals the product of the determinants of the two matrices (see Ex. 5.8). Assuming this lemma, we have, from Eq. 5.22,

$$|G'| = |L| \cdot |G| \cdot |\tilde{L}| = |G| \cdot |L| \cdot |\tilde{L}| \cdot = |G| \cdot |L\tilde{L}| = |G|, \quad (5.24)$$

because of Eq. 5.23 and $|E| = 1$.

So far we have assumed the components of a matrix to be all real numbers, but this is an unnecessary restriction. If we allow complex numbers for the components of a matrix, it is necessary to define the *Hermitian conjugate* of a matrix. This is derived from the matrix by taking its transpose and forming the complex conjugate of each element. H is the Hermitian conjugate of G if $H_{mn} = G_{nm}{}^*$. A matrix is said to be a *Hermitian matrix* if it is equal to its own Hermitian conjugate; it is the generalization of the symmetrical matrix. We shall not have many occasions in this text to make use of matrices with complex coefficients, but their study is important in more advanced work.

Exercise I.5

5.1. Find the matrix sum of A and B where

$$A = \begin{matrix} 1 & -3 & 2 \\ 3 & 0 & 3 \\ -2 & -3 & -1 \end{matrix} \qquad B = \begin{matrix} 0 & 3 & -1 \\ -3 & 1 & -3 \\ 1 & 3 & 2 \end{matrix}$$

5.2. Find the matrix products AB and BA of the matrices in Ex. 5.1.
5.3. Find the symmetric and antisymmetric parts of A and B above.
5.4. Find the symmetric and antisymmetric parts of the following matrices:

$$\begin{matrix} 1 & 4 & 0 \\ 2 & 2 & 0 \\ 0 & 0 & 3 \end{matrix} \qquad \text{and} \qquad \begin{matrix} 0 & -1 & 0 \\ 9 & 0 & 0 \\ 0 & 0 & 1 \end{matrix}$$

5.5. Find the determinant and the cofactors of the elements of the following matrices:

$$\begin{matrix} 1 & 1 & 0 \\ 1 & 1 & 0 \\ 0 & 0 & 2 \end{matrix} \qquad \begin{matrix} 1 & 1 & 1 \\ 1 & 1 & 0 \\ 1 & 0 & 2 \end{matrix} \qquad \begin{matrix} 0 & 1 & 1 \\ 1 & 1 & 0 \\ 1 & 0 & 2 \end{matrix}$$

5.6. Write out the reciprocal matrices to those in Ex. 5.5 that are not singular matrices.

5.7. Take each of the three matrices of Ex. 5.5 as the matrix G in Eq. 5.19 and find their eigenvalues. Write out the matrices in normal form, and set up the equation of the quadratic surface representing each matrix in its normal form.

5.8. Find the determinant of each of the matrices A and B, and of AB and BA of Ex. 5.1. Compare the determinant of the matrix products with the product of the two determinants.

5.9. Check the theorem that $|A|\cdot|B| = |AB|$ using the matrices of Ex. 5.5.

5.10. Find the Hermitian conjugates of the matrices

$$\begin{pmatrix} 0 & i \\ -i & 0 \end{pmatrix} \qquad \begin{pmatrix} 0 & i \\ i & 0 \end{pmatrix} \qquad \begin{pmatrix} z & x+iy \\ x-iy & -z \end{pmatrix}$$

Which of these matrices are Hermitian?

5.11. Use Eq. 5.19 to diagonalize the matrices

$$\begin{pmatrix} 0 & 1 \\ -1 & 0 \end{pmatrix} \qquad \begin{pmatrix} 0 & 1 \\ 1 & 0 \end{pmatrix}$$

5.12. Use Eq. 5.19 to diagonalize the matrices of Ex. 5.10. Note that the Hermitian matrices have real eigenvalues.

5.13. Show that, if the matrix G in Eq. 5.14 had any antisymmetrical part, it would not make any difference to the quadratic surface associated with the matrix.

5.14. Prove that the matrix product of the unit dyads $(\mathbf{i}_m\mathbf{i}_q)(\mathbf{i}_s\mathbf{i}_n)$ is equal to $(\mathbf{i}_m\mathbf{i}_n)\delta_{qs}$, and hence justify the dot product notation in the equation following Eq. 5.12 and in the proof of Eq. 5.13.

I.6 Vector Calculus in Three Dimensions.

A scalar function of position has a well-defined value at every point in space. If the coordinate system is changed by a rotation like Eq. 4.3:

$$x_m = \sum_n l_{mn} x_n' \quad \text{and} \quad x_m' = \sum_n l_{nm} x_n, \tag{6.1}$$

the scalar function may change its form. The value of the scalar function $f(x_1, x_2, x_3)$ originally associated with the point (x_1, x_2, x_3) is now associated with the new coordinates x_1', x_2', x_3' given by Eq. 6.1. The scalar can be expressed as a function of the x'-coordinates instead of the x-coordinates, and this is done by substituting x_m from the first of Eqs. 6.1 into the function $f(x_1, x_2, x_3)$ and deriving the new function from the identity

$$f'(x_1', x_2', x_3') = f(x_1, x_2, x_3)$$

and

$$f(x_1, x_2, x_3) = f\left(\sum_n l_{1n} x_n', \ \sum_n l_{2n} x_n', \ \sum_n l_{3n} x_n'\right). \tag{6.2}$$

If it happens that the new function has the same form as the original one, it is called a *scalar invariant function*. A simple example of such a scalar invariant function is

$$f(x_1, x_2, x_3) = x_1^2 + x_2^2 + x_3^2,$$

which transforms according to Eq. 6.1 into

$$f'(x_1', x_2', x_3') = x_1'^2 + x_2'^2 + x_3'^2, \tag{6.3}$$

so that the two functions are not only equal (by definition) but also have the same functional form.

A vector function of position has well-defined components at every point in space: it is an ordered set of three functions of position. In elementary physics the vector function associates a directed magnitude with each point in space, the three components of the directed magnitude being the three component functions of the vector function. If the coordinate system is rotated according to Eq. 6.1, the vector function goes through a transformation at every point according to Eq. 4.3. Thus if we write the original vector function in the form*:

$$\mathbf{F} = (F_1(\mathbf{r}), F_2(\mathbf{r}), F_3(\mathbf{r})),$$

where $\mathbf{r}$ is the *position vector* $\mathbf{r} = (x_1, x_2, x_3)$, then the same vector function after rotation according to Eq. 4.3 is

$$\mathbf{F} = (F_1'(\mathbf{r}'), F_2'(\mathbf{r}'), F_3'(\mathbf{r}')),$$

where

$$F_m'(\mathbf{r}') = \sum_n l_{nm} F_n(\mathbf{r}) \tag{6.4}$$

and $\mathbf{r}'$ is the position vector with components x_1', x_2', x_3'. To find the forms of the three new functions F_m' we substitute Eq. 6.1 into the original functions, thus:

$$F_m'(\mathbf{r}') = \sum_n l_{nm} F_n(\sum_s l_{1s} x_s', \quad \sum_s l_{2s} x_s', \quad \sum_s l_{3s} x_s'). \tag{6.5}$$

If it happens that the functions F_m' are the same form as the functions F_m, the vector function is called a *vector invariant function*. A simple example of a vector function that is invariant for a particular class of rotations is the velocity of a point on a rotating rigid body:

$$\mathbf{F} = (\omega x_2, \ -\omega x_1, 0) \quad \text{where} \quad \omega \text{ is a constant.}$$

The three component functions are $F_1 = \omega x_2, F_2 = -\omega x_1, F_3 = 0$. If we use the transformation for rotation through any angle θ around the z-axis for the components l_{mn} in Eq. 6.5, we find easily that

$$F_1' = \omega x_2', \quad F_2' = -\omega x_1', \quad F_3' = 0,$$

so that the vector function has exactly the same form as before the rotation, independently of the angle of rotation θ.

A particularly important class of vector invariant functions is found if we use the three component functions that are the three partial derivatives of a scalar invariant function of position. This vector is called the *gradient*. If the scalar invariant function is Φ, the gradient of Φ is written:

$$\mathbf{grad}\,\Phi \quad \text{or} \quad \nabla\Phi = (\partial\Phi/\partial x_1, \partial\Phi/\partial x_2, \partial\Phi/\partial x_3). \tag{6.6}$$

* $F_1(\mathbf{r})$, etc., are abbreviations for $F_1(x_1, x_2, x_3)$, etc., functions of three variables each.

Under the rotation of coordinates, Eq. 6.1, the scalar invariant function remains unchanged, as in Eq. 6.2: $\Phi'(x') = \Phi(x)$, while the transformation rule, Eq. 6.5, yields

$$(\mathbf{grad}\,\Phi)_m' = \sum_n l_{nm}\partial\Phi/\partial x_n = \sum_n (\partial x_n/\partial x_m')(\partial\Phi/\partial x_n),$$

where the last step can be proved from the first of Eqs. 6.1. Summing the partial differentials gives $\partial\Phi/\partial x_m'$ for the right side of the last equation, and then using the scalar invariance of Φ we have finally

$$(\mathbf{grad}\,\Phi)' = (\partial\Phi'/\partial x_1',\quad \partial\Phi'/\partial x_2',\quad \partial\Phi'/\partial x_3'). \tag{6.7}$$

Since Φ is a scalar invariant function, it has the same form in both coordinate systems, so that Φ' is the same function of x' as Φ is of x and $\partial\Phi'/\partial x'$ is the same functional form as $\partial\Phi/\partial x$; therefore from Eq. 6.7, $(\mathbf{grad}\,\Phi)'$ is the same functional form in each component as $\mathbf{grad}\,\Phi$; the gradient of a scalar invariant function is therefore a vector invariant.*

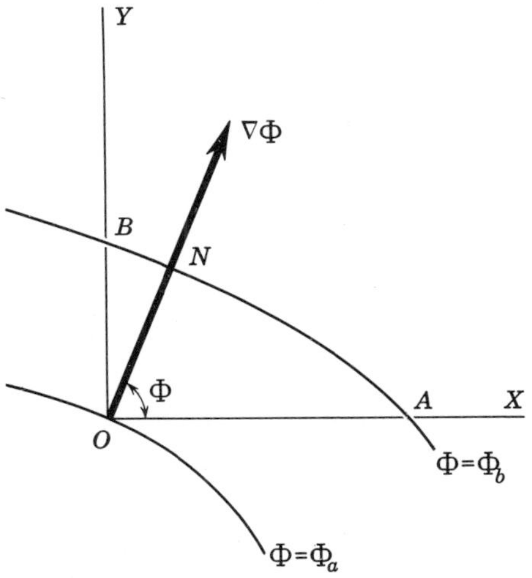

FIG. 6.1. The vector gradient of a scalar function Φ. By definition:

$$\partial\Phi/\partial x = \text{Lim.}(\Phi_b-\Phi_a)/OA, \quad \partial\Phi/\partial y = \text{Lim.}(\Phi_b-\Phi_a)/OB$$

and

$$|\nabla\Phi|^2 = (\partial\Phi/\partial x)^2+(\partial\Phi/\partial y)^2 = \text{Lim.}(\Phi_b-\Phi_a)^2(1/OA^2+1/OB^2)$$
$$= \text{Lim.}(\Phi_b-\Phi_a)^2/ON^2 = (\partial\Phi/\partial n)^2$$

* If Φ is scalar but not invariant, then $\nabla\Phi$ is vector but not invariant. The scalar function Φ may be invariant only for a special group of transformations, called the symmetry group of the function. $\nabla\Phi$ is then a vector invariant for the same group of transformations. The theory of groups is an important branch of mathematics with extensive applications, but we shall not have time in this course to do more than mention this point of contact with it.

A simple geometrical interpretation of the gradient is as follows: the direction of $\mathbf{grad}\ \Phi$ is the direction of the maximum rate of change or the slope of the function Φ, and the magnitude of $\mathbf{grad}\ \Phi$, namely,

$$\mathbf{grad}\ \Phi = [(\partial\Phi/\partial x_1)^2+(\partial\Phi/\partial x_2)^2+(\partial\Phi/\partial x_3)^2]^{1/2}, \tag{6.8}$$

is equal to the maximum slope of the function Φ.

In unit vector notation it is convenient to write

$$\mathbf{grad}\ \Phi = \mathbf{i}_1\partial\Phi/\partial x_1+\mathbf{i}_2\partial\Phi/\partial x_2+\mathbf{i}_3\partial\Phi/\partial x_3. \tag{6.9}$$

The total variation of a function Φ can be written as the sum of its partial variations:

$$d\Phi = (\partial\Phi/\partial x_1)dx_1+(\partial\Phi/\partial x_2)dx_2+(\partial\Phi/\partial x_3)dx_3 \tag{6.10}$$

and this can be written in the useful vector form

$$d\Phi = \mathbf{grad}\ \Phi\cdot d\mathbf{r} = d\mathbf{r}\cdot\mathbf{grad}\ \Phi \tag{6.11}$$

where $\mathbf{dr}$ is the *displacement vector* (dx_1, dx_2, dx_3).

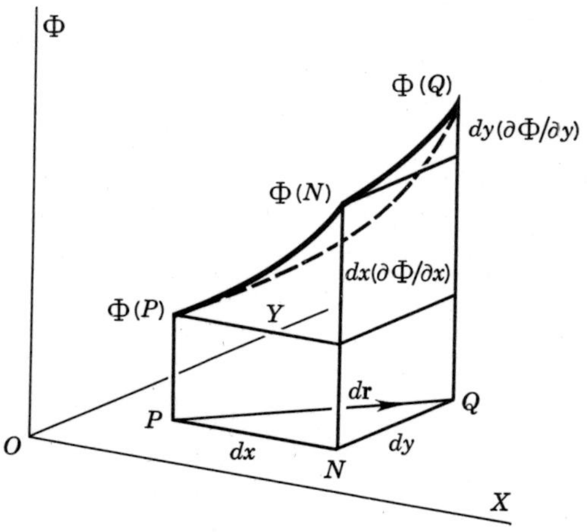

FIG. 6.2. Equation 6.10 illustrated in two dimensions. The total increment in Φ is $\Phi(Q)-\Phi(P)$, and this is the sum of two parts equal, respectively, to $dx\cdot(\partial\Phi/\partial x)$ and $dy\cdot(\partial\Phi/\partial y)$

If we interpret Φ as a potential function for a force $\mathbf{F}$, we have, by elementary mechanics, work done = change in potential energy,

$$-d\Phi = F_1dx_1+F_2dx_2+F_3dx_3, \tag{6.12}$$

which, by comparison with Eq. 6.11, yields

$$F_1 = -\partial\Phi/\partial x_1 \text{ etc.} \quad \text{or} \quad \mathbf{F} = -\mathbf{grad}\ \Phi. \tag{6.13}$$

From a different point of view, Eq. 6.9 may be thought of as instructing us to perform a certain operation or set of operations on the function Φ, the operations being defined symbolically by the *del operator*,

$$\nabla \equiv \sum_n \mathbf{i}_n(\partial/\partial x_n), \tag{6.14}$$

where we have to sum over all three values of the subscript. The operator is linear in the sense that

$$\nabla(\Phi+\Psi) = \nabla\Phi+\nabla\Psi. \tag{6.15}$$

It also obeys the familiar distributive rule,

$$\nabla(\Phi\Psi) = \Psi\,\nabla\Phi+\Phi\,\nabla\Psi. \tag{6.16}$$

We can take the scalar product between the operator form and some vector function, writing

$$\nabla\cdot\mathbf{F} = \sum_n \mathbf{i}_n(\partial/\partial x_n)\cdot\sum_k \mathbf{i}_k F_k. \tag{6.17}$$

So long as the unit vectors are constant in direction throughout space, as they are in ordinary Cartesian coordinates, we can take $\mathbf{i}_k$ and the scalar product sign to the left through the differentiations, and write

$$\nabla\cdot\mathbf{F} = \sum_{n,k}\sum \mathbf{i}_n\cdot\mathbf{i}_k\partial F_k/\partial x_n = \sum_k \partial F_k/\partial x_k. \tag{6.18}$$

This is of course a scalar quantity; it is called the *divergence* of $\mathbf{F}$. We can also take the vector product of the del operator and any vector function thus:

$$\nabla\times\mathbf{V} = \begin{vmatrix} \mathbf{i}_1 & \mathbf{i}_2 & \mathbf{i}_3 \\ \partial/\partial x_1 & \partial/\partial x_2 & \partial/\partial x_3 \\ V_1 & V_2 & V_3 \end{vmatrix} \tag{6.19}$$

in accordance with Eq. 2.12. We note that the first row must be used as basis for computing the determinant, to keep the operators in their proper places. For reasons that will appear later, this function is called the *curl* of $\mathbf{V}$.

Successive applications of the del operator occur in many problems, and we list some of the specific results below.

The *Laplacian operator* is $\nabla\cdot\nabla$ or ∇^2, where

$$\nabla\cdot\nabla\Phi = \partial^2\Phi/\partial x_1{}^2+\partial^2\Phi/\partial x_2{}^2+\partial^2\Phi/\partial x_3{}^2. \tag{6.20}$$

The *double curl* is

$$\nabla\times\nabla\times\mathbf{V} = \nabla(\nabla\cdot\mathbf{V})-\nabla\cdot\nabla\mathbf{V}. \tag{6.21}$$

(Compare this with the ordinary vector formula $\mathbf{A} \times \mathbf{B} \times \mathbf{C} = \mathbf{B}(\mathbf{A} \cdot \mathbf{C}) - \mathbf{C}(\mathbf{A} \cdot \mathbf{B})$.)
The *divergence curl*, to be compared with Eq. 2.13, is

$$\nabla \cdot \nabla \times \mathbf{V} = \begin{vmatrix} \partial/\partial x_1 & \partial/\partial x_2 & \partial/\partial x_3 \\ \partial/\partial x_1 & \partial/\partial x_2 & \partial/\partial x_3 \\ V_1 & V_2 & V_3 \end{vmatrix} = 0. \tag{6.22}$$

Note that the first row must be used as basis for computing the determinant: two identical rows make it vanish identically.

We can find the dyadic product of the del operator and a vector function, and so write out a matrix form as follows:

$$\nabla \mathbf{V} = \sum_{j,k}\sum \partial V_j/\partial x_k \, \mathbf{i}_k \mathbf{i}_j, \tag{6.23}$$

or in matrix notation:

$$||\nabla \mathbf{V}|| = \begin{pmatrix} \partial V_1/\partial x_1 & \partial V_2/\partial x_1 & \partial V_3/\partial x_1 \\ \partial V_1/\partial x_2 & \partial V_2/\partial x_2 & \partial V_3/\partial x_2 \\ \partial V_1/\partial x_3 & \partial V_2/\partial x_3 & \partial V_3/\partial x_3 \end{pmatrix}. \tag{6.24}$$

The physical usefulness of these operators will appear later in the course.

Exercise I.6

6.1. Show that any scalar function of the distance from the origin is a scalar invariant function for rotations about any axis through the origin.

6.2. Show that the vector function

$$f(r)(0, \omega z, -\omega y)$$

is an invariant function for rotations about the x-axis; $f(r)$ being any scalar function of the distance from the origin and ω a constant.

6.3. Show that the vector $\mathbf{r}f(r)$ is a vector invariant for rotations about any axis through the origin.

6.4. Prove that the scalar product of two vector invariant functions is a scalar invariant function.

6.5. Prove that the divergence of a vector invariant function is a scalar invariant function.

6.6. Prove that the curl of a vector invariant function is a vector invariant function under rotations of coordinates.

6.7. Prove that the curl of a vector invariant function reverses sign as a result of coordinate inversions. If these transformations are included, the curl of a vector invariant is therefore a pseudo-vector invariant function.

6.8. Prove that the Laplacian of a scalar invariant is a scalar invariant.

6.9. Prove that the divergence curl of a vector invariant is a pseudo-scalar invariant.

6.10. Prove the following formulas, in which $\mathbf{r}$ is the position vector (x, y, z), $\mathbf{r}_1$ is the unit vector in the direction of $\mathbf{r}$: $\mathbf{r}_1 = (x/r, y/r, z/r)$, r is the magnitude of $\mathbf{r}$, n is a constant scalar, and $\mathbf{c}$ a constant vector.

(a) $\nabla r^n = nr^{n-1}\mathbf{r}_1$ (b) $\nabla^2 r^n = n(n-1)r^{n-2}$
(c) $\nabla \cdot \mathbf{r} = 3$ (d) $\nabla \times \mathbf{r} = 0$
(e) $\nabla(\mathbf{r}\cdot\mathbf{c}) = \mathbf{c}$ (f) $\nabla \times (r\mathbf{c}) = \mathbf{r}_1 \times \mathbf{c}$
(g) $\nabla \cdot (r\mathbf{c}) = \mathbf{r}_1 \cdot \mathbf{c}$ (h) $\nabla \cdot (r^n\mathbf{r}) = (n+3)r^n$

6.11. Verify Eq. 6.21 in detail.
6.12. Prove that $\nabla(\mathbf{A}\cdot\mathbf{B}) = (\mathbf{A}\cdot\nabla)\mathbf{B} + (\mathbf{B}\cdot\nabla)\mathbf{A} + \mathbf{A}\times(\nabla\times\mathbf{B}) + \mathbf{B}\times(\nabla\times\mathbf{A})$.
6.13. Prove that $\nabla\cdot(\mathbf{A}\times\mathbf{B}) = \mathbf{B}\cdot\nabla\times\mathbf{A} - \mathbf{A}\cdot\nabla\times\mathbf{B}$.
6.14. Prove that $\nabla\times(\mathbf{A}\times\mathbf{B}) = \mathbf{A}(\nabla\cdot\mathbf{B}) - \mathbf{B}(\nabla\cdot\mathbf{A}) + (\mathbf{B}\cdot\nabla)\mathbf{A} - (\mathbf{A}\cdot\nabla)\mathbf{B}$.
6.15. Prove that $(\mathbf{A}\cdot\nabla)\mathbf{A} = \frac{1}{2}\nabla(\mathbf{A}\cdot\mathbf{A}) - \mathbf{A}\times\nabla\times\mathbf{A}$.

I.7 Vector Integration. We have seen in §2 that the scalar product of a force times a displacement represents the work done by the force during the displacement. Now suppose the force is given as some vector function of position, and its point of application is moved along some definite curve; how shall we compute the total work done? To answer this question we may divide the curve into a large number of small elements, take the scalar product of the force at the center of each element times the vector element, and add up all the results. The smaller the elements, and the more numerous, the more exact this procedure will be. If there exists a limiting result as every element is reduced indefinitely in length, it is defined as the line integral of the force along the curve. Formally we write

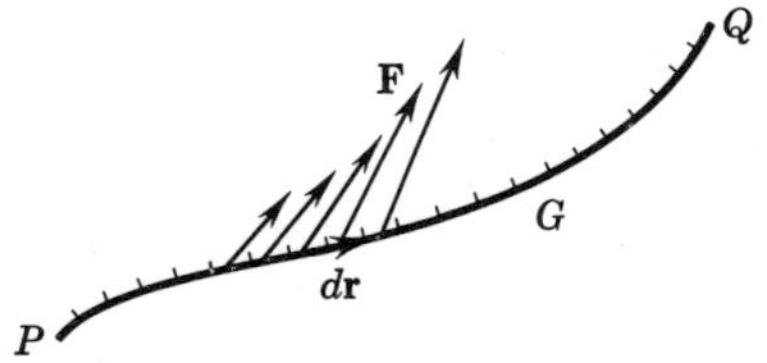

FIG. 7.1. The line integral in Eq. 7.1

$$\int_C \mathbf{F}\cdot d\mathbf{r} = \text{Lim} \sum \mathbf{F}\cdot d\mathbf{r}, \tag{7.1}$$

where C indicates the particular curve along which the integral is evaluated, $\mathbf{F}$ is the force, and $d\mathbf{r}$ is the element of displacement on the curve. Before proceeding to the limit we may expand the scalar product:

$$\mathbf{F}\cdot d\mathbf{r} = F_1 dx + F_2 dy + F_3 dz. \tag{7.2}$$

Now each component of $\mathbf{F}$ is a function of all three variables x, y, z of position on the curve. But the equation of the curve may be taken as two relations between the coordinates, and this means that any two of the coordinates can be expressed as functions of the third. Thus the values of $F_1(x, y, z)$ on the curve C can be expressed as a function of

x only, $F_1(x, y, z) = f_1(x)$ on C; similarly $F_2(x, y, z) = f_2(y)$ on C, and $F_3(x, y, z) = f_3(z)$ on C. The integral then becomes just the sum of three ordinary integrals between different limits in different variables:

$$\int_C \mathbf{F}\cdot d\mathbf{r} = \int_{x_a}^{x_b} f_1(x)dx + \int_{y_a}^{y_b} f_2(y)dy + \int_{z_a}^{z_b} f_3(z)dz. \tag{7.3}$$

In one important special case this detailed procedure turns out to be unnecessary. We saw in Eq. 6.13 that, if a potential function exists, the force is equal to the gradient of a potential, $\mathbf{F} = -\mathbf{grad}\ \Phi$, and the work done becomes

$$\int_C \mathbf{F}\cdot d\mathbf{r} = -\int_C \mathbf{grad}\ \Phi\cdot d\mathbf{r} = -\int_C d\Phi = \Phi_a - \Phi_b, \tag{7.4}$$

from Eq. 6.11; where a and b indicate the beginning and the end of the curve C. The line integral in this case therefore depends only on the end points of the curve, not on the path itself. The work done is equal to the potential difference between the ends and is independent of the path.

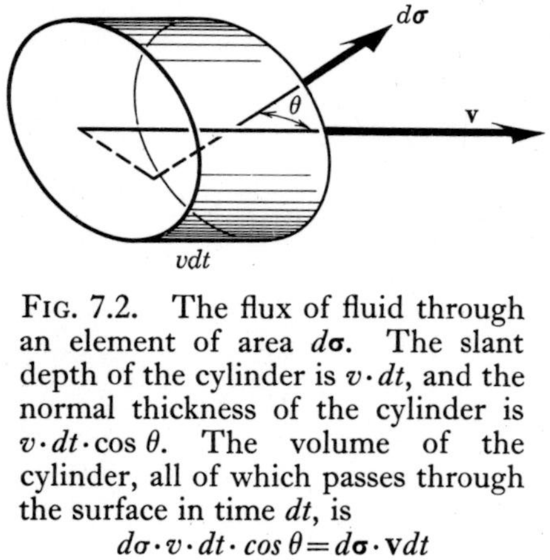

FIG. 7.2. The flux of fluid through an element of area $d\boldsymbol{\sigma}$. The slant depth of the cylinder is $v\cdot dt$, and the normal thickness of the cylinder is $v\cdot dt\cdot \cos\theta$. The volume of the cylinder, all of which passes through the surface in time dt, is

$d\sigma\cdot v\cdot dt\cdot \cos\theta = d\boldsymbol{\sigma}\cdot \mathbf{v}dt$

Consider the velocity vector of any fluid motion as a function of position and time. Imagine some surface drawn through space, and ask what volume of the fluid passes through this surface in some small interval of time dt. First consider any small element $d\boldsymbol{\sigma}$ of the surface, then the volume flux through $d\boldsymbol{\sigma}$ is evidently the scalar product of the velocity $\mathbf{V}$ and the surface element $d\boldsymbol{\sigma}$ multiplied by dt: $\mathbf{V}\cdot d\boldsymbol{\sigma}dt$. The volume flux per unit time is then simply $\mathbf{V}\cdot d\boldsymbol{\sigma}$. The total flux passing through the whole surface is obtained by adding the contributions from every element, and by taking the elements small enough to avoid variations of $\mathbf{V}$ over the area of any one element. The limiting result is defined as the surface integral of $\mathbf{V}$ over the surface and written

$$\iint_S \mathbf{V}\cdot d\boldsymbol{\sigma} = \operatorname{Lim} \sum \mathbf{V}\cdot d\boldsymbol{\sigma}. \tag{7.5}$$

We may expand the scalar product into its component terms:

$$\iint_S \mathbf{V}\cdot d\boldsymbol{\sigma} = \iint V_1(x, y, z)d\sigma_1 + \iint V_2(x, y, z)d\sigma_2 + \iint V_3(x, y, z)d\sigma_3.$$

The surface is determined by one equation between the three variables x, y, z, and this permits any one variable to be expressed in terms of the other two. Noting that $d\sigma_1 = dydz$, $d\sigma_2 = dzdx$, $d\sigma_3 = dxdy$, we then have three ordinary integrals in two dimensions on the right side of the last equation:

$$\iint_S \mathbf{V}\cdot d\boldsymbol{\sigma} = \iint V_1(y, z)dydz + \iint V_2(z, x)dxdz + \iint V_3(x, y)dxdy, \tag{7.6}$$

with appropriate limits.

If the vector $\mathbf{V}$ can be expressed as the curl of another vector function, the surface integral transforms into a line integral and we have Stokes' theorem:

$$\iint_S \nabla\times\mathbf{F}\cdot d\boldsymbol{\sigma} = \int_C \mathbf{F}\cdot d\mathbf{r}, \tag{7.7}$$

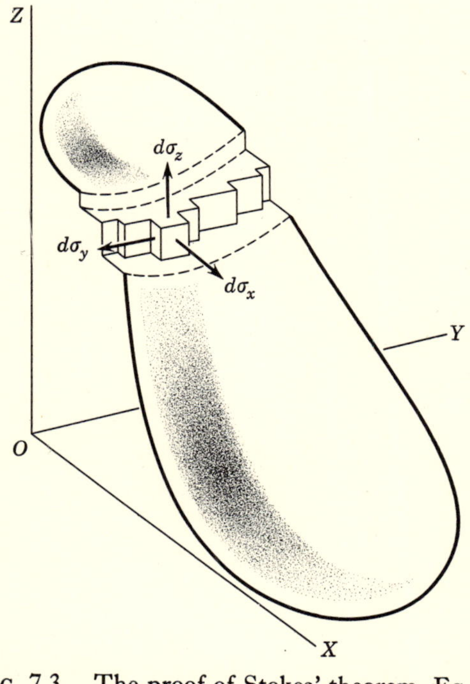

FIG. 7.3. The proof of Stokes' theorem, Eq. 7.7

where C represents the periphery curve of the surface S and the element $d\mathbf{r}$ is directed around C in such a way that the right-hand screw rule would give the direction of $d\boldsymbol{\sigma}$, the surface element in its neighborhood. To demonstrate this theorem we fill the surface S with small elements parallel to one or other of the coordinate planes and evaluate the line integral around any such element, proving that each is equal to the surface integral over that element. Then we note that the sum of all the surface integrals is equal to the total surface integral, at least in the limit when the number of elements is indefinitely great, whereas the sum of the line integrals is equal to the line integral around the outside periphery because the line integrals on adjacent loops cancel each other.

Stokes' theorem leads to an important corollary: If $\mathbf{F}$ is well-behaved and irrotational, $\nabla \times \mathbf{F} = 0$, then there exists a potential function Φ such that $\mathbf{F} = -\mathbf{grad}\,\Phi$. Thus, putting $\nabla \times \mathbf{F}$ zero in the integrand of Eq. 7.7 leads to zero line integral of $\mathbf{F}$ around any closed loop or periphery, and this implies through Eq. 7.4 that $\Phi_a = \Phi_b$ when the curve closes on itself, so there exists a single-valued function Φ defined at every point.

If the surface integral of Eq. 7.5 is taken over a closed surface, *i.e.* one that encloses a definite volume, we have another theorem called *Gauss' theorem*:

$$\iint_S \mathbf{V} \cdot d\boldsymbol{\sigma} = \iiint \nabla \cdot \mathbf{V}\, d\tau, \tag{7.8}$$

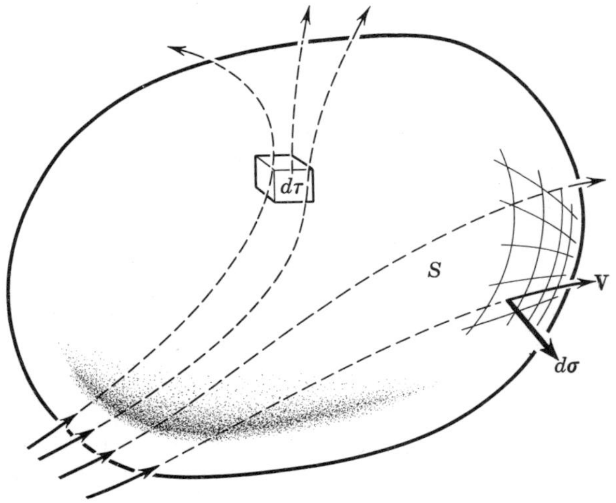

FIG. 7.4. Equation 7.8 illustrated. The lines of flow in the neighborhood of the volume element $d\tau$ have been drawn to represent the existence of a divergence in the element

where the volume integral is an ordinary triple integral of the divergence of the vector function, over the volume τ enclosed by the surface S. This theorem is proved by filling the volume with small elements whose sides are parallel to the three coordinate planes, and proving that Eq. 7.8 is true for each such element; then noting that the surface integrals over adjacent elements cancel on their common surfaces because their respective surface element vectors $d\boldsymbol{\sigma}$ are opposite in direction. Thus the sum of all the surface integrals equals the surface integral over the outermost surface only, as required to prove Gauss' theorem.

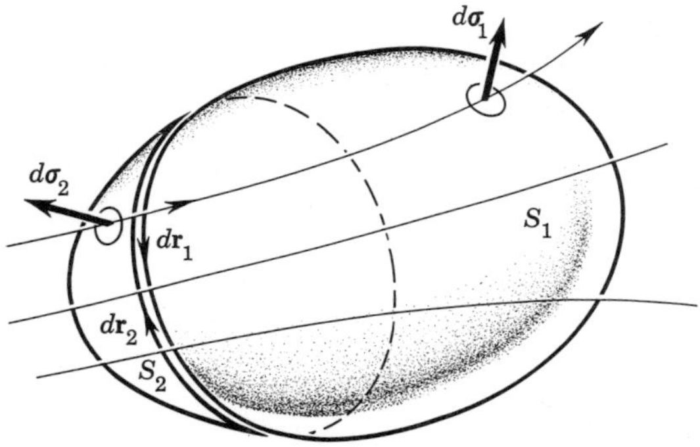

FIG. 7.5. The surface integral $\int \mathbf{v}\cdot d\boldsymbol{\sigma}$ can be split into two parts, one over the surface S_1 and the other over S_2. If the function $\mathbf{v}$ is solenoidal, the two parts are equal and opposite. By reversing the sign of $d\boldsymbol{\sigma}$ and $d\mathbf{r}$ on S_2, we can prove that the surface integral depends only on the periphery

Gauss' theorem also leads to an important corollary: If the vector function $\mathbf{V}$ is solenoidal everywhere in a certain region, *i.e.* if $\nabla\cdot\mathbf{V}=0$, the surface integral of $\mathbf{V}$ depends only on the perimeter of the surface. Thus, putting $\nabla\cdot\mathbf{V}=0$ in Eq. 7.8 yields a surface integral that is zero over the closed surface; then we may divide this closed surface into two parts separated by any arbitrary perimeter, reverse the sign of $d\boldsymbol{\sigma}$ on one of these two parts to conform with the right-hand screw rule, and see at once that to make the total surface integral vanish, as we know it does, the two parts must have equal surface integrals. We see that this is true for any pair of surfaces having a common perimeter; hence the corollary is proved. Applications of these theorems and corollaries appear specially in hydrodynamics.

Exercise I.7

7.1. The vector function representing the gravitational force $\mathbf{f}$ between two masses m_1 and m_2 is

$$\mathbf{f} = -Gm_1m_2\,\mathbf{r}_1/r^2,$$

where G is the gravitational constant, $\mathbf{r}_1$ is the unit vector in the direction of the line of centers of the two masses, and r is the distance between the two centers. Find the line integral $\int \mathbf{f} \cdot d\mathbf{r}$ when one of the masses is held fixed and the other moved from a point A to a point B, the two sets of coordinates relative to the first mass being (r_A, θ_A, ϕ_A) and (r_B, θ_B, ϕ_B).

7.2. Find the surface integral $\iint \mathbf{f} \cdot d\boldsymbol{\sigma}$ in the above problem, over the surface of a sphere of radius a with its center on the fixed mass.

7.3. A vector function $\mathbf{f}$ is defined by

$$\mathbf{f} = (-y/r^2, x/r^2, 0) \quad \text{at the point } (x, y, z),$$

where $r^2 = x^2 + y^2 + z^2$. Find the line integral $\int \mathbf{f} \cdot d\mathbf{r}$ around a closed circular path of radius a, in the $x-y$ plane and with its center at the origin of coordinates.

7.4. A vector function is defined by $\mathbf{f} = (x^n, y^n, z^n)$, where n is some constant, at the point (x, y, z). Find the line integral $\int \mathbf{f} \cdot d\mathbf{r}$ along a straight line joining the origin to the point (1, 1, 0).

7.5. Find the line integral of the same vector function as in Ex. 7.4 around the closed paths: (a) a circle of radius 1 in the $x-y$ plane with its center at the origin, and (b) a circle of radius 1 in the plane $z=1$, with its center on the z-axis.

7.6. Using the same vector function as in Ex. 7.4, find the flux integral $\iint \mathbf{f} \cdot d\boldsymbol{\sigma}$ over the following surfaces:

(a) The unit circle in the $x-y$ plane with center at the origin.
(b) The unit circle in the plane $z=1$ and center on the z-axis.
(c) The unit circle in the $y-z$ plane with center at the origin.
(d) The unit circle in the plane $x=1$ and center on the x-axis.
(e) The unit sphere with center at the origin.

7.7. Check Stokes' theorem for each of the surface integrals in Ex. 7.6.

7.8. Check Gauss' theorem in Ex. 7.6(e). Consider the validity of Gauss' theorem in this problem when the constant n is negative, by actually evaluating both the volume and surface integrals.

I.8 Curvilinear Coordinates. So far our differential operators have been expressed entirely in terms of Cartesian coordinates. In practice, other systems of coordinates are often more convenient. Referring back to Eq. 4.11, we may introduce more general coordinates, which will here be symbolized by q_m, through the definitions

$$q_m = q_m(x_1, x_2, x_3) \quad (m = 1, 2, 3). \tag{8.1}$$

These three functions are equivalent to linear local transformations

$$dq_m = \sum_n (\partial q_m/\partial x_n) dx_n, \tag{8.2}$$

where the coefficients $(\partial q_m/\partial x_n)$ are not restricted to satisfy the conditions for a pure rotation—cf. Eqs. 4.13 and 4.6.

Consider an element of volume in the q-diagram, a q-cell, at some point P bounded by arcs dq_1, dq_2, and dq_3. These increments of q are not in general equal to the lengths of the edges as measured in the

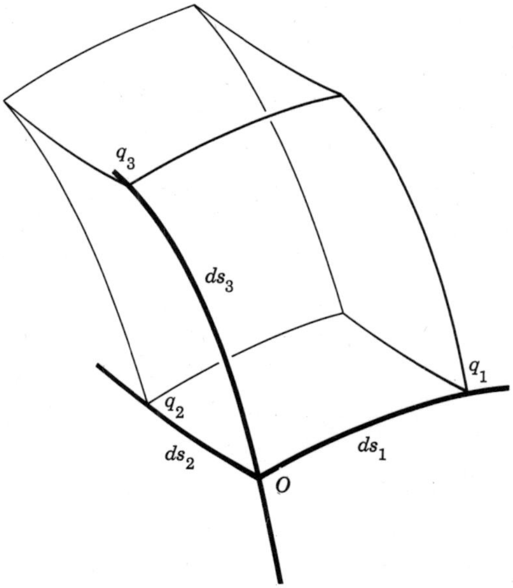

FIG. 8.1. A volume element in orthogonal curvilinear coordinates; its dimensions are $ds_1 = h_1 dq_1$, $ds_2 = h_2 dq_2$, $ds_3 = h_3 dq_3$

Cartesian system. If the curvilinear system is orthogonal, we may rotate the Cartesian system until its axes coincide with the arcs at the point of interest. To do this, let the rotation be written as the linear transformation

$$ds_m = \sum_n l_{mn} dx_n \tag{8.3}$$

and adjust the angle of rotation so that the new Cartesian elements ds_m are parallel to the arcs:

$$ds_m = h_m dq_m. \tag{8.4}$$

Comparing this with Eq. 8.2 fixes the transformation coefficients:

$$l_{mn} = h_m(\partial q_m/\partial x_n). \tag{8.5}$$

The coefficients l_{mn} must satisfy the Eq. 4.6 because they represent a pure rotation, and from this we can show that

$$1/h_m{}^2 = \sum_n (\partial q_m/\partial x_n)^2. \tag{8.6}$$

The length of the arc dq_m is therefore $h_m dq_m$, the functions h_m being determined from Eq. 8.6 and the functions in Eq. 8.1. In simple cases it is easier to write down the lengths of the arcs from geometrical intuition than to prove them analytically in the above fashion. For example, in spherical polar coordinates, we have

$$\left.\begin{array}{lll} dq_1 = dr, & ds_1 = dr, & \text{so that } h_1 = 1 \\ dq_2 = d\theta, & ds_2 = rd\theta, & \text{so that } h_2 = r \\ dq_3 = d\phi, & ds_3 = r\sin\theta\, d\phi & \text{so that } h_3 = r\sin\theta \end{array}\right\}. \tag{8.7}$$

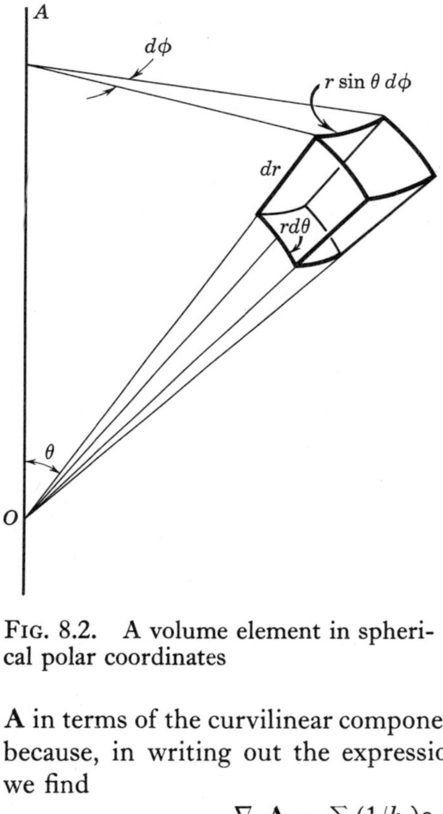

FIG. 8.2. A volume element in spherical polar coordinates

We have taken the rotated Cartesian system s_m tangent to the curvilinear system at the point of interest, so that if the unit vectors $\mathbf{e}_j$ are tangent to the curvilinear system, we have the usual Cartesian expressions for differential operators:

$$\nabla = \sum_j \mathbf{e}_j \partial/\partial s_j. \tag{8.8}$$

or, from Eq. 8.4,

$$\nabla = \sum_j (1/h_j)\mathbf{e}_j \partial/\partial q_j. \tag{8.9}$$

Expressing a vector function $\mathbf{A}$ in terms of the unit vectors $\mathbf{e}_j$:

$$\mathbf{A} = \sum_j A_j \mathbf{e}_j \tag{8.10}$$

and, operating on this with the gradient operator in the form Eq. 8.9, we can try to work out both the divergence and curl of $\mathbf{A}$ in terms of the curvilinear components. Now we run into difficulties because, in writing out the expression of the divergence for example, we find

$$\nabla\cdot\mathbf{A} = \sum_j (1/h_j)\mathbf{e}_j(\partial/\partial q_j)\cdot\sum_k A_k\mathbf{e}_k \tag{8.11}$$

and note that the unit vectors $\mathbf{e}_k$ are no longer constants as they were in the Cartesian case: the curvilinear coordinate lines change directions and so the unit vectors tangent to them also change directions as we move about in space. We therefore need to find the partial derivatives of the unit vectors. This complication can be avoided by appealing to Gauss' theorem, which must evidently be true no matter what coordinate system is being used, as it is an invariant relation. We write this theorem down for the small element of volume under discussion:

$$d\tau = ds_1 ds_2 ds_3 = h_1 h_2 h_3 dq_1 dq_2 dq_3.$$

The divergence integral in Gauss' theorem is then just $\nabla \cdot \mathbf{A} ds_1 ds_2 ds_3$. The surface integral in Gauss' theorem is the flux of $\mathbf{A}$ over the six elements of surface of the volume $d\tau$ and is easily seen to be

$$dq_1 \partial(A_1 ds_2 ds_3)/\partial q_1 + dq_2 \partial(A_2 ds_3 ds_1)/\partial q_2 + dq_3 \partial(A_3 ds_1 ds_2)/\partial q_3.$$

We now use Eq. 8.4 in the derivatives in this expression, noting that the h_j's are variables, while the dq_j's are constant increments for the given volume element. Thus for the surface integral in Gauss' theorem, we get

$$dq_1 dq_2 dq_3 [\partial(A_1 h_2 h_3)/\partial q_1 + \partial(A_2 h_3 h_1)/\partial q_2 + \partial(A_3 h_1 h_2)/\partial q_3].$$

Equating this to the volume or divergence integral written previously and canceling the volume element on both sides, we find at once

$$\nabla \cdot \mathbf{A} = (1/h_1 h_2 h_3)[\partial(A_1 h_2 h_3)/\partial q_1 + \partial(A_2 h_3 h_1)/\partial q_2 + \partial(A_3 h_1 h_2)/\partial q_3], \tag{8.12}$$

which is the desired expression for the divergence in terms of the general curvilinear coordinates.

The curl of $\mathbf{A}$ in curvilinear coordinates can be found by using Stokes' theorem on a small element of area bounded by four curvilinear coordinate lines. The result can be expressed most conveniently in the form of a determinant analogous to Eq. 6.19:

$$\nabla \times \mathbf{A} = (1/h_1 h_2 h_3) \begin{vmatrix} h_1 \mathbf{e}_1 & h_2 \mathbf{e}_2 & h_3 \mathbf{e}_3 \\ \partial/\partial q_1 & \partial/\partial q_2 & \partial/\partial q_3 \\ h_1 A_1 & h_2 A_2 & h_3 A_3 \end{vmatrix}. \tag{8.13}$$

Applying Eq. 8.9 to **grad** Φ and equating this to the $\mathbf{A}$ in Eq. 8.11 give the Laplacian operator in curvilinear coordinates:

$$\nabla \cdot \nabla \Phi = \frac{1}{h_1 h_2 h_3} \left[\frac{\partial}{\partial q_1} \left(\frac{h_2 h_3}{h_1} \frac{\partial \Phi}{\partial q_1} \right) + \frac{\partial}{\partial q_2} \left(\frac{h_3 h_1}{h_2} \frac{\partial \Phi}{\partial q_2} \right) + \frac{\partial}{\partial q_3} \left(\frac{h_1 h_2}{h_3} \frac{\partial \Phi}{\partial q_3} \right) \right]. \tag{8.14}$$

Several interesting and useful theorems concerning transformations to curvilinear coordinates follow.

In general we can solve Eq. 8.1 to give the x's as functions of the q's: formally

$$x_m = x_m(q_1, q_2, q_3), \tag{8.15}$$

which can be differentiated:

$$dx_m = \sum_n (\partial x_m/\partial q_n) dq_n. \tag{8.16}$$

Writing this into the right side of Eq. 8.2, we have

$$dq_m = \sum_{s,n}\sum (\partial q_m/\partial x_n)(\partial x_n/\partial q_s)dq_s. \tag{8.17}$$

This yields the identity

$$\sum_n (\partial q_m/\partial x_n)(\partial x_n/\partial q_s) = \delta_{ms}. \tag{8.18}$$

Comparing this with the rule for matrix multiplication, we see that the matrix whose components are $\partial q_m/\partial x_n$ is the reciprocal of the matrix whose components are $\partial x_m/\partial q_n$.

Let us write $\mathbf{e}_m$ for the unit vectors associated with the increments ds_m. Recalling that Eq. 8.3 is the analog of Eq. 4.3, we write the analog of Eq. 4.1 as follows:

$$\sum_n ds_n \mathbf{e}_n = \sum_m dx_m \mathbf{i}_m$$

and use Eq. 8.4 to give

$$\sum_n h_n dq_n \mathbf{e}_n = \sum_m dx_m \mathbf{i}_m. \tag{8.19}$$

From this we can show that

$$h_n \mathbf{e}_n = \sum_m (\partial x_m/\partial q_n)\mathbf{i}_m. \tag{8.20}$$

To verify this, multiply Eq. 8.20 on both sides by dq_n, sum all values of n, and note that $dx_m = \sum_n (\partial x_m/\partial q_n)dq_n$.

From Eq. 8.20, we may now write

$$h_1 h_2 h_3(\mathbf{e}_1 \cdot \mathbf{e}_2 \times \mathbf{e}_3) = |\partial x_m/\partial q_n|, \tag{8.21}$$

where the right-hand side is the notation for the determinant of all the component partial derivatives. This result follows at once from Eq. 2.13 if we consider Eq. 8.20 as equating $(\partial x_m/\partial q_1)$ to the mth component of the vector $\mathbf{A} \equiv h_1\mathbf{e}_1$, etc. This determinant is called the *Jacobian* of the transformation and is often written

$$|\partial x_m/\partial q_n| = J(x; q). \tag{8.22}$$

Because the **e**-vectors are orthogonal unit vectors, we have $\mathbf{e}_1 \cdot \mathbf{e}_2 \times \mathbf{e}_3 = 1$, and Eqs. 8.21 and 8.22 give

$$J(x; q) = h_1 h_2 h_3. \tag{8.23}$$

From Eq. 8.4 and Eq. 8.23, we therefore have, for an element of volume,

$$d\tau = ds_1 ds_2 ds_3 = J(x; q)dq_1 dq_2 dq_3. \tag{8.24}$$

From Eq. 8.19, we can also write

$$\mathbf{i}_m = \sum_n h_n \mathbf{e}_n (\partial q_n / \partial x_m), \tag{8.25}$$

and therefore

$$\mathbf{i}_1 \cdot \mathbf{i}_2 \times \mathbf{i}_3 = |h_n \partial q_n / \partial x_m| = h_1 h_2 h_3 |\partial q_n / \partial x_m|. \tag{8.26}$$

Since the left side of this is unity, we have, from Eq. 8.23,

$$J(q; x) = |\partial q_n / \partial x_m| = 1/J(x; q). \tag{8.27}$$

Every transformation of coordinates possesses its appropriate Jacobian. Thus, in general, Eq. 8.24 is equivalent to

$$dx_1 dx_2 dx_3 = J(x; q) dq_1 dq_2 dq_3 \tag{8.28}$$

and Eq. 8.27 yields

$$dq_1 dq_2 dq_3 = J(q; x) dx_1 dx_2 dx_3. \tag{8.29}$$

A pure rotation, in which no change in volume occurs, has a Jacobian equal to unity.

The square length of the displacement whose components are dq_1, dq_2, and dq_3 is given by using the Cartesian components of the length given in Eq. 8.16, thus:

$$ds^2 = \sum_p dx_p{}^2 = \sum\sum\sum_{m,n,p} (\partial x_p / \partial q_n)(\partial x_p / \partial q_m) dq_n dq_m$$

or

$$ds^2 = \sum\sum_{m,n} g_{mn} dq_m dq_n, \tag{8.30}$$

where

$$g_{mn} \equiv \sum_p (\partial x_p / \partial q_m)(\partial x_p / \partial q_n). \tag{8.31}$$

The set of components, g_{mn}, is called the *metric tensor* of the q-coordinate system. All intrinsic properties of a coordinate system can be expressed in terms of this tensor. The branch of mathematics known as differential geometry takes off from this point.

So far in this section we have been thinking of a single fixed set of points and assigning different sets of coordinates to this one set of points. Thus Eq. 8.28 is the relation between two different coordinate measures of one particular volume element. It is equally possible, and sometimes very useful, to interpret the mathematical formalism the other way around. Thus we can think of Eq. 8.1 as a set of instructions telling us to move a point from the position (x_1, x_2, x_3) to the position (q_1, q_2, q_3); the two positions being measured in the same coordinate system. With this re-interpretation, all the equations of this section refer to the motions of points in a fixed coordinate system. Eq. 8.29

now tells us that the volume occupied by a given set of points is increased by their motion in the ratio $J(q; x)$:

$$d\tau(q) = J(q; x)d\tau(x). \tag{8.32}$$

Consider the expression

$$\begin{aligned} dq^2 &= \sum_m dq_m dq_m = \sum\sum\sum_{m,n,p} (\partial q_m/\partial x_n)(\partial q_m/\partial x_p)dx_p dx_n \\ &= \sum\sum_{p,n} g^{pn} dx_p dx_n, \end{aligned} \tag{8.33}$$

where

$$g^{pn} \equiv \sum_m (\partial q_m/\partial x_n)(\partial q_m/\partial x_p). \tag{8.34}$$

Note that these components are not the same as Eq. 8.31. Eq. 8.33 indicates that the length of a line element carried with the points in their motion depends not only on its original length, but also on its direction. The general theory of finite strains in continuous media is based on this formalism.

Exercise I.8

8.1. Derive Eq. 8.6 from Eqs. 8.5 and 4.6.

8.2. Prove that $h_m{}^2 = \sum_n (\partial x_n/\partial q_m)^2$.

8.3. Show that, for a pure rotation from one set of Cartesian coordinates to another set of Cartesian coordinates, $\partial q_m/\partial x_n = \partial x_n/\partial q_m$.

8.4. Carefully distinguish, in words, the meanings of the partial derivatives $\partial q_m/\partial x_n$ and $\partial x_n/\partial q_m$, respectively.

8.5. Discuss the comparison between Eqs. 8.18 and 4.6.

8.6. Discuss the distinction between Eqs. 8.31 and 8.34. Use Eq. 8.18 to prove that $\sum_n g_{mn}\, g^{np} = \delta_{mp}$.

8.7. Use Eq. 8.19 to verify Eq. 8.25.

8.8. Taking spherical polar coordinates defined by Eq. 8.1 in the form

$$\begin{aligned} q_1 &= r = (x^2+y^2+z^2)^{1/2}, \\ q_2 &= \theta = \cos^{-1}[z/(x^2+y^2+z^2)^{1/2}], \\ q_3 &= \phi = \tan^{-1}(y/x), \end{aligned}$$

find the scale factors h_m from Eq. 8.6 as functions of r, ϕ, and θ.

8.9. Taking cylindrical polar coordinates defined by Eq. 8.1 in the form

$$q_1 = \rho = (x^2+y^2)^{1/2},\ q_2 = \phi = \tan^{-1}(y/x),\ q_3 = z,$$

use Eq. 8.6 to find the scale factors h_m as functions of ρ, ϕ, and z.

8.10. Prove the following formulas for cylindrical polar coordinates:

(a) $\nabla \cdot \mathbf{A} = (1/r)(\partial/\partial r)(rA_r) + (1/r)\partial A_\phi/\partial\phi + \partial A_z/\partial z.$

(b) $\nabla \times \mathbf{A} = \mathbf{e}_r[(1/r)\partial A_z/\partial\phi - \partial A_\phi/\partial z] + \mathbf{e}_\phi[\partial A_r/\partial z - \partial A_z/\partial r] + \mathbf{e}_z[(1/r)(\partial/\partial r)(rA_\phi) - (1/r)\partial A_r/\partial\phi].$

(c) $\nabla^2 f = (1/r)(\partial/\partial r)(r\partial f/\partial r) + (1/r^2)\partial^2 f/\partial\phi^2 + \partial^2 f/\partial z^2.$

8.11. Prove the following formulas for spherical polar coordinates:

(a) $$\nabla\cdot\mathbf{A} = (1/r^2)(\partial/\partial r)(r^2A_r)+(1/r\sin\theta)(\partial/\partial\theta)(\sin\theta A_\theta) + (1/r\sin\theta)\partial A_\phi/\partial\phi.$$

(b) $$\nabla\times\mathbf{A} = (\mathbf{e}_r/r\sin\theta)[(\partial/\partial\theta)(A_\phi\sin\theta)-\partial A_\theta/\partial\phi] + (\mathbf{e}_\theta/r)[(1/\sin\theta)\partial A_r/\partial\phi-(\partial/\partial r)(rA_\phi)] + (\mathbf{e}_\phi/r)[(\partial/\partial r)(rA_\theta)-\partial A_r/\partial\theta].$$

(c) $$\nabla^2 f = (1/r^2)(\partial/\partial r)(r^2\partial f/\partial r)+(1/r^2\sin\theta)(\partial/\partial\theta)(\sin\theta\,\partial f/\partial\theta) + (1/r\sin\theta)^2\partial^2 f/\partial\phi^2.$$

8.12. Apply the results of Ex. 8.10 to the unit vectors $\mathbf{e}_r$, $\mathbf{e}_\phi$, and $\mathbf{e}_z$ in place of $\mathbf{A}$ to find the derivatives of the unit vectors in cylindrical polar coordinates:

$$\nabla\cdot\mathbf{e}_r = 1/r, \qquad \nabla\cdot\mathbf{e}_\phi = \nabla\cdot\mathbf{e}_z = 0,$$
$$\nabla\times\mathbf{e}_r = \nabla\times\mathbf{e}_z = 0, \qquad \nabla\times\mathbf{e}_\phi = \mathbf{e}_z/r.$$

8.13. Apply Ex. 8.11(a) and (b) to prove the following formulas·

$$\nabla\cdot\mathbf{e}_r = 2/r, \quad \nabla\cdot\mathbf{e}_\theta = (1/r)\cot\theta, \quad \nabla\cdot\mathbf{e}_\phi = 0,$$
$$\nabla\times\mathbf{e}_r = 0, \quad \nabla\times\mathbf{e}_\theta = \mathbf{e}_\phi/r, \quad \nabla\times\mathbf{e}_\phi = (1/r)\cot\theta\,\mathbf{e}_r-\mathbf{e}_\theta/r,$$

all in spherical polar coordinates.

8.14. Use the results of Ex. 8.11 (c) and Ex. 8.13 to prove that in spherical polar coordinates

$$\nabla^2\mathbf{A} = L_r\mathbf{e}_r+L_\theta\mathbf{e}_\theta+L_\phi\mathbf{e}_\phi,$$

where

$$L_r = \nabla^2A_r-2A_r/r^2-(2/r^2\sin\theta)[(\partial/\partial\theta)(\sin\theta\,A_\theta)+\partial A_\phi/\partial\phi],$$
$$L_\theta = \nabla^2A_\theta-(1/r\sin\theta)^2[A_\theta-2\sin^2\theta\,\partial A_r/\partial\theta+2\cos\theta\,\partial A_\phi/\partial\phi],$$
$$L_\phi = \nabla^2A_\phi-(1/r\sin\theta)^2[A_\phi-2\sin\theta\,\partial A_r/\partial\phi-2\cos\theta\,\partial A_\theta/\partial\phi].$$

8.15. Prove from Eq. 8.25 that, for the vectors $\mathbf{e}_n$ to be mutually orthogonal, it is necessary that the transformations of Eq. 8.1 satisfy the relations:

$$\sum_p\{(\partial q_p/\partial x_m)(\partial q_p/\partial x_n)\div\sum_s(\partial q_p/\partial x_s)^2\} = \delta_{mn}.$$

References

The following texts contain useful material for collateral reading with the contents of this chapter.

The Nature of Physical Reality, H. Margenau, McGraw-Hill Book Co., Inc., New York, 1950.

Philosophy of Science, Philipp Frank, Prentice-Hall, Inc., Englewood Cliffs, N.J., 1957.

The Mathematics of Physics and Chemistry, H. Margenau and G. M. Murphy, 2nd Ed., D. Van Nostrand Co., Inc., Princeton, N.J., 1956.

Higher Mathematics for Engineers and Physicists, I. S. and E. S. Sokolnikoff, McGraw-Hill Book Co., Inc., New York, 1941.

Methods of Theoretical Physics, P. M. Morse and H. Feshbach, McGraw-Hill Book Co., Inc., New York, 1953.

Introduction to Theoretical Physics, Leigh Page, D. Van Nostrand Co., Inc., Princeton, N.J., 1952.

Theoretical Physics: Mechanics, F. W. Constant, Addison-Wesley Publishing Co., Inc., Reading, Mass., 1954.

Vector and Tensor Analysis, A. P. Wills, Prentice-Hall, Inc., Englewood Cliffs, N.J., 1931.

A Short Table of Integrals, B. O. Peirce, Ginn and Co., Boston, Mass., 1929.

Tables of Functions, E. Jahnke and F. Emde, Dover Publications, Inc., New York, 1945.

American Institute of Physics Handbook, McGraw-Hill Book Co., Inc., New York, 1957.

Chapter II

THE CONTINUUM THEORY OF MATTER

Chapter II

THE CONTINUUM THEORY OF MATTER

In this chapter we discuss the most classical part of physics, *viz.* those mechanical properties of matter that can be observed without resource to microscopic investigation. To set up this continuum theory of matter we need only the concepts familiar from a study of Newton's laws of motion: space, time, velocity, force, mass or inertia, energy, and the ability of matter to transmit stress. It is fashionable to belittle this continuum theory of matter because it is possible to go so much further with the atomic hypothesis. However, it is worth while to emphasize that the atomic theory of matter demands acceptance of all the same concepts as does the continuum theory, except only the ability of matter to transmit stress; the atomic theory "explains" this ability only by assuming that empty space between atoms can transmit stress. This assumption in turn hinges on the mathematical theory of fields: gravitational, electromagnetic, and nuclear force fields, which again are continuum theories. Merely as preparation for the more difficult task of understanding modern field theory, the study of the continuum theory of matter is therefore pedagogically important; and as a reasonable theory of observed facts, it is actually in a more secure logical position than the atomic theory, in that its basic postulates are intuitively more self-evident. The reason for accepting the atomic theory as superior is that it can embrace a wider range of phenomena. The classical atomic theory of matter will be taken up in Chapter III.

II.9 The Vibrating String. The motion of an elastic string under tension is the prototype of problems in the continuum theory of matter. Let the string be constrained to move in a horizontal plane (to eliminate the effect of gravitation): the x-axis is taken as the undisturbed configuration of the string. The string has a mass ρ per unit length and is under tension S. It is surrounded by a medium that exerts a frictional force R per unit length per unit transverse velocity of the string. The displacement of the string, y, from the x-axis, is assumed to be so small that the tension S is independent of the motion, and the motion of any element of the string is always perpendicular to the x-axis. The equation of motion for any element of length dx at position x under these limitations is then easily shown to be

$$dx \cdot S \cdot \partial^2 y/\partial x^2 = dx \cdot R \cdot \partial y/\partial t + \rho dx \cdot \partial^2 y/\partial t^2, \tag{9.1}$$

where the left side represents the outward force due to the tension, and the right side gives the resistance due to friction and inertia. Provided there is no discontinuity in any of the functions or their derivatives appearing in this equation, we can cancel the element dx from the equation:

$$(S/\rho)\partial^2 y/\partial x^2 - (R/\rho)\partial y/\partial t - \partial^2 y/\partial t^2 = 0, \tag{9.2}$$

valid at all points on the string.

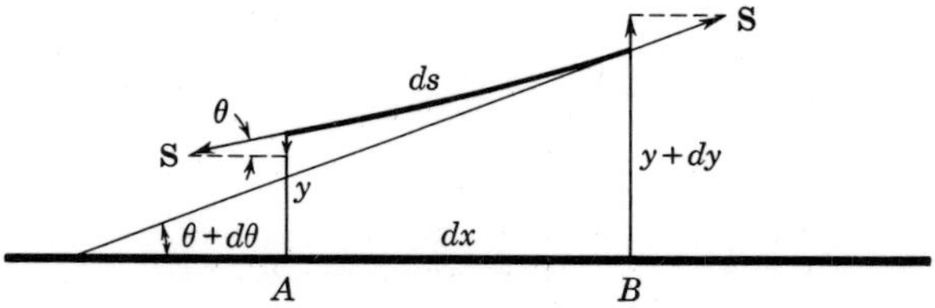

FIG. 9.1. An element of string displaced laterally while under tension. When the displacement is small enough, its length ds is practically equal to dx, and $\tan\theta = dy/dx = \sin\theta$. The net restoring force is $\mathbf{s}\, dx\, d^2y/dx^2$

The solution of this equation by the method of separation of the variables is straightforward. Let

$$y(x, t) = X(x)T(t) \tag{9.3}$$

and substitute this into Eq. 9.2. We find:

$$(S/\rho)d^2X/dx^2 = -k^2X, \quad \text{and} \quad (R/\rho)dT/dt + d^2T/dt^2 = -k^2T, \tag{9.4}$$

where k^2 is an arbitrary constant. Assuming k^2 to be real and positive, we have immediately

$$X(x) = X_0 \sin(2\pi x/\lambda) \quad \text{or} \quad X_0 \cos(2\pi x/\lambda), \tag{9.5}$$

where X_0 is an arbitrary amplitude and

$$1/\lambda = (k/2\pi)\sqrt{\rho/S}. \tag{9.6}$$

We also have

$$T(t) = \exp(-Rt/2\rho)\sin\omega t \quad \text{or} \quad \exp(-Rt/2\rho)\cos\omega t, \tag{9.7}$$

where

$$\omega = (k^2 - R^2/4\rho^2)^{1/2}. \tag{9.8}$$

Linear combinations of these alternative solutions are also solutions, and one linear combination gives the following typical solution:

$$y(x, t) = y_0 \exp(-Rt/2\rho) \sin(2\pi x/\lambda - \omega t), \tag{9.9}$$

which represents a propagated wave motion with phase velocity

$$V = \lambda\omega/2\pi = (\omega/k)\sqrt{S/\rho}, \tag{9.10}$$

multiplied by a time *damping factor* $\exp(-Rt/2\rho)$ that reduces the whole wave amplitude by a factor $1/e$ in a *characteristic time* $\tau = 2\rho/R$.

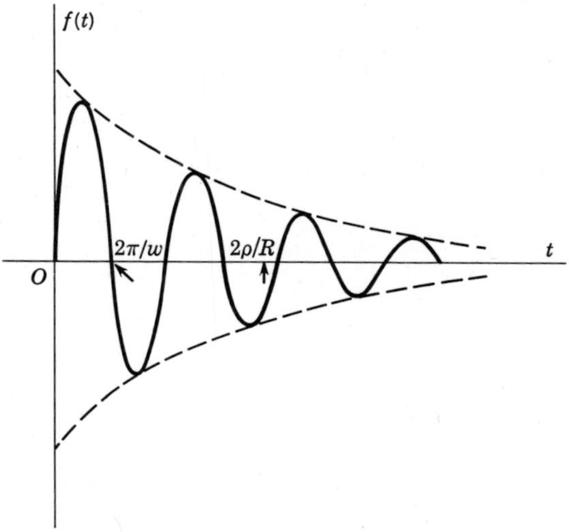

FIG. 9.2. Equation 9.9 at $x=0$

This does not exhaust the possible solutions of Eq. 9.2 because it was assumed that the constant k^2 was real and positive in Eq. 9.4. Actually k^2 was stated there to be quite arbitrary. There exists another physically interesting set of solutions, of the form:

$$y(x, t) = f(x)e^{i\omega t} \tag{9.11}$$

where ω is real, either positive or negative. Writing Eq. 9.11 into Eq. 9.2 yields

$$d^2f/dx^2 + K^2f = 0,$$

where

$$K^2 = \rho\omega^2/S - iR\omega/S. \tag{9.12}$$

This equation is similar to the first of Eqs. 9.4, except that k^2 is now replaced by K^2S/ρ, which is complex. The solution is

$$f = \exp(\pm iKx). \tag{9.13}$$

From Eq. 9.12, we have

$$K = (2\pi/\lambda_0)(1-iR/\omega\rho)^{1/2},$$

where (9.14)

$$\lambda_0 = (2\pi/\omega)\sqrt{S/\rho}.$$

When $R/\omega\rho \ll 1$, this may be expanded in series. Neglecting terms of higher order than $(R/\omega\rho)^2$, we have:

$$K = (2\pi/\lambda_0)[1+\tfrac{1}{2}(R/2\omega\rho)^2 - \tfrac{1}{2}iR/\omega\rho] \quad (9.15)$$

and therefore, in this case, using negative K in the exponent of Eq. 9.13:

$$f(x) = e^{-x/l}\exp(-2\pi ix/\lambda), \quad (9.16)$$

where

$$1/\lambda = (1/\lambda_0)[1+\tfrac{1}{2}(R/2\omega\rho)^2] \quad (9.17)$$

and

$$l = 2\sqrt{\rho S}/R.$$

Combining this with the time factor, as in Eq. 9.11, we have for the complete solution:

$$y(x, t) = y_0 e^{-x/l}\exp[i(\omega t - 2\pi x/\lambda)]. \quad (9.18)$$

This solution represents a wave propagated in the direction of positive x, having an amplitude that diminishes with increasing x, the distance in which the amplitude diminishes by the ratio $1/e$ being $l = 2\sqrt{\rho S}/R$.

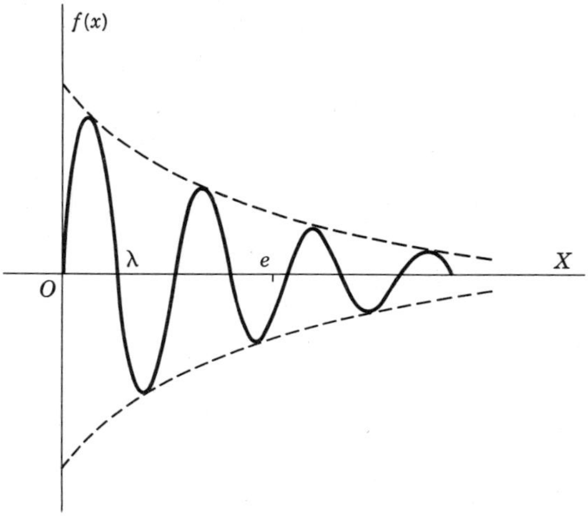

FIG. 9.3. Equation 9.18 at $t=0$. There is an Argand diagram at every point on the x-axis, perpendicular to this axis, and the curve sketched here is a projection of a screw onto the plane of the diagram

The velocity of propagation is

$$V = \omega\lambda/2\pi = V_0/[1+\tfrac{1}{2}(R/2\omega\rho)^2], \quad V_0 = \sqrt{S/\rho}. \tag{9.19}$$

We may use this solution only for values of x to the right of the source of the disturbance. To the left of the source point we must return to Eq. 9.13 and choose the positive sign for K in the exponent, and with this choice we can derive the alternative solution:

$$y(x, t) = y_0 e^{x/l} \exp\,[i(\omega t + 2\pi x/\lambda)], \tag{9.20}$$

which represents a wave propagated in the direction of negative x, and damping with increasing negative values of x. These two solutions together, each starting out from the source at the origin, evidently describe the resulting motion of the string when the point of the string at the origin is constrained to move periodically:

$$x = 0: \qquad y(0, t) = y_0 e^{i\omega t}. \tag{9.21}$$

The proper solution is not continuous across the source point—different solutions are required on either side.

Exercise II.9

9.1. Derive Eq. 9.1 from first principles of mechanics.

9.2. Find the linear combination of the expressions in Eqs. 9.5 and 9.7 required to form the solutions displayed in Eq. 9.9. Prove quite generally that the sum of any two solutions of Eq. 9.2 is also a solution.

9.3. An elastic string is fixed at the two ends, $x=0$ and $x=L$; use these boundary conditions to find the possible wavelengths from Eq. 9.6 and the possible frequencies from Eq. 9.8.

9.4. Discuss the physical meaning of the critical condition $R=2k\rho$ in Eq. 9.8. Discuss also the condition when $R\to 0$.

9.5. Find the velocity of propagation of the wave on the string as a function of frequency.

9.6. Find the group velocity of waves on the string.

9.7. Derive Eq. 9.20 for propagation in the negative direction.

9.8. Note that Eq. 9.11 is a solution of Eq. 9.2 whether ω is a positive or a negative quantity, with the same function $f(x)$ in either case. This ambiguity of the sign of ω carries through to Eqs. 9.18 and 9.19. Form appropriate linear combinations of all these solutions including those with negative values of ω to obtain a real function instead of the exponential form given in the text. Find the wave that has a maximum value at $x=0$ and $t=0$; obtain both the forward and the backward propagating waves.

9.9. The solutions presented in this section have all presumed small damping; consider instead the solutions when damping is critical (see Ex. 9.4). Prove that while no standing waves can exist in this case, propagated waves are still produced no matter how great the damping may be.

9.10. An elastic string is stretched with tension T and surrounded by a material that exerts a restoring lateral force on each element of the string proportional to the lateral displacement of the element. Prove that the differential equation of lateral motion of the string is

$$\partial^2 y/\partial t^2 = (T/\rho)\partial^2 y/\partial x^2 - (K/\rho)y,$$

where the restoring force due to the surrounding material on an element of string of length dx is Kdx per unit lateral displacement, and ρ is the linear density of the string. This form of differential equation is called the *Klein-Gordon equation.*

9.11. Find the solutions of the Klein-Gordon equation, Ex. 9.10, that have the form: $y(x, t) = u(x)e^{i\omega t}$ and show that they represent waves propagated with speed

$$c = \sqrt{T/\rho}/\sqrt{1-K/\rho\omega^2}.$$

Discuss qualitatively the dispersion of these waves.

9.12. If the string in the last two exercises has length L and is clamped at both ends, show that the spectrum of frequencies of its normal modes of motion (standing waves) is $\omega_n = \sqrt{T/\rho}\ \sqrt{K/T+(n\pi/L)^2}$ where n is any non-zero integer.

II.10 Vibrating Membranes.

We consider the vibrations of a flexible membrane stretched under uniform tension S and clamped around some plane periphery of specified shape. Figure 10.1 shows an element $dxdy$ of the membrane displaced a distance u normally from its equilibrium (flat) position; u is a function of time and of position (x, y) in the membrane. The net force in the direction of the normal displacement, due to the tensions acting across the y-edges, is approximately

$$S[(\partial u/\partial x)_{x+dx} - (\partial u/\partial x)_x]dy = S\,\partial^2 u/\partial x^2\ dxdy. \tag{10.1}$$

The approximations involved here require that u be very small and S unchanged by the motion. The net force across the x-edges is similarly

$$S[(\partial u/\partial y)_{y+dy} - (\partial u/\partial y)_y]dx = S\,\partial^2 u/\partial y^2\ dxdy. \tag{10.2}$$

Adding these and equating them to the mass times acceleration of the element leads, after canceling the area $dxdy$ on both sides, to the equation

$$\partial^2 u/\partial t^2 = (S/\rho)\{\partial^2 u/\partial x^2 + \partial^2 u/dy^2\}, \tag{10.3}$$

where ρ is the mass per unit area of the membrane.

Consider first a rectangular membrane clamped at the lines $x=0$, $x=X$, $y=0$, $y=Y$. We shall seek only solutions in which the variables separate, and these we can guess must have the form

$$u(x, y, t) = u_0 e^{\pm i\omega t} e^{\pm ikx} e^{\pm ijy}, \tag{10.4}$$

where ω, k, and j are constants to be found. The ambiguous signs in the exponents are intended to remind us that each factor in Eq. 10.4 represents any linear combination of sine and cosine functions. Substi-

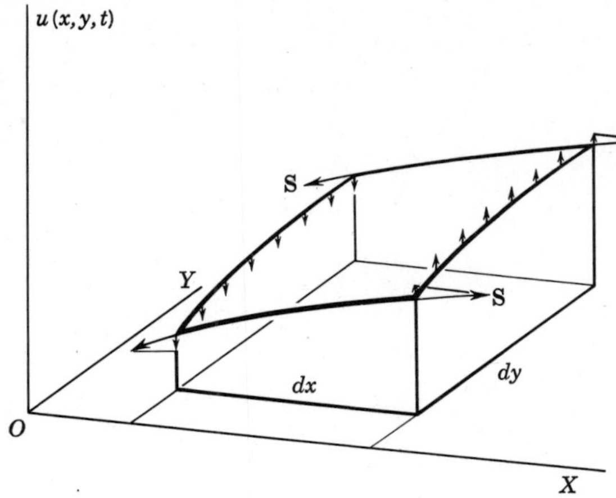

FIG. 10.1. An element of vibrating membrane. The vertical components of force are equal to $\mathbf{s}(\partial u/\partial x)$ evaluated at x on the left edge of the element, and at $x+dx$ on the right edge. Their difference is the restoring force. Similar forces act on the edges parallel to the x-axis, given by $\mathbf{s}(\partial u/\partial y)$

tution of the form Eq. 10.4 into Eq. 10.3 yields the following relation between the constants:

$$\omega^2 = (S/\rho)(k^2+j^2). \qquad (10.5)$$

Further progress requires a discussion of the boundary conditions, at the periphery of the membrane, which state that u must be permanently zero there. This suggests the following solution:

$$u(x, y, t) = u_0 \exp(\pm i\omega_{nm}t) \sin(n\pi x/X) \sin(m\pi y/Y) \qquad (10.6)$$

where, from Eq. 10.5,

$$\omega_{nm}{}^2 = (S/\rho)(n^2/X^2+m^2/Y^2)\pi^2, \qquad (10.7)$$

n and m being non-zero integers to ensure the vanishing of u at the boundaries X and Y. The frequencies ω_{nm} are the normal modes of vibration of the membrane. There are $n-1$ nodes on the x-side and $m-1$ nodes on the y-side.

In general, the actual motion of the membrane will not coincide with any one of these normal modes. Instead we may build up the motion by forming linear combinations of the normal modes, in the manner of Fourier series. Let us examine the solution formed as follows:

$$u(x, y, t) = \sum_{n,m}\sum A_{nm} \cos(\omega_{nm}t) \sin(n\pi x/X) \sin(m\pi y/Y), \qquad (10.8)$$

where A_{nm} are constant amplitudes. We try to fit this to some initial shape of the membrane at time $t=0$, for example:

$$u(x, y, 0) = f(x, y) = \sum_{n,m}\sum A_{nm} \sin(n\pi x/X) \sin(m\pi y/Y). \quad (10.9)$$

Multiply both sides of this equation by the factor

$$\sin(n'\pi x/X) \sin(m'\pi y/Y),$$

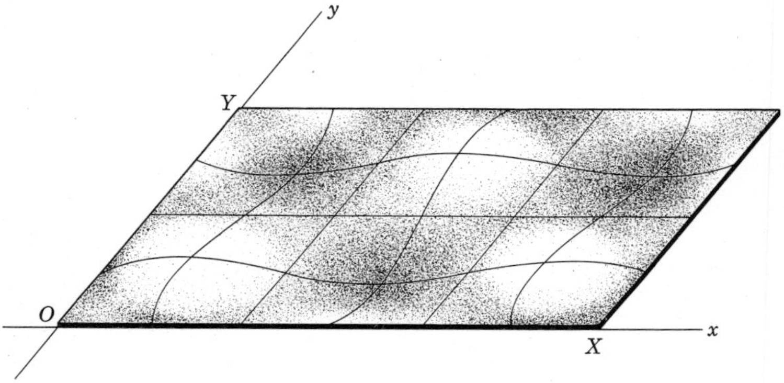

FIG. 10.2. A normal mode of a vibrating rectangular membrane, Eq. 10.7, with $n=3$ and $m=2$

where n' and m' are integers, and integrate the resulting product with respect to x and y over the area of the membrane:

$$\int_0^X \int_0^Y f(x, y) \sin(n'\pi x/X) \sin(m'\pi y/Y) dx dy$$

$$= \sum_{n,m}\sum A_{nm} \int_0^X \sin(n'\pi x/X) \sin(n\pi x/X) dx$$

$$\int_0^Y \sin(m'\pi y/Y) \sin(m\pi y/Y) dy$$

$$= \tfrac{1}{4} XYA_{n'm'}, \quad (10.10)$$

an equation that gives the amplitude $A_{n'm'}$ for any normal mode n', m' when we know the function $f(x, y)$ for the initial shape at $t = 0$.

Consider now a membrane clamped in a circular periphery of radius a. Evidently the Cartesian form of equation used for the rectangular membrane is awkward for a discussion of circular boundaries, and it is more convenient to transform the equation to plane polar coordinates. Thus, instead of Eq. 10.3, we now have to solve (cf. Ex. 8.10):

$$\partial^2 u/\partial t^2 = (S/\rho)\{(1/r)(\partial/\partial r)(r\partial u/\partial r) + (1/r^2)\partial^2 u/\partial\theta^2\}. \quad (10.11)$$

Again we look only for solutions whose variables separate, and so represent standing waves only, as we shall verify later. Thus let

$$u(r, \theta, t) = R(r)F(\theta)T(t). \tag{10.12}$$

Substituting this into Eq. 10.11 and dividing by RFT, we find

$$(1/Rr)(d/dr)(rdR/dr)+(1/r^2F)d^2F/d\theta^2 = (\rho/ST)d^2T/dt^2. \tag{10.13}$$

The left-hand side of this equation is independent of t, because all the functions and operations on this side depend only on r and θ; while the right side is independent of r and θ. Each side must therefore be equal to the same constant, which we shall write $-\omega^2\rho/S$ so that we have the two independent equations derived from Eq. 10.13:

$$d^2T/dt^2 = -\omega^2T \tag{10.14}$$

and

$$(r/R)(d/dr)(rdR/dr)+(1/F)d^2F/d\theta^2+\rho r^2\omega^2/S = 0, \tag{10.15}$$

with ω an arbitrary constant. The general solution of Eq. 10.14 is

$$T(t) = A \exp(\pm i\omega t). \tag{10.16}$$

Eq. 10.15 separates again into independent parts depending, respectively, on r and θ alone; the part depending on θ is

$$d^2F/d\theta^2 = -m^2F, \tag{10.17}$$

where m^2 is the arbitrary separation constant. The solution of this is

$$F(\theta) = C \exp(\pm im\theta). \tag{10.18}$$

Now the function u, and therefore also the function F, is required by the physics of the problem to be a single valued function of the angle θ, so $F(\theta)$ must be periodic in θ with period 2π. This requires that the constant m be restricted to integral values in Eq. 10.18:

$$m = 0, 1, 2, \cdots . \tag{10.19}$$

Putting Eq. 10.18 into Eq. 10.15, we find for the part of that equation depending only on r:

$$(1/r)(d/dr)(rdR/dr)+(\rho\omega^2/S-m^2/r^2)R = 0. \tag{10.20}$$

To reduce this we make the following transformation:

$$x = \omega r\sqrt{\rho/S}, \quad J(x) = R(r) \tag{10.21}$$

and obtain the *Bessel equation*:

$$(1/x)(d/dx)(xdJ/dx)+(1-m^2/x^2)J = 0. \tag{10.22}$$

Solving this equation by series expansion, one finds the *Bessel functions of the first kind*, represented by the series:

$$J_m(x) = \sum_{k=0}^{\infty} \frac{(-1)^k x^{m+2k}}{2^{m+2k} k!(m+k)!}. \tag{10.23}$$

There is another set of series solutions, called *Bessel functions of the second kind*, which have an infinity at the origin. These solutions cannot occur in the problem under discussion, but would have to be included in a discussion of a ring-shaped (annular) membrane in which the origin is not within the boundary.

Gathering together the various steps we have been through in setting up the solution of Eq. 10.11, we have finally:

$$u(r, \theta, t) = A \exp(\pm i\omega t) \exp(\pm im\theta) J_m(r\omega\sqrt{\rho/S}), \tag{10.24}$$

where A is an arbitrary amplitude. The only boundary condition to be imposed by the periphery is that this function be zero at all times when $r=a$. This can be satisfied only by making the Bessel function fit with one of its zeros at $r=a$:

$$J_m(a\omega\sqrt{\rho/S}) = 0. \tag{10.25}$$

Actually the zeros of the Bessel functions are tabulated in most mathematical handbooks. We give the following samples here simply as a basis for a few numerical exercises. The boundary condition requires simply that $a\omega\sqrt{\rho/S} = x_{mk}$ for some value of m and k, and this restricts the frequencies to have one of the following values:

$$w_{mk} = (x_{mk}/a)\sqrt{S/\rho}, \tag{10.26}$$

which is then the spectrum of the normal modes.

TABLE OF VALUES OF x_{mk} where $J_m(x_{mk}) = 0$

k	$m = 0$	$m = 1$	$m = 2$	$m = 3$	$m = 4$
1	2.405	0	0	0	0
2	5.520	3.832	5.136	6.380	7.588
3	8.653	7.015	8.417	9.761	11.065
4	11.792	10.174	11.620	13.015	14.372

Imagine the membrane to start off at time $t=0$ from a flat shape, having an initial velocity $\partial u/\partial t$, some arbitrarily given function of position:

$$\partial u(r, \theta, t)/\partial t_{t=0} = G(r, \theta); \quad u(r, \theta, 0) = 0. \tag{10.27}$$

We try to form this initial velocity pattern out of linear combinations of the normal modes in Eq. 10.24. In so doing we must write $\sin \omega_{mk}t$

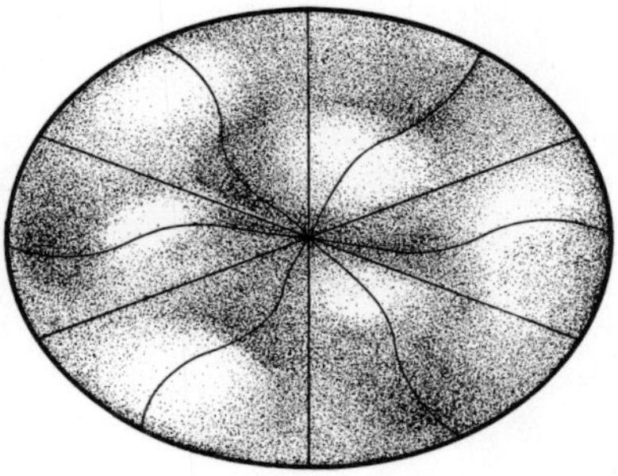

FIG. 10.3. A vibrating circular membrane with $m=3$ and $k=3$

in place of $\exp(i\omega t)$ to satisfy the second of Eqs. 10.27. Therefore, we have

$$u(r, \theta, t) = \sum_{m,k}\sum \sin \omega_{mk}t\,[B_{km} \cos m\theta + D_{km} \sin m\theta]J_m(r\omega_{mk}\sqrt{\rho/S}), \tag{10.28}$$

where B_{km} and D_{km} are arbitrary constants to be adjusted to satisfy the first of Eqs. (10.27). Thus we wish to differentiate Eq. 10.28 with respect to time:

$$G(r, \theta) = \sum_{m,k}\sum \omega_{mk}(B_{km} \cos m\theta + D_{km} \sin m\theta)J_m(r\omega_{mk}\sqrt{\rho/S}). \tag{10.29}$$

We can solve this equation for the constants B_{km} and D_{km} by making use of the following orthogonality relations:

$$\int_0^a J_m(r\omega_{mj}\sqrt{\rho/S})J_m(r\omega_{mk}\sqrt{\rho/S})rdr = \delta_{jk}C_{mj}, \tag{10.30}$$

where

$$C_{mj} = -\tfrac{1}{2}a^2J_{m-1}(a\omega_{mj}\sqrt{\rho/S})J_{m+1}(a\omega_{mj}\sqrt{\rho/S}).$$

In fact, multiplying Eq. 10.29 by $\cos m'\theta J_{m'}(r\omega_{m'k'}\sqrt{\rho/S})$ and integrating over the area of the membrane, we find

$$B_{k'm'} = (1/\pi\omega_{m'k'}C_{m'k'}) \int_0^{2\pi}\int_0^a G(r, \theta) \cos m'\theta\; J_{m'}(r\omega_{m'k'}\sqrt{\rho/S})rdrd\theta, \tag{10.31}$$

which can in principle be evaluated for any desired function $G(r, \theta)$.

Again multiplying Eq. 10.29 by $\sin m'\theta J_{m'}(r\omega_{m'k'}\sqrt{\rho/S})$ and integrating over the area of the membrane, we find

$$D_{k'm'} = (1/\pi\omega_{m'k'}C_{m'k'}) \int_0^{2\pi}\int_0^a G(r, \theta) \sin m'\theta \, J_{m'}(r\omega_{m'k'}\sqrt{\rho/S}) r dr d\theta. \tag{10.32}$$

In discussing the one-dimensional motion of the vibrating string, we were able to find simple solutions representing propagated wave motion. In fact, some elementary discussions describe standing waves in a string as the resultant of two oppositely progressing waves—the incident and the reflected wave. In the one-dimensional case, the propagated wave of small enough amplitude is such that the shape of the wave is unchanged during transmission. Propagated wave solutions also exist for the two-dimensional problem like the flexible membrane discussed in the present section, but their mathematical form is quite complicated. An essential feature of the propagated wave in two dimensions is that its form does change during propagation, leaving a developing wake behind it. For this reason it is not possible to construct standing waves from the superposition of two propagating waves.

Exercise II.10

10.1. Prove that, if one is free to design the dimensions of a rectangular membrane, the lowest two frequencies of its normal modes, ω_1 and ω_2, necessarily satisfy the restriction $\omega_1 < \omega_2 < 2\omega_1$.

10.2. Use Eq. 10.7 to design a rectangular membrane in which $S = 1.00 \times 10^6$ dynes per cm, and $\rho = 1.00$ g per cm^2 so that its lowest two frequencies are 300 and 400 cycles per sec. Find its next two frequencies. Sketch the nodes for each mode of vibration.

10.3. Prove that, if one is free to design a circular membrane, its entire spectrum of normal modes is determined by its lowest frequency, and there is no freedom of choice of two frequencies as in a rectangular one.

10.4. Use Eq. 10.26 to design a circular membrane in which $S = 1.00 \times 10^6$ dynes per cm, and $\rho = 1.00$ g per cm^2, its lowest frequency being 300 cycles per sec. Find the next three frequencies in its spectrum and sketch the nodes for each.

10.5. Consider an elastic beam whose Young's modulus is Y, with a uniform cross section having a radius of gyration K; let the beam have a uniform mass ρ per unit length. Let it be displaced laterally a distance y at position x from one end. Prove that the bending moment, *i.e.* the torque exerted across a section at position x, is

$$M = YK^2\partial^2 y/\partial x^2.$$

Considering any small element of the beam, show that its equation of motion is

$$\partial^2 y/\partial t^2 = K^2(Y/\rho)\partial^4 y/\partial x^4,$$

where gravitation has been neglected; taking boundary conditions

$$x = 0,\ y = y'' = 0; \qquad x = L,\ y = y'' = 0,$$

show that solutions of the form

$$y = u(x) \sin \omega_n t$$

exist with a spectrum of frequencies given by

$$\omega_n = K(Y/\rho)^{1/2} (n\pi/L)^2,$$

where n is any integer.

10.6. A stretched membrane is embedded in a medium that exerts a lateral restoring force on each element of the membrane, proportional to the area of the element and to the lateral displacement. Show that the equation of motion for lateral displacement is the Klein-Gordon Equation:

$$\partial^2 u/\partial t^2 = (S/\rho)(\partial^2 u/\partial x^2 + \partial^2 u/\partial y^2) - (K/\rho)u,$$

where S is the tension in the membrane, $Kdxdy$ is the restoring force on an element $dxdy$ per unit lateral displacement, ρ the density per unit area of the membrane, and x, y the position in the membrane.

10.7. Show that the equation of Ex. 10.6 has solutions of the form of Eq. 10.4 with $\omega^2 = (S/\rho)(k^2 + j^2) + K/\rho$.

10.8. If the membrane of Ex. 10.6 is clamped in a rectangle of sides X and Y, show that its normal modes have frequencies ω_{nm} given by

$$\omega_{nm}{}^2 = (S/\rho)[(n\pi/X)^2 + (m\pi/Y)^2] + K/\rho,$$

where n and m are any two non-zero integers.

10.9. If the membrane of Ex. 10.6 is clamped in a circular frame of radius a, show that its normal modes have frequencies

$$\omega_{mk} = (S/\rho)^{1/2} \sqrt{(x_{mk}/a)^2 + K/S},$$

where $m = 0, 1, 2, \cdots$ and $k = 1, 2, 3, \cdots$; x_{mk} are the zeros of the Bessel function $J_m(x)$.

II.11 Description of Strains in Continuous Media.

Consider two particles of the medium at neighboring points A and B when the medium is unstrained. Any strain of the medium moves the particles, that at A to A', and that at B to B'. Let $d\mathbf{r}$ be the vector difference between the positions of A and B, and let $\mathbf{u}$ be the vector AA', $\mathbf{u} + d\mathbf{u}$ the vector BB'. The *relative strain* between A and B is defined as the vector $d\mathbf{u}$.

The motion of A due to the strain, *i.e.* the vector $\mathbf{u}$, may be expressed as a function of the initial position of A. This means that each component of the vector $\mathbf{u}$ is a function of position:

$$u_j = u_j(x_1, x_2, x_3). \qquad (11.1)$$

To each of these functions we may apply the method of Eq. 6.11 and write

$$du_j = d\mathbf{r}\cdot\nabla u_j \; (j = 1, 2, 3). \tag{11.2}$$

These three equations can be combined by multiplying each by the corresponding unit vector $\mathbf{i}_j$ and adding the equations together. The result is

$$d\mathbf{u} = d\mathbf{r}\cdot\nabla\mathbf{u}, \tag{11.3}$$

where $\nabla\mathbf{u}$ is a dyadic expression like that in Eq. 6.24.

To see the physical meaning of the components of the dyadic $\nabla\mathbf{u}$ we consider some particularly simple strains and evaluate the corresponding dyadics explicitly.

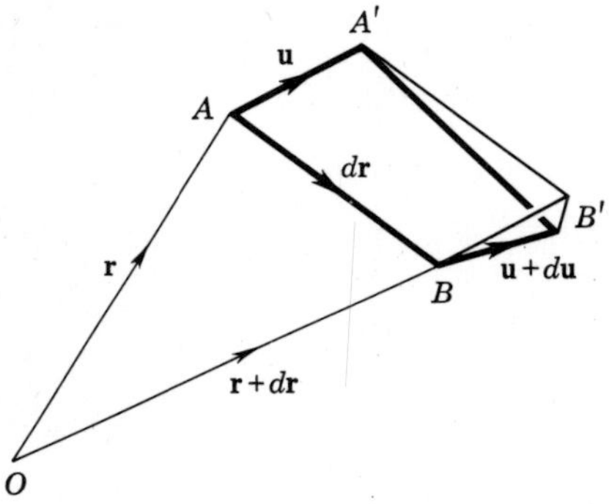

FIG. 11.1. The symbols used in describing strains

(A) A simple elongation in the x_1-direction, with no other change.

In this case the vector $\mathbf{u}$ is everywhere along the x_1-direction, so that $d\mathbf{u}$ is also everywhere along the x_1-direction:

$$d\mathbf{u} = du_1\mathbf{i}_1. \tag{11.4}$$

Moreover, the elongation is the same at all points in any plane perpendicular to the x_1-direction, so the partial derivatives with respect to x_2 and x_3 are everywhere zero. The only surviving term in the dyadic therefore comes from the derivative $\partial u_1/\partial x_1$, and in fact the complete dyadic in this case is

$$\nabla\mathbf{u} = \begin{matrix} \partial u_1/\partial x_1 & 0 & 0 \\ 0 & 0 & 0 \\ 0 & 0 & 0 \end{matrix} \tag{11.5}$$

Similarly, if the elongation were in the x_2-direction, or in the x_3-direction, we would find the corresponding diagonal element present in the dyadic. Note that

$$\partial u_1/\partial x_1 = (L-L_0)/L_0, \tag{11.6}$$

which is the *relative elongation* due to the strain.

(B) A shear in the x_2-direction in the x_1-x_2 plane.

In this case the motion is in the x_2-direction so the only non-zero component of $\mathbf{u}$ is u_2, and the motion is independent of x_3 and of x_2.

The only derivative that does not vanish is therefore $\partial u_2/\partial x_1$, which is then the only non-zero element of the dyadic $\nabla\mathbf{u}$ describing the strain. In general, we infer that the element $\partial u_j/\partial x_k$ of the general dyadic represents a motion in the j-k plane of an element dx_k toward the j-direction through an angle whose tangent is $\partial u_j/\partial x_k$. When this angle is sufficiently small, we may equate the tangent to the angle in radians; thus, when they are small, the non-diagonal elements of the dyadic $\nabla\mathbf{u}$ are the shear angles.

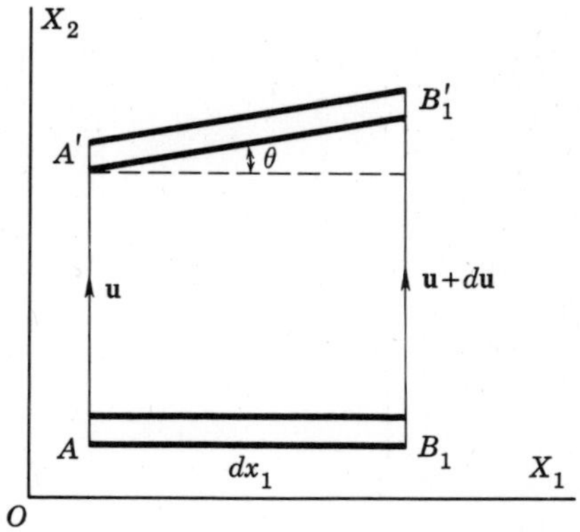

FIG. 11.2. A combined shear and rotation. The shear is $\frac{1}{2}\tan\theta = \frac{1}{2}\partial u_2/\partial x_1 = \frac{1}{2}\theta$; the rotation is $\frac{1}{2}\theta$

When both shears $\partial u_2/\partial x_1$ and $\partial u_1/\partial x_2$ are present, it is conventional to take the sum of the two angles $\partial u_2/\partial x_1 + \partial u_1/\partial x_2$ as the angle of the shear. The components of the shear dyadic are thus one-half the *shear angles*.

An important observed feature of small strains in actual materials is their simple additivity, and this feature permits us to use the matrix algebra set up in the previous chapter. Thus the complete dyadic $\nabla\mathbf{u}$ represents the combined effect of all the strains represented by the individual components: its matrix is the matrix sum of all the matrices like Eq. 11.5 with the various components in their appropriate places.

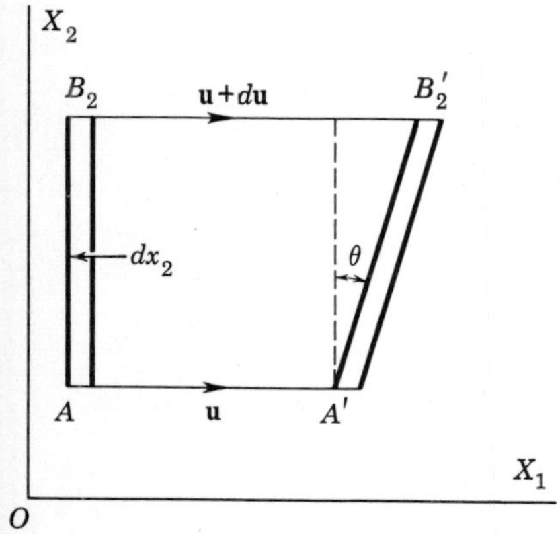

FIG. 11.3. A combined shear and rotation. The shear is $\frac{1}{2}\tan\theta = \frac{1}{2}\partial u_1/\partial x_2 = \frac{1}{2}\theta$; the rotation is $-\frac{1}{2}\theta$

The most general matrix $\nabla\mathbf{u}$ that we can set up includes terms that would not ordinarily be called strains at all, but represent rather mere rotations without distortion or strain of the material. For example, if it should happen that $\partial u_2/\partial x_1 = -\partial u_1/\partial x_2$ so that the matrix is anti-symmetrical, the angle of shear of dx_1 toward the 2-direction is opposite to that of dx_2 toward the 1-direction, and the whole material is simply rotated in the 1-2 plane. To separate the rotation from the strain, it is convenient to introduce the transpose of the matrix $\nabla\mathbf{u}$ and form the two matrices, (*i*) the symmetrical part and (*ii*) the antisymmetrical part, thus:

$$\Phi = \tfrac{1}{2}(\nabla\mathbf{u}+\widetilde{\nabla\mathbf{u}}) \quad \text{and} \quad \Theta = \tfrac{1}{2}(\nabla\mathbf{u}-\widetilde{\nabla\mathbf{u}}). \tag{11.7}$$

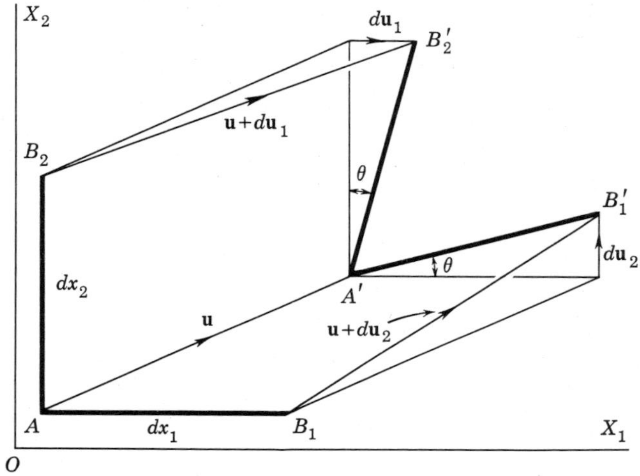

FIG. 11.4. A shear without rotation. The shear is
$\tfrac{1}{2}(\partial u_1/\partial x_2+\partial u_2/\partial x_1) = \tfrac{1}{2}\tan\theta+\tfrac{1}{2}\tan\theta = \theta$

The matrix $\nabla\mathbf{u}$ is the sum of these two:

$$\nabla\mathbf{u} = \Phi+\Theta. \tag{11.8}$$

Here Φ is symmetrical and represents pure strain, whereas Θ is antisymmetrical and represents a rotation. The explicit components of Φ are

$$\begin{matrix} \partial u_1/\partial x_1 & \tfrac{1}{2}(\partial u_1/\partial x_2+\partial u_2/\partial x_1) & \tfrac{1}{2}(\partial u_1/\partial x_3+\partial u_3/\partial x_1) \\ \tfrac{1}{2}(\partial u_2/\partial x_1+\partial u_1/\partial x_2) & \partial u_2/\partial x_2 & \tfrac{1}{2}(\partial u_2/\partial x_3+\partial u_3/\partial x_2). \\ \tfrac{1}{2}(\partial u_3/\partial x_1+\partial u_1/\partial x_3) & \tfrac{1}{2}(\partial u_3/\partial x_2+\partial u_2/\partial x_3) & \partial u_3/\partial x_3 \end{matrix} \tag{11.9}$$

The relative strain vector due to this pure strain dyadic is

$$d\mathbf{u}_{\text{strain}} = d\mathbf{r}\cdot\Phi, \tag{11.10}$$

in place of Eq. 11.3.

Because the matrix Φ is symmetrical it is possible to find a coordinate system relative to which it is in diagonal form. The method of diagonalization has already been given in Eq. 5.19. This fact means that any pure strain can be reduced by a suitable choice of coordinates to a sum of three pure elongations (or compressions). Suppose that we have found the diagonal form and write out the dyadic as follows:

$$\Phi = (\partial u_1/\partial x_1)\mathbf{i}_1\mathbf{i}_1 + (\partial u_2/\partial x_2)\mathbf{i}_2\mathbf{i}_2 + (\partial u_3/\partial x_3)\mathbf{i}_3\mathbf{i}_3. \tag{11.11}$$

The sum of these three components (omitting the dyads) equals the invariant sum of the diagonal elements of the matrix in any coordinate system and is called its *Trace*, so

$$\text{Trace}\,\Phi = \nabla\cdot\mathbf{u}. \tag{11.12}$$

We can see the physical meaning of this invariant quantity as follows: consider any small element of volume before strain $\delta V = \delta x_1\delta x_2\delta x_3$ and compare it with the new volume of the same material after strain:

$$\delta V' = (\delta x_1 + du_1)(\delta x_2 + du_2)(\delta x_3 + du_3).$$

Neglecting quantities of second or higher order in the elongations du_j, we have:

$$\begin{aligned}\delta V' - \delta V &= du_1\delta x_2\delta x_3 + du_2\delta x_3\delta x_1 + du_3\delta x_1\delta x_2\\ &= (\partial u_1/\partial x_1 + \partial u_2/\partial x_2 + \partial u_3/\partial x_3)\delta x_1\delta x_2\delta x_3,\end{aligned}$$

so that finally:

$$(\delta V' - \delta V)/\delta V = \nabla\cdot\mathbf{u}. \tag{11.13}$$

The divergence of the strain vector is thus the relative volume dilation due to the strain. Incidentally, it vanishes if the strain is a pure shear.

While Eq. 11.10 gives the relative displacements due to pure strain, the complete relative displacement vector of Eq. 11.2 includes a part due to rotation, *viz.*

$$d\mathbf{u}_{\text{rot}} = d\mathbf{r}\cdot\Theta. \tag{11.14}$$

Explicitly the components of Θ are

$$\begin{matrix} 0 & \frac{1}{2}(\partial u_2/\partial x_1 - \partial u_1/\partial x_2) & \frac{1}{2}(\partial u_3/\partial x_1 - \partial u_1/\partial x_3)\\ \frac{1}{2}(\partial u_1/\partial x_2 - \partial u_2/\partial x_1) & 0 & \frac{1}{2}(\partial u_3/\partial x_2 - \partial u_2/\partial x_3).\\ \frac{1}{2}(\partial u_1/\partial x_3 - \partial u_3/\partial x_1) & \frac{1}{2}(\partial u_2/\partial x_3 - \partial u_3/\partial x_2) & 0 \end{matrix} \tag{11.15}$$

If for convenience we write $\boldsymbol{\alpha}$ for the vector $\frac{1}{2}\nabla\times\mathbf{u}$, the matrix Θ is

$$\Theta = \begin{pmatrix} 0 & \alpha_3 & -\alpha_2\\ -\alpha_3 & 0 & \alpha_1\\ \alpha_2 & -\alpha_1 & 0 \end{pmatrix} \quad \text{where} \quad \boldsymbol{\alpha} = \tfrac{1}{2}\nabla\times\mathbf{u} \tag{11.16}$$

It is easy to prove that the displacement due to the rotation is also

$$d\mathbf{u}_{\text{rot}} = d\mathbf{r}\cdot\Theta = -d\mathbf{r}\times\boldsymbol{\alpha}. \tag{11.17}$$

This expression shows that $\boldsymbol{\alpha}$ is a vector directed along the axis of the rotation, having a length equal to the magnitude of the angle of the rotation. If the rotation were to occur in unit time, $\boldsymbol{\alpha}$ would then be the vector angular velocity of the rotation and $\mathbf{u}$ would be the linear velocity of the particles; this re-interpretation of Eq. 11.16 is often useful.

The exact theory of finite strains is based on the techniques of differential geometry mentioned at the end of §8. Thus Eq. 8.32 gives the exact ratio between volumes before and after strain. In the notation of the present section, it is clear that the new coordinates q of Eq. 8.32 are now given by

$$q_m = x_m + u_m, \tag{11.18}$$

so the Jacobian of the motion is the determinant:

$$J(q;x) = |\partial(x_m+u_m)/\partial x_n|$$
$$= \begin{vmatrix} 1+\partial u_1/\partial x_1 & \partial u_2/\partial x_1 & \partial u_3/\partial x_1 \\ \partial u_1/\partial x_2 & 1+\partial u_2/\partial x_2 & \partial u_3/\partial x_2 \\ \partial u_1/\partial x_3 & \partial u_2/\partial x_3 & 1+\partial u_3/\partial x_3 \end{vmatrix}. \tag{11.19}$$

If the coordinate system is rotated until the strain matrix is in its normal form, the nondiagonal elements vanish, and we have for this determinant simply the product of its diagonal elements:

$$J(q;x) = (1+\partial u_1/\partial x_1)(1+\partial u_2/\partial x_2)(1+\partial u_3/\partial x_3). \tag{11.20}$$

It is now easy to show that, if one adopts the same approximations as in the theory of small strains, the above result reduces to the same as Eq. 11.13. Thus we have on this approximation

$$J(q;x) = 1+\nabla\cdot\mathbf{u}+\text{second- and higher-order terms}. \tag{11.21}$$

Then, from Eq. 8.32, $J(q;x) = \delta V(q)/\delta V(x)$, we see at once that

$$\nabla\cdot\mathbf{u} = [\delta V(q)-\delta V(x)]/\delta V(x), \tag{11.22}$$

in agreement with Eq. 11.13.

Exercise II.11

11.1. Find, and describe in words, all the individual component elongations, shears, and rotations represented by the matrix of strains:

$$(1/100)\cdot\begin{pmatrix} 1 & 4 & 2 \\ 2 & 0 & 3 \\ -2 & 3 & -1 \end{pmatrix}.$$

What is the volume dilation in this strain? Diagonalize the pure shear part of this strain matrix.

11.2. The relative displacement vector $\mathbf{u}$ is defined as a function of position $\mathbf{r}$ by the equation

$$\mathbf{u} = k\mathbf{r} + \mathbf{a} \times \mathbf{r}$$

where k is a constant scalar and $\mathbf{a}$ is a constant vector. Find the matrix $\nabla\mathbf{u}$ and separate it into pure strain and pure rotational parts. Describe the situation in words.

11.3. Verify that the result of Ex. 11.2 is consistent with Eq. 11.16 by proving quite generally that

$$\nabla \times (\mathbf{a} \times \mathbf{r}) = 2\mathbf{a}$$

when $\mathbf{a}$ is a constant vector.

11.4. Two shears in the same plane each 0.10 radian are superposed. Assume the additivity principle for the strain matrix and use trigonometrical tables to calculate the error in estimating the resulting shear to be 0.20 radian.

11.5. The elongation in each principal direction is 10%. Find the second- and third-order terms neglected in Eq. 11.22 for the volume dilation.

II.12 Description of Stresses in Continuous Media.

If external forces are acting on the medium across its boundaries, the medium is said to be under stress. The stress is transmitted through the medium to all points. If we imagine a section cut at any point, the material to one side of this section exerts forces on the material to the other side. A pull is positive and a push is negative by convention. The force per unit area across a plane normal to the j-direction has three components S_{jk} $k=1, 2, 3$; and for each value of $k, j=1, 2,$ or 3, so there are nine components in all. These nine components can be written as a matrix. A convenient notation for this is to write (X_x, Y_x, Z_x) for the force per unit area across a section normal to the x-direction, and on this notation the whole matrix of the stresses is written as

$$S = \begin{matrix} X_x & Y_x & Z_x \\ X_y & Y_y & Z_y \\ X_z & Y_z & Z_z \end{matrix} . \tag{12.1}$$

Given this matrix at any point in the medium, it is possible to find the stress—force per unit area—across any section in an arbitrary direction at that point in the material. If the normal to the section is represented by the unit vector $\mathbf{n}$, this force is found to be

$$\mathbf{f} = \mathbf{n} \cdot S \tag{12.2}$$

which equation expresses $\mathbf{f}$ as a linear vector function of $\mathbf{n}$.

The nondiagonal elements of the stress matrix are called the *shear* components. We can prove that the shear components of stress are symmetrical:

$$X_y = Y_x, \quad X_z = Z_x, \quad Y_z = Z_y. \tag{12.3}$$

To do this, consider a small rectangular element of the material with its sides parallel to the coordinate planes. Compute the torque on this element due to the stresses X_y and Y_x when the stress is uniform (not a

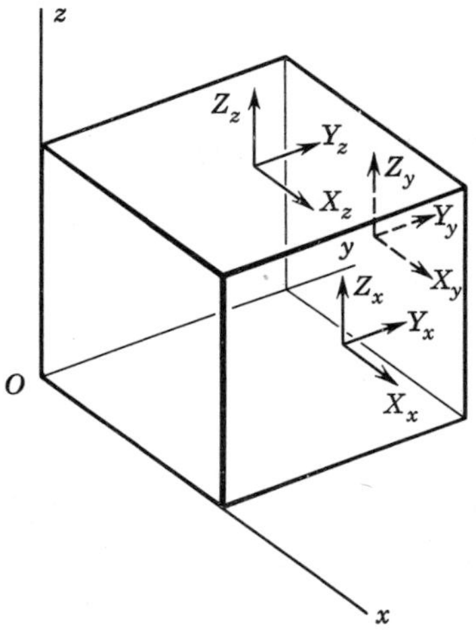

FIG. 12.1. The convention for positive stress components. Arrows show directions of forces acting on the cubic element of material due to the material outside the cube

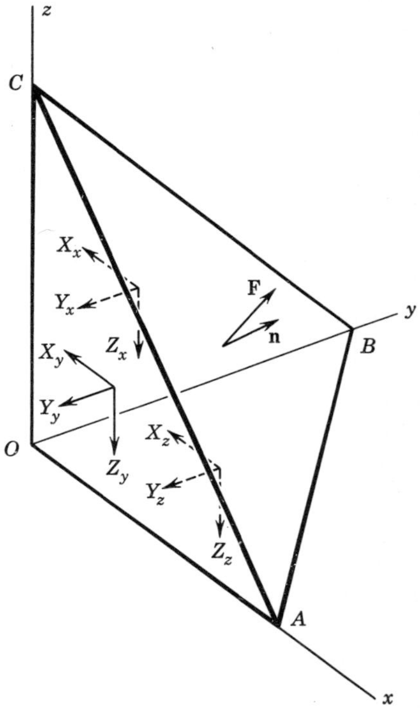

FIG. 12.2. Equation 12.2 illustrated. The force **F** per unit area acting on the triangle ABC is balanced by the stresses over the other three faces

function of position). The ratio of this torque to the rotational inertia of the element gives the angular acceleration of the element, which must remain finite no matter how small the volume of the element may be. If X_y were not exactly equal to Y_x, the angular acceleration would become indefinitely great for indefinitely small volume elements. The forces on the faces of the element vary as the square of its linear dimension—hence the torque varies as the cube; because the density is constant, the mass varies as the cube of the linear dimension. Therefore the rotational inertia varies as the fifth power of the linear dimension and the acceleration as the inverse square.

We have seen in the previous section that a rotation is represented by an antisymmetrical strain matrix, but here we must understand that it is not caused by any antisymmetrical stress matrix. It is, in fact, caused by stresses that are functions of position—*i.e.* by a nonuniform symmetrical stress matrix. The resultant force on any finite volume integrates to give a net torque. This will be shown in detail at the end of this section.

We can calculate the work done by the stresses as they cause strains, and hence the energy of the strained material. Consider a small unstrained cube of material having sides L, and take three of its faces as fixed coordinate planes. The stress in the material, represented by the matrix S, produces forces on the other three faces, according to Eq. 12.2, equal to

$$\mathbf{F}_n = L^2\mathbf{i}_n \cdot S. \tag{12.4}$$

The displacements of these three faces, relative to the fixed sides, are given by Eq. 11.10 as

$$d\mathbf{u}_n = L\mathbf{i}_n \cdot \Phi. \tag{12.5}$$

If the strain Φ is changed slightly, the change being represented by $\delta\Phi$, the small displacements of the three faces are $\delta d\mathbf{u}_n = L\mathbf{i}_n \cdot \delta\Phi$. The differential work done by the stresses on the three faces is the scalar product of the force and displacement, $\mathbf{F}_n \cdot \delta d\mathbf{u}_n$ or

$$\delta W = (L^2\mathbf{i}_n \cdot S)\cdot(L\mathbf{i}_n \cdot \delta\Phi). \tag{12.6}$$

If we write out the matrices S and Φ in dyadic notation:

$$S = \sum_{m,p}\sum S_{mp}\mathbf{i}_m\mathbf{i}_p \quad \text{and} \quad \Phi = \sum_{m,p}\sum \Phi_{mp}\mathbf{i}_m\mathbf{i}_p,$$

this works out to give

$$\delta W = L^3 \sum_{n,m}\sum S_{nm}\,\delta\Phi_{nm}, \tag{12.7}$$

or, in the other notation:

$$\delta W/L^3 = X_x\delta\Phi_{11} + Y_y\delta\Phi_{22} + Z_z\delta\Phi_{33} + 2X_y\delta\Phi_{12} + 2Z_x\delta\Phi_{31} + 2Y_z\delta\Phi_{23}, \tag{12.8}$$

because of the symmetry of both Φ and S. The first three terms are the work done on the material in volume expansion, the others the work done by the shear stresses. When the stresses are known as functions of the strain components—*viz.* the elastic relations to be discussed in the next section—the differential work can be integrated as a function of the final strains.

If we neglect internal friction (*i.e.* viscosity) we can assume that the work done by the stresses is in fact a single-valued function of the strain. This means that if we think of $U = W/L^3$ as a function of the nine variables Φ_{nm}, the order of any two partial differentiations is commutable: $\partial^2 U/\partial\Phi_{11}\partial\Phi_{12} = \partial^2 U/\partial\Phi_{12}\partial\Phi_{11}$ etc., or in general:

$$\partial^2 U/\partial\Phi_{nm}\partial\Phi_{pq} = \partial^2 U/\partial\Phi_{pq}\partial\Phi_{nm}. \tag{12.9}$$

From Eq. 12.7, we have $S_{nm} = \partial U/\partial\Phi_{nm}$. Therefore Eq. 12.9 implies

$$\partial S_{nm}/\partial\Phi_{pq} = \partial S_{pq}/\partial\Phi_{nm}. \tag{12.10}$$

These are called the *elastic identities* and lead to important simplification of the elasticity relations discussed in the next section.

Since the stress matrix S is symmetrical, we can find a coordinate system in which it is diagonal, with no apparent shear components. This coordinate system is called the *principal coordinate system* for that particular stress system. In the principal coordinates the stress matrix is

$$S = \begin{bmatrix} S_{11} & 0 & 0 \\ 0 & S_{22} & 0 \\ 0 & 0 & S_{33} \end{bmatrix} = S_{11}\mathbf{i}_1\mathbf{i}_1 + S_{22}\mathbf{i}_2\mathbf{i}_2 + S_{33}\mathbf{i}_3\mathbf{i}_3. \tag{12.11}$$

Consider again the small parallelepiped with sides parallel to the principal coordinate planes. The net force acting on this small element due to the stresses on its six faces is

$$\begin{aligned} \mathbf{f}d\tau &= dx(\partial S_{11}/\partial x)\mathbf{i}_1 dydz + dy(\partial S_{22}/\partial y)\mathbf{i}_2 dzdx + dz(\partial S_{33}/\partial z)\mathbf{i}_3 dxdy \\ &= dxdydz[\mathbf{i}_1\partial S_{11}/\partial x + \mathbf{i}_2\partial S_{22}/\partial y + \mathbf{i}_3\partial S_{33}/\partial z]. \end{aligned} \tag{12.12}$$

The net force per unit volume on the element due to nonuniform stress is therefore

$$\mathbf{f} = \nabla\cdot S. \tag{12.13}$$

Equating the net force on a small element of material to the rate of change of its momentum, we have the equation of motion:

$$dxdydz\nabla\cdot S = (d/dt)(\rho dxdydz\; d\mathbf{u}/dt). \tag{12.14}$$

Following the given element of material in its motion, the mass $\rho dxdydz$ remains constant, so that this factor can be taken out of the differentiation sign on the right side of this equation; providing that no discontinuities occur in the element, we can then cancel $dxdydz$ and write the equation as

$$\nabla\cdot S = \rho d^2\mathbf{u}/dt^2, \tag{12.15}$$

valid at any point in the material. In studying hydrodynamics we shall be obliged to take into account the distinction between d/dt and $\partial/\partial t$; the former representing the rate of change in a volume element moving with the local velocity of the fluid, the latter that in a stationary volume element. If the velocities are not too great we may neglect this distinction and write Eq. 12.15 in the approximate form

$$\nabla \cdot S = \rho \partial^2 \mathbf{u} / \partial t^2, \tag{12.16}$$

for the equation of motion at any point in the material. After we have found the elastic relations between stress and strain, we can eliminate

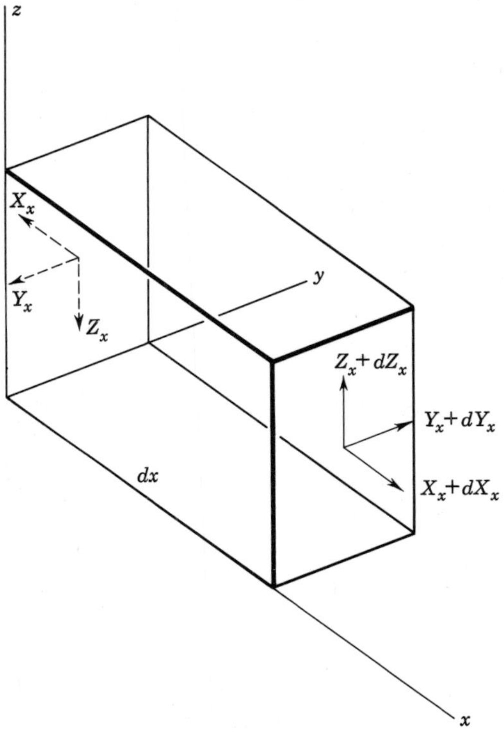

FIG. 12.3. Equation 12.13 illustrated. $dX_x = dx(\partial X_x/\partial x)$, *etc.*

the stress components from the equation of motion and derive a differential equation in the strain vector components alone.

The foregoing considerations apply to an infinitesimally small volume of material. The net force on a finite region of material under strain must be due entirely to the stresses acting on its boundaries, for the forces across internal surfaces must cancel by Newton's third law.

If we compute the net force on the finite region by integrating the body force over the region, we obtain:

$$\text{Net force on } V = \iiint_V \nabla \cdot S d\tau. \tag{12.17}$$

This transforms by Gauss' theorem to yield:

$$\text{Net force on } V = \iint S \cdot d\boldsymbol{\sigma} = \iint S \cdot \mathbf{n} d\sigma, \tag{12.18}$$

which, by comparison with Eq. 12.2, is indeed the integrated effect of the forces acting across the boundaries of the region.

We proved above that the stress tensor is necessarily symmetrical and stated that rotational motion (antisymmetrical "strain") can be produced only by spatial variations of the symmetrical stress, not by any antisymmetrical stress. We can now give this statement an analytical expression. The surface force at any boundary element is $S \cdot d\boldsymbol{\sigma}$ and the torque exerted by this force about an arbitrary origin, is $\mathbf{r} \times S \cdot d\boldsymbol{\sigma}$ so the net torque on the whole region V due to the surface stresses is

$$\text{Torque on } V = \iint \mathbf{r} \times S \cdot d\boldsymbol{\sigma} = -\iint d\boldsymbol{\sigma} \cdot S \times \mathbf{r}. \tag{12.19}$$

This transforms by Gauss' theorem into the volume integral:

$$\text{Torque on } V = -\iiint_V \nabla \cdot (S \times \mathbf{r}) d\tau. \tag{12.20}$$

The next step is not obvious, but it can be proved as an exercise that the last expression of Eq. 12.20 is identical with the following:

$$\text{Torque on } V = \iiint_V \mathbf{r} \times (\nabla \cdot S) d\tau. \tag{12.21}$$

Thus we have found that the net torque on the whole region vanishes unless the divergence of S is not zero; if S were a uniform stress there would be no torque. Eq. 12.21 merely states that the torque due to the surface forces—which is what we started out with in Eq. 12.19—is identical with the torque due to the body forces $\nabla \cdot S$. This is as it should be if the theory is to be self-consistent.

Exercise II.12

12.1. Prove Eq. 12.2. Hint: consider a small tetrahedron of material with three of its faces lying in the coordinate planes and the fourth face normal to the unit vector $\mathbf{n}$. Set up the equilibrium of this tetrahedron under the action of the stresses on its faces.

12.2. Carry out the detailed proof of Eq. 12.3 as outlined in the text.

12.3. Check the derivation of Eq. 12.21 from Eq. 12.20 using the symmetry of S.

12.4. Consider a rigid body rotation and write

$$\partial \mathbf{u}/\partial t = \boldsymbol{\omega} \times \mathbf{r}$$

in Eq. 12.16. Then, by comparison with Eq. 12.21, show that the angular momentum density can be represented by the vector

$$\mathbf{L} = \rho \mathbf{r} \times (\boldsymbol{\omega} \times \mathbf{r}).$$

12.5. Show that the angular momentum density can be written in the form

$$\mathbf{L} = K \cdot \boldsymbol{\omega}$$

where K is the dyadic $K = \rho(r^2 E - \mathbf{rr})$ and E is the identity dyadic.

12.6. A rigid cylinder with its axis along the z-axis of coordinates and radius equal to R is clamped on the axis. Show that under equilibrium conditions a tangential force applied to the surface of the cylinder equal to S dynes per cm^2 in the direction of increasing angle θ will produce a stress described by the following matrix:

$$(R^2 S/r^2)\begin{bmatrix} -\sin 2\theta & \cos 2\theta & 0 \\ \cos 2\theta & \sin 2\theta & 0 \\ 0 & 0 & 0 \end{bmatrix}$$

where $r^2 = x^2 + y^2$, $x = r\cos\theta$, $y = r\sin\theta$.

12.7. Find the divergence of the stress matrix in the last exercise at any point x, y inside the cylinder. Compare this result with Gauss' theorem $\iiint \nabla \cdot S d\tau = \iint d\boldsymbol{\sigma} \cdot S$ and explain the apparent discrepancy.

II.13 Elasticity: Hooke's law. In first approximation, for small enough strains, Hooke's law, in its simplest form, states that stress is proportional to strain. In its most general form, the law states that each component of the stress matrix is a linear function of all the strain components. We write out the nine components of stress in some arbitrary order:

$$S_{11}, S_{22}, S_{33}, S_{23}, S_{31}, S_{12}, S_{32}, S_{13}, S_{21}$$

and renumber the subscripts from 1 to 9:

$$S_1, S_2, S_3, S_4, S_5, S_6, S_7, S_8, S_9;$$

we do exactly the same thing with the strain components, and then write out the general form of Hooke's law in the convenient notation:

$$S_m = \sum_n c_{mn}\Phi_n \quad (m, n = 1, 2, 3 \cdots 9). \tag{13.1}$$

This means that there are 81 *moduli of elasticity*, c_{mn}. Actually, of course, because both S and Φ are symmetrical, there are only six different components of each matrix, and therefore at most 36 different moduli of elasticity. Referring back to the elastic identities, Eq. 12.10, we see that there are many identities between the moduli of the type $c_{61}=c_{16}$; in fact, the totality of identities, Eq. 12.10, results in the moduli being a symmetrical set:

$$c_{nm} = c_{mn}. \tag{13.2}$$

The 36 moduli therefore reduce to 21 different moduli, *i.e.* the six diagonal elements of the set plus one-half the remaining thirty. The general scheme may then be written out in the x-y-z notation as follows:

$$\left.\begin{aligned}
X_x &= c_{11}\Phi_{11}+c_{12}\Phi_{22}+c_{13}\Phi_{33}+2c_{14}\Phi_{23}+2c_{15}\Phi_{31}+2c_{16}\Phi_{12}\\
Y_y &= c_{12}\Phi_{11}+c_{22}\Phi_{22}+c_{23}\Phi_{33}+2c_{24}\Phi_{23}+2c_{25}\Phi_{31}+2c_{26}\Phi_{12}\\
Z_z &= c_{13}\Phi_{11}+c_{23}\Phi_{22}+c_{33}\Phi_{33}+2c_{34}\Phi_{23}+2c_{35}\Phi_{31}+2c_{36}\Phi_{12}\\
Y_z &= c_{14}\Phi_{11}+c_{24}\Phi_{22}+c_{34}\Phi_{33}+2c_{44}\Phi_{23}+2c_{45}\Phi_{31}+2c_{46}\Phi_{12}\\
Z_x &= c_{15}\Phi_{11}+c_{25}\Phi_{22}+c_{35}\Phi_{33}+2c_{45}\Phi_{23}+2c_{55}\Phi_{31}+2c_{56}\Phi_{12}\\
X_y &= c_{16}\Phi_{11}+c_{26}\Phi_{22}+c_{36}\Phi_{33}+2c_{46}\Phi_{23}+2c_{56}\Phi_{31}+2c_{66}\Phi_{12}
\end{aligned}\right\}. \tag{13.3}$$

If the material has cubic symmetry with respect to the axes x, y, z, the elastic moduli are further reduced in number; thus the moduli between any pressure component and the corresponding extension must be the same for all three symmetry axes, the shear stress over a symmetry plane produces only a shear strain in that plane, and the moduli for all three symmetry planes must be the same. The scheme of Eq. 13.3 then reduces to

$$\begin{aligned}
X_x &= c_{11}\Phi_{11}+c_{12}\Phi_{22}+c_{12}\Phi_{33}\\
Y_y &= c_{12}\Phi_{11}+c_{11}\Phi_{22}+c_{12}\Phi_{33}\\
Z_z &= c_{12}\Phi_{11}+c_{12}\Phi_{22}+c_{11}\Phi_{33}\\
Y_z &= 2c_{44}\Phi_{23}\\
Z_x &= 2c_{44}\Phi_{31}\\
X_y &= 2c_{44}\Phi_{12}
\end{aligned} \tag{13.4}$$

There are now at most three different elastic moduli other than zero, for materials having cubic symmetry of structure.

Since $2\Phi_{23}$ is the total angle of shear in the y-z plane, it is evident that c_{44} is the coefficient of rigidity for shear in any one of the three symmetry planes: in the usual notation,

$$c_{44} = \mu. \tag{13.5}$$

To interpret the other two coefficients, we may consider first an arrangement set up to measure Young's modulus, Y: the only stress is now X_x, a tension in one of the symmetry axes, and Eqs. 13.4 reduce to

$$X_x = Y(L-L_0)/L_0 = Y\Phi_{11}, \quad Y_y = Z_z = 0. \tag{13.6}$$

From symmetry, we have $\Phi_{22}=\Phi_{33}$, and all shear components are zero. Comparing Eqs. 13.6 and 13.4, we have

$$Y\Phi_{11} = c_{11}\Phi_{11}+2c_{12}\Phi_{22}, \quad 0 = c_{12}\Phi_{11}+\Phi_{22}(c_{11}+c_{12}). \quad (13.7)$$

The last equation solves for Φ_{22}:

$$\Phi_{22} = -c_{12}\Phi_{11}/(c_{11}+c_{12}).$$

Writing this into the first of Eqs. 13.7, we have Young's modulus

$$Y = c_{11}-2c_{12}{}^2/(c_{11}+c_{12}). \quad (13.8)$$

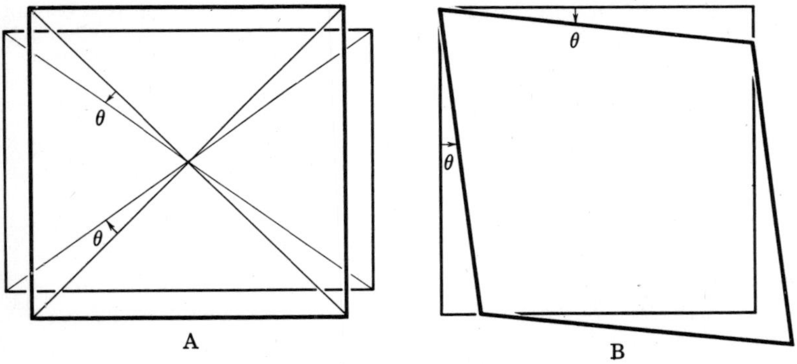

FIG. 13.1. The two kinds of shear strain in a cubic material (A) Principal axes of strain coincide with the axes of symmetry of the crystal, (B) Principal axes of strain at 45° from the axes of symmetry of the crystal. The appropriate elastic constants are $\frac{1}{2}(c_{11}-c_{12})$, for case A, and c_{44} for case B

Secondly, consider an arrangement set up to measure the bulk modulus K, in which the three diagonal stresses are equal pressures:

$$X_x = Y_y = Z_z = -p = K(V-V_0)/V_0 \quad (13.9)$$

and the three extensions are

$$\Phi_{11} = \Phi_{22} = \Phi_{33} = (L-L_0)/L_0 = (V-V_0)/3V_0. \quad (13.10)$$

All shear components are again zero. The first three equations in 13.4 all become the same, *viz.*:

$$K(V-V_0)/V_0 = (c_{11}+2c_{12})(V-V_0)/3V_0,$$

yielding for the bulk modulus

$$K = (c_{11}+2c_{12})/3. \quad (13.11)$$

If the stresses were measured with respect to axes that do not coincide with the symmetry axes of the crystal, the elastic relations would become very complicated: pressure could produce shear, shear stress could produce changes in volume, and so on. In order to handle such complications, one has to study the theory of space groups and the systematic classification of crystal symmetries. The elastic equations must remain invariant under the group of rotations characteristic of the crystal symmetry, and this restricts the possible relations, thus simplifying the problem. We shall not go into the details here. If, however, the material is completely isotropic, *i.e.* its properties are entirely independent of direction, the elastic relations become extremely simple. In fact, the relations between stress and strain also become independent of direction and are, therefore, independent of all rotations of the coordinate system. To see what this requires mathematically, we can multiply each line of Eq. 13.4 by the appropriate dyad $\mathbf{i}_1\mathbf{i}_1$, $\mathbf{i}_2\mathbf{i}_2$, etc., so that on adding the lines all together, the left side becomes the dyadic stress. The result then condenses into the following equation:

$$S = c_{12}(\mathrm{Tr}\cdot\Phi)E + 2c_{44}\Phi + (c_{11} - c_{12} - 2c_{44})(D\Phi), \qquad (13.12)$$

where $D\Phi \equiv \Phi_{11}\mathbf{i}_1\mathbf{i}_1 + \Phi_{22}\mathbf{i}_2\mathbf{i}_2 + \Phi_{33}\mathbf{i}_3\mathbf{i}_3$ is the diagonal part only of the strain matrix Φ. For isotropy, we need this equation to be invariant for all coordinate rotations. Now S, E, and Φ are tensors, and the coefficients c_{12}, $\mathrm{Tr}\cdot\Phi$, and $2c_{44}$ are scalar invariants, so that the first three parts of the equation transform as tensors under all rotations; but the last part, *viz.* $D\Phi$, is only part of a tensor—the diagonal part—and this will not transform in the same way as a complete tensor. The invariance of Eq. 13.12 therefore requires that we get rid of the term $D\Phi$ so that the whole equation can be a true tensor equation. This can be done by equating the coefficient of the offending term to zero:

$$2c_{44} = c_{11} - c_{12}. \qquad (13.13)$$

The elastic moduli of a pure isotropic material must then satisfy this identity, thus reducing the number of independent moduli to two. The ratio

$$A = 2c_{44}/(c_{11} - c_{12}) \qquad (13.14)$$

is called the anisotropy factor. When A is unity, the material is isotropic; otherwise not. The observed anisotropy factor for lead is 3.9, and for aluminum it is 1.2. The two independent elastic moduli for any isotropic material can be taken either μ and K, Y and K, or μ and Y. Another useful modulus is Poisson's ratio σ, the ratio of the lateral contraction to the elongation in Young's modulus experiment.

We can simplify the expression in Eq. 13.12 for Hooke's law as it applies to isotropic materials, and write it in the form:

$$S = (K-2\mu/3)(\nabla\cdot\mathbf{u})E+2\mu\Phi. \tag{13.15}$$

We can now use the law of elasticity given in Eq. 13.4 and Eq. 12.8 to find the energy per unit volume in strained material as a function of the strain components, the material being a cubic crystal. From the two equations quoted, we have, for the variation in energy due to any infinitesimal change in strain components,

$$\begin{aligned}\delta U = {} & \delta\Phi_{11}[c_{11}\Phi_{11}+c_{12}(\Phi_{22}+\Phi_{33})] \\ & +\delta\Phi_{22}[c_{11}\Phi_{22}+c_{12}(\Phi_{11}+\Phi_{33})] \\ & +\delta\Phi_{33}[c_{11}\Phi_{33}+c_{12}(\Phi_{11}+\Phi_{22})] \\ & +4c_{44}(\Phi_{23}\delta\Phi_{23}+\Phi_{31}\delta\Phi_{31}+\Phi_{12}\delta\Phi_{12}).\end{aligned} \tag{13.16}$$

To integrate this increment from all zero strains to any arbitrary final strain components, we can start by increasing Φ_{11} from zero to its final value while keeping all the other components zero; then we can keep Φ_{11} at its final value, increase Φ_{22} from zero to its final value keeping the other components zero, and so on, until each component has reached its final value. In this way we find for the total energy density as a function of the final strain components:

$$\begin{aligned}U = {} & \tfrac{1}{2}c_{11}(\Phi_{11}^2+\Phi_{22}^2+\Phi_{33}^2) \\ & +c_{12}(\Phi_{11}\Phi_{22}+\Phi_{22}\Phi_{33}+\Phi_{33}\Phi_{11}) \\ & +2c_{44}(\Phi_{12}^2+\Phi_{23}^2+\Phi_{31}^2).\end{aligned} \tag{13.17}$$

Exercise II.13

13.1. Derive Eq. 13.15 in detail from Eq. 13.12.

13.2. Repeat the discussion of this whole section suitably simplified by assuming isotropic material from the outset, finally deriving Eq. 13.15 directly.

13.3. Prove that Eq. 13.17 follows from Eq. 13.16 whether we start as in the text by increasing Φ_{11}, or in some other way, *e.g.* by first increasing Φ_{22}. Discuss this result in terms of the conservation of energy.

13.4. Show that the energy density due to a pure compression of a cubic crystal is $U=\frac{1}{2}K[(V-V_0)/V_0]^2$.

13.5. Show that the energy density due to a shear strain through an angle θ in any one plane of an isotropic material is $U=\frac{1}{2}\mu\theta^2$.

13.6. Show that the energy density in a cubic crystal under the kind of strain produced in the Young's modulus experiment is

$$U = \tfrac{1}{2}Y[(L-L_0)/L_0]^2.$$

Why does Poisson's ratio not appear in this result?

13.7. Find the work required to compress isothermally by 10% one cubic centimeter of several cubic crystals. Hint: see Table 4.1 of *Solid State Physics* by Charles Kittel.

II.14 Wave Propagation in Elastic Media. We can combine the equation of motion, Eq. 12.16, and the law of elasticity, Eq. 13.12, for nonisotropic crystals or Eq. 13.15 for isotropic materials, eliminate the stress components, and so derive a differential equation for the displacement **u** as a function of position and time. In isotropic materials the equation has two types of solution, one corresponding to the propagation of longitudinal waves and the other to transverse waves. In nonisotropic materials the equation is much more complicated; even in crystals having simple cubic symmetry there are usually three kinds of wave propagation for any given direction of propagation with three different velocities of propagation; it is only for the principal directions in the crystal that the waves are reduced to purely longitudinal and transverse types. In this section, we restrict the discussion to the propagation of waves in an isotropic elastic medium.

From Eq. 13.15, we have

$$\nabla\cdot S = (K-2\mu/3)\nabla\nabla\cdot\mathbf{u}+2\mu\nabla\cdot\Phi. \tag{14.1}$$

From Eq. 11.9, we can show that

$$2\nabla\cdot\Phi = \nabla\cdot\nabla\mathbf{u}+\nabla\nabla\cdot\mathbf{u}. \tag{14.2}$$

Using the identity

$$\nabla\nabla\,\cdot\; = \nabla\cdot\nabla+\nabla\times\nabla\times \tag{14.3}$$

in Eqs. 14.1 and 14.2, we see that

$$\nabla\cdot S = (K+4\mu/3)\nabla\cdot\nabla\mathbf{u}+(K+\mu/3)\nabla\times\nabla\times\mathbf{u}. \tag{14.4}$$

In case the strain is irrotational, $\nabla\times\mathbf{u}=0$, Eq. 14.4 becomes

$$\nabla\cdot S = (K+4\mu/3)\nabla\cdot\nabla\mathbf{u}, \tag{14.5}$$

while in case the strain is solenoidal, $\nabla\cdot\mathbf{u}=0$, we return to Eqs. 14.1 and 14.2 to derive

$$\nabla\cdot S = \mu\nabla\cdot\nabla\mathbf{u}. \tag{14.6}$$

Using these two alternatives in Eq. 12.16 to eliminate the stress matrix, we have for the

irrotational case: $$\rho\partial^2\mathbf{u}/\partial t^2 = (K+4\mu/3)\nabla\cdot\nabla\mathbf{u}, \tag{14.7}$$

and for the

solenoidal case: $$\rho\partial^2\mathbf{u}/\partial t^2 = \mu\nabla\cdot\nabla\mathbf{u}. \tag{14.8}$$

In taking the two alternative cases, solenoidal and irrotational, we are making use of a theorem to the effect that a certain class of vector functions, to which the vector **u** belongs, can always be expressed as the sum of two vector functions, one of which is solenoidal and the other is irrotational. The proof of this theorem is outlined very briefly here; it is usually known as *Helmholtz' theorem*.

Given the vector function $\mathbf{u}$ we can form two functions, a scalar f and a vector $\mathbf{g}$, where

$$\left.\begin{aligned} f &= (1/4\pi)\iiint (1/r)\nabla\cdot\mathbf{u}\,d\tau \\ \mathbf{g} &= (1/4\pi)\iiint (1/r)\nabla\times\mathbf{u}\,d\tau \end{aligned}\right\}, \tag{14.9}$$

r being the distance from the volume element $d\tau$ to the point at which the function, f or $\mathbf{g}$, is being defined. We then define the two vectors

$$\mathbf{v} = \nabla\times\mathbf{g} \quad \text{and} \quad \mathbf{w} = \nabla f \tag{14.10}$$

and prove that

$$\mathbf{u} = \mathbf{v}+\mathbf{w}. \tag{14.11}$$

From Eq. 14.10, however, it is clear that, identically

$$\nabla\cdot\mathbf{v} = 0 \quad \text{and} \quad \nabla\times\mathbf{w} = 0,$$

so that Eq. 14.11 actually has yielded $\mathbf{u}$ as the sum of a solenoidal and an irrotational vector function, as assumed. The proof of Eq. 14.11 from Eqs. 14.9 and 14.10 is left as an exercise.

Returning to the wave equations, 14.7 and 14.8, we shall first consider plane wave solutions. The restriction to plane waves implies that we seek only such solutions as are independent of two of the coordinates, say y and z, being functions only of x and t. Thus we have always a number of the partial derivatives of all the components of $\mathbf{u}$ zero:

$$\partial u_j/\partial y = \partial u_j/\partial z = 0 \quad (j = 1, 2, 3). \tag{14.12}$$

Let us discuss irrotational waves first. Applying Eq. 14.7, we have

$$\partial^2 u_j/\partial t^2 = V_1^2\partial^2 u_j/\partial x^2 \quad (j = 1, 2, 3), \tag{14.13}$$

where

$$V_1^2 = (K+4\mu/3)/\rho.$$

The condition for irrotational waves, $\nabla\times\mathbf{u}=0$, reduces by Eq. 14.12 to

$$\partial u_2/\partial x = \partial u_3/\partial x = 0. \tag{14.14}$$

Therefore, Eq. 14.13 becomes

$$\partial^2 u_2/\partial t^2 = \partial^2 u_3/\partial t^2 = 0, \tag{14.15}$$

showing that the two components transverse to the x-axis undergo no accelerated changes, or are nonoscillatory. The oscillatory component now satisfies the equation

$$\partial^2 u_1/\partial t^2 = V_1^2\partial^2 u_1/\partial x^2. \tag{14.16}$$

The general solution to this equation is

$$u_1(x, t) = F(x \pm V_1 t), \tag{14.17}$$

where F is an arbitrary function of the variable $x \pm V_1 t$. For example, we could choose

$$u_1(x, t) = u_0 \cos [2\pi(x - V_1 t)/\lambda], \tag{14.18}$$

representing a longitudinal wave transmitted with speed V_1 having an arbitrary wavelength λ and amplitude u_0.

The condition for a solenoidal wave is that $\nabla \cdot \mathbf{u} = 0$, and, when combined with Eq. 14.12, this means that

$$\partial u_1/\partial x = 0 \tag{14.19}$$

and the wave equation 14.8 becomes

$$\partial^2 u_j/\partial t^2 = V_2^2 \partial^2 u_j/\partial x^2, \quad \text{where} \quad V_2^2 = \mu/\rho. \tag{14.20}$$

Using the condition for plane waves equation (14.12), this reduces to two equations $j=2$ and $j=3$, the equation with $j=1$ being removed because of Eq. 14.19. Thus:

$$\left.\begin{aligned} \partial^2 u_2/\partial t^2 &= V_2^2 \partial^2 u_2/\partial x^2 \\ \partial^2 u_3/\partial t^2 &= V_2^2 \partial^2 u_3/\partial x^2 \end{aligned}\right\}. \tag{14.21}$$

The general solutions of these equations are again

$$u_2 = F_2(x \pm V_2 t) \quad \text{and} \quad u_3 = F_3(x \pm V_2 t)$$

and these two functions can be chosen quite generally and independently of each other. The choice of solution depends on boundary conditions. Thus we might have, for example,

$$\left.\begin{aligned} u_2 &= u_0 \cos [2\pi(x - V_2 t)/\lambda - \phi] \\ u_3 &= u_0 \cos [2\pi(x - V_2 t)/\lambda] \end{aligned}\right\}, \tag{14.22}$$

where ϕ is an arbitrary phase angle. The direction of the resultant vector $\mathbf{u}$ makes an angle θ with the y-direction given by $\tan \theta = u_3/u_2$, and at any given position, say $x=0$, this angle is a function of time;

$$\begin{aligned} \cot \theta &= \cos (2\pi V_2 t/\lambda + \phi)/ \cos (2\pi V_2 t/\lambda) \\ &= \cos \phi - \sin \phi \tan (2\pi V_2 t/\lambda). \end{aligned} \tag{14.23}$$

The vector rotates around the x-axis, going through the complete circle in the period of the wave; at any given instant in time, the end of the vector traces out a corkscrew pattern in space. We say that the plane of polarization of the wave is rotating as the wave is propagated.

So far we have been considering the oscillatory motion of the material at any point, as represented by the behavior of the vector displacement $\mathbf{u}$. In practice it is sometimes of greater interest to study the oscillatory behavior of other physical parameters, and here we shall consider briefly the propagation of pressure waves in an elastic material. Under a purely hydrostatic compression, we can simply write the stress matrix as

$$S = -pE \tag{14.24}$$

and Eq. 13.15 yields simply

$$-p = K\nabla \cdot \mathbf{u}. \tag{14.25}$$

Taking the divergence of Eq. 14.7, rearranging the order of some of the differentiations, and making use of Eq. 14.25, we find

$$\partial^2 p/\partial t^2 = V_1^2 \nabla \cdot \nabla p. \tag{14.26}$$

Let us consider spherical pressure waves, in which the wave fronts are concentric spheres around the source at the origin. In this case, p is a function only of r and t, and it is convenient to express Eq. 14.26 in spherical polar coordinates. The wave equation (14.26) is now

$$\partial^2 p/\partial t^2 = V_1^2 (1/r^2)(\partial/\partial r)(r^2 \partial p/\partial r). \tag{14.27}$$

To solve this we make the substitution

$$p = q/r \tag{14.28}$$

and derive

$$\partial^2 q/\partial t^2 = V_1^2 \partial^2 q/\partial r^2. \tag{14.29}$$

This equation is precisely of the same form as that for plane waves (cf. Eq. 14.16) and has for its general solution

$$q = q(r - V_1 t), \tag{14.30}$$

an arbitrary function of the variable $r - V_1 t$. The solution of the pressure wave equation (see Eq. 14.28) is, therefore,

$$p(r, t) = (1/r) q(r - V_1 t). \tag{14.31}$$

In particular, for example, we may have

$$p(r, t) = (1/r) \sin [2\pi(r/\lambda - \nu t)], \tag{14.32}$$

where

$$\nu\lambda = V_1$$

and either ν or λ may be taken as arbitrary.

The factor $(1/r)$ in the solution ensures the conservation of energy in accordance with the inverse square law during propagation. From Eqs. 13.17 and 13.9, the energy density in a pressure wave is $\frac{1}{2}p^2/K$, and from Eq. 14.31 this is

$$U = q^2/2Kr^2. \tag{14.33}$$

If we follow the progress of the wave, keeping the function q at a constant value, the energy density will decrease as the inverse square of distance; the total energy in a spherical shell propagating out with the speed V_1 is therefore conserved.

Exercise II.14

14.1. Compare Eq. 14.8 and Eq. 9.2. Given a flexible string of cross section 0.100 mm, density 1.00 g per cm^3, find the tension needed to cause transverse waves to propagate with the same speed that transverse waves propagate in a continuous three-dimensional medium having the same mass density as the string and a rigidity modulus of $\mu = 8.2 \times 10^{11}$ dynes per cm^2.

14.2. In the two-dimensional membrane of §10, find the solutions of the wave equation corresponding to rectilinear wave fronts.

14.3. At the end of §10 it was stated that we could not build up standing waves in a two-dimensional membrane by superposing propagated waves as one can with the standing waves in a string. Explain carefully why the solutions found in Eq. 14.32 cannot be used for this purpose.

14.4. Express Eq. 14.26 in terms of cylindrical polar coordinates. Substitute a few trial solutions in the equation. Discuss the difficulties.

14.5. Prove that Eq. 14.17 is a solution of Eq. 14.16. Prove that any wave form whatever propagates without change in wave form in this problem.

14.6. Find the velocity of propagation of changes in the angular velocity of an element in elastic material. Hint: see Eq. 11.16 and take the curl of the appropriate wave equation.

14.7. Verify by direct substitution that Eq. 14.32 is a solution of Eq. 14.27.

14.8. Consider a long elastic rod of uniform cross section subject to a varying tension at one end. Develop the theory of propagation of extension waves along the rod and find the velocity of their propagation in terms of the density of the material and its Young's modulus. Hint: review Ex. 13.6 and develop the analog of Eq. 14.25.

II.15 Hydrokinematics. Imagine a volume in space through which fluid is moving. The rate at which the mass of the fluid inside this volume is increasing is equal to the source rate of mass within the volume minus the rate at which mass flows out through the surface of the volume. Expressed mathematically this statement becomes:

$$\iiint(\partial\rho/\partial t)d\tau = \iiint R d\tau - \iint \rho\mathbf{v}\cdot d\boldsymbol{\sigma}, \tag{15.1}$$

where ρ is the density of the fluid, $\mathbf{v}$ is its velocity, and R is the source rate of mass per unit volume; each of these quantities is to be treated as a function of position and time. The surface integral is taken over the entire surface of the volume and we can make use of Gauss' theorem, Eq. 7.8, to rewrite it as the volume integral $-\iiint\nabla\cdot(\rho\mathbf{v})d\tau$. We now let the volume of integration shrink indefinitely around any point of interest and, provided that all integrands involve only functions that are continuous in the neighborhood of the point, we can cancel the volume element and integral signs from the equation, obtaining instead a differential equation valid at the point of interest; the result is the equation of continuity:

$$\partial\rho/\partial t = R - \nabla\cdot(\rho\mathbf{v}). \tag{15.2}$$

The following special cases are of interest:

the equation of continuity with zero source,

$$\partial\rho/\partial t = -\nabla\cdot(\rho\mathbf{v}), \tag{15.3}$$

the equation of continuity for steady flow conditions where $\partial\rho/\partial t = 0$,

$$\nabla\cdot(\rho\mathbf{v}) = 0, \tag{15.4}$$

the equation of continuity for an incompressible fluid of uniform density

$$(\rho = \text{constant}), \quad \nabla\cdot\mathbf{v} = 0. \tag{15.5}$$

The last equation states that in a uniform incompressible fluid the velocity vector is solenoidal. From the last paragraph of §7 we see therefore that the velocity vector defines a flux integral through any loop, depending only on the periphery. We may select any closed loop and draw lines through its periphery everywhere tangent to the velocity vector **v**. So long as **v** is a definite function of position—non-turbulent motion—these lines cannot cross each other, and they form a tube of flow. A cross section can be made through such a tube of flow at any place, and so long as **v** is a solenoidal vector, the flux through the section must exactly balance the flux through any other section.

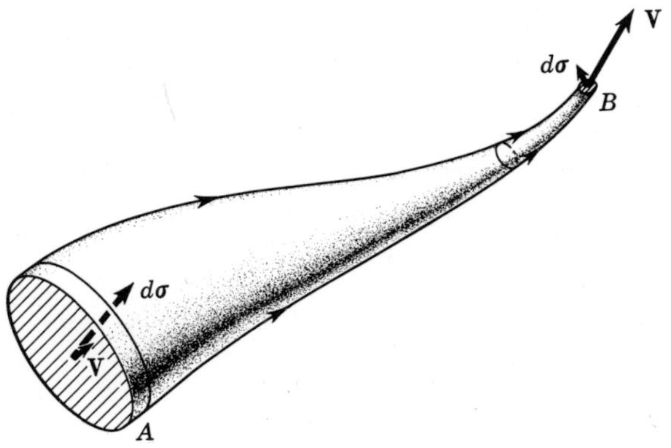

FIG. 15.1. A tube of flow in an incompressible fluid. At A the integral $\int \mathbf{v}\cdot d\boldsymbol{\sigma}$ equals the integral $\int \mathbf{v}\cdot d\boldsymbol{\sigma}$ at B

Consider a fluid in which there is a very strong source density in the immediate neighborhood of some point that we may take as the origin of coordinates. It is convenient to divide the region in two parts, one that is as small as we can make it and still contain the entire source, and the other consisting of the whole of space outside the source. Outside the source R is everywhere zero, and Eqs. 15.3, 15.4, and 15.5 apply, depending on the conditions assumed. Within the source volume we are not safe in going over to the differential equations, because the source function is in danger of behaving discontinuously at the origin. Instead we must go back to the original integral equation (15.1). Here we can assume steady conditions, $\partial\rho/\partial t = 0$, and equate the volume integral of the source rate to the total source strength as an arbitrary constant in the problem. We then have the total flow into the outer region equal to the total source strength inside the source volume. This forms a boundary condition on the problem in the outer region, where there are no sources.

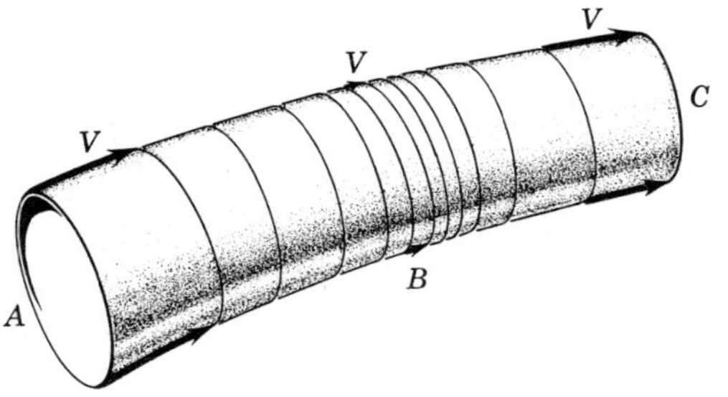

FIG. 15.2. A tube of flow in a compressible fluid. The flux integral is not uniform down the tube. The decreased kinetic energy in the region B is compensated for by an increase in elastic energy. If the flow velocity is equal to the velocity of propagation of sound in the fluid, the compression can remain stationary in the tube

The foregoing considerations apply whether there exists a velocity potential or not, *i.e.* whether or not the velocity is irrotational. We shall not discuss rotational motion however until later; here we shall assume

$$\nabla \times \mathbf{v} = 0. \tag{15.6}$$

In this case, we know that a velocity potential does exist, *cf.* §7, so that

$$\mathbf{v} = -\nabla\Phi. \tag{15.7}$$

From this and Eq. 15.5, the equation of continuity for an incompressible fluid, we have

$$\nabla\cdot\nabla\Phi = 0, \tag{15.8}$$

showing that the velocity potential for an incompressible fluid is a solution of Laplace's equation. The solutions of Laplace's equation will be discussed in detail in Chapter IV.

We discuss here the concept of convective derivatives. Consider each component of the fluid velocity to be a function of position and time, written in the implicit form $v_j(x, y, z, t)$, $j=1, 2, 3$. Expand the variation dv_j in terms of its partial derivatives thus:

$$dv_j = (\partial v_j/\partial t)dt + (\partial v_j/\partial x)dx + (\partial v_j/\partial y)dy + (\partial v_j/\partial z)dz. \quad (15.9)$$

Divide this equation by dt and write $dx/dt = v_1$, etc., which means that the variations dx, dy, and dz refer to the displacement of a particular piece of fluid material moving with the velocity (v_1, v_2, v_3), and, by the same token, the derivative dv_j/dt that appears on the left of the equation is the acceleration of this piece of material; it is called the *convective acceleration*, as distinct from the partial derivative with respect to time, the point-acceleration. The equation now reads:

$$dv_j/dt = \partial v_j/\partial t + \sum_k v_k(\partial v_j/\partial x_k), \quad (15.10)$$

where j and k both can take the values 1, 2, 3, and x_1, x_2, x_3 are written in place of x, y, z, respectively. The three equations obtained here with the three values of j are combined into one vector equation:

$$d\mathbf{v}/dt = \partial \mathbf{v}/\partial t + \sum_k v_k(\partial \mathbf{v}/\partial x_k) \quad (15.11)$$

and a natural notation for this is

$$d\mathbf{v}/dt = \partial \mathbf{v}/\partial t + \mathbf{v}\cdot\nabla\mathbf{v}. \quad (15.12)$$

We reiterate that the derivative $d\mathbf{v}/dt$ refers to a given element of material of the fluid followed in its motion through a small time interval, while the partial derivative $\partial\mathbf{v}/\partial t$ refers to a fixed position in space, so that we are comparing the velocities of two different pieces of material at the two close instants of time. The full derivative $d\mathbf{v}/dt$ is the convective derivative.

The convective derivative can be evaluated for any function of position and time, whether scalar or vector. Consider the density of the fluid as a scalar function of position and time. Its convective derivative is simply

$$d\rho/dt = \partial\rho/\partial t + \mathbf{v}\cdot\nabla\rho. \quad (15.13)$$

It is clear that for an incompressible fluid, the convective derivative of the density must be identically zero—we are following a given piece of material whose density cannot change. This is true even if the fluid is actually a mixture in which the density is not the same at all points; a

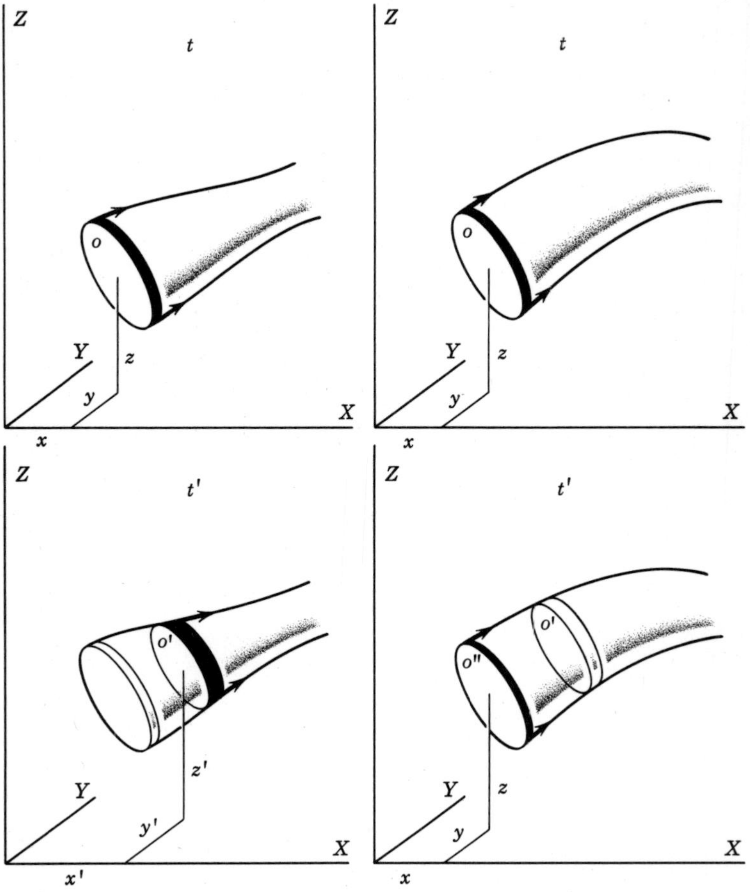

FIG. 15.3. The left-hand pair of diagrams illustrate the convective derivative: $d\rho/dt = \text{Lim.}(\rho' - \rho)/(t' - t)$. The right-hand pair of diagrams illustrate the partial derivative with respect to time: $\partial\rho/\partial t = \text{Lim.}(\rho'' - \rho)/(t' - t)$

given element of incompressible fluid remains of the same fixed density at all times. Thus we arrive at the theorem:

$$\partial\rho/\partial t = -\mathbf{v}\cdot\nabla\rho \tag{15.14}$$

for an incompressible fluid, even when the density is not uniform in space because of heterogeneities in the fluid material. This equation then gives the rate at which density at a fixed point changes in time due to the space-variations in the density around the fixed point. If we now combine Eq. 15.14 with Eq. 15.3, the equation of continuity with zero source, we find that $-\mathbf{v}\cdot\nabla\rho = -\nabla\cdot(\rho\mathbf{v})$ and therefore $\rho\nabla\cdot\mathbf{v} = 0$; thus we have for the motion of an incompressible, possibly heterogeneous fluid, a velocity that is necessarily a solenoidal vector:

$$\nabla \cdot \mathbf{v} = 0. \tag{15.15}$$

This statement is considerably stronger than that implied in Eq. 15.5 which was derived only for a uniform incompressible fluid.

Exercise II.15

15.1. Show that $f=1/r$ where $r^2=x^2+y^2$ is a solution of Laplace's equation. Prove this both by expressing the Laplacian operator in Cartesian coordinates and in spherical polar coordinates.

15.2. Use Eq. 8.9 and the results of Ex. 8.8 in spherical polar coordinates to find the velocity vector corresponding to the velocity potential $\Phi=1/r$.

15.3. Find the total mass flow through any sphere centered on the origin due to the flow described in Ex. 15.2.

15.4. What total source rate must be provided at the origin to maintain the steady flow described in the last two exercises?

15.5. Prove that $\Phi=-\frac{1}{2}\ln(x^2+y^2)$ is a solution of Laplace's equation in three dimensions using (a) Cartesian coordinates and (b) cylindrical polar coordinates.

15.6. Find the velocity vector when the velocity potential is the function Φ defined in Ex. 15.5.

15.7. Find the mass of fluid passing through a cylinder per unit length, coaxial with the polar axis of coordinates in Ex. 15.6.

15.8. What source must be provided to maintain the steady flow described in the last two exercises?

15.9. Find the operator $\mathbf{v}\cdot\nabla$ appearing in Eq. 15.12 when the velocity function is that found in Ex. 15.2 and Ex. 15.6; use the appropriate polar coordinates in each case.

15.10. Find the convective time derivatives of the velocity vectors in Ex. 15.2 and Ex. 15.6. Note that the partial time derivatives are identically zero in each case because the flow is steady.

II.16 Hydrodynamical Equations of Motion. A fluid is a continuous elastic medium in the sense of §12. The discussion leading up to Eq. 12.15 shows that the equation of motion can be written

$$\nabla \cdot S = \rho d\mathbf{v}/dt, \tag{16.1}$$

where S is the stress tensor in the fluid, and $d\mathbf{v}/dt$ is the convective derivative of the material velocity. In studying elastic media possessing rigidity we neglected the difference between $d\mathbf{v}/dt$ and $\partial\mathbf{v}/\partial t$. This involves assuming that $\mathbf{v}$ is rather small, as we can see from Eq. 15.12, and therefore the product of $\mathbf{v}$ and its space derivatives can be neglected. In fluid motion this would be a very bad neglect. Another difference between fluid motion and motion of a solid is in the possible form of the stress matrix. In an ideal fluid with no viscosity there is no shear stress component in any coordinate system, and, as we know from elementary hydrostatics, the stress is always a pure pressure equal in all directions. Therefore, in this case we must have

$$S = -Ep, \tag{16.2}$$

E being, as before, the identity matrix. In particular we have

$$\nabla \cdot S = -\nabla p \tag{16.3}$$

and the general equation of motion for the ideal fluid is

$$-\nabla p = \rho d\mathbf{v}/dt = \rho \partial \mathbf{v}/\partial t + \rho \mathbf{v} \cdot \nabla \mathbf{v}. \tag{16.4}$$

To discuss the special case of irrotational motion we cite the following lemma (see Ex. 6.15),

$$\mathbf{v} \times \nabla \times \mathbf{v} = (\nabla \mathbf{v}) \cdot \mathbf{v} - \mathbf{v} \cdot \nabla \mathbf{v} = \tfrac{1}{2} \nabla v^2 - \mathbf{v} \cdot \nabla \mathbf{v}. \tag{16.5}$$

When $\nabla \times \mathbf{v} = 0$, this yields

$$\mathbf{v} \cdot \nabla \mathbf{v} = \tfrac{1}{2} \nabla v^2. \tag{16.6}$$

Putting this into Eq. 16.4, we have the equation for irrotational motion:

$$\nabla p + \tfrac{1}{2} \rho \nabla v^2 + \rho \partial \mathbf{v}/\partial t = 0. \tag{16.7}$$

If the fluid is uniform and incompressible, ρ is a constant; when the motion also is steady, $\partial \mathbf{v}/\partial t = 0$, we deduce from Eq. 16.7 the equation of Bernoulli:

$$\nabla(p/\rho + \tfrac{1}{2} v^2) = 0. \tag{16.8}$$

Integrating this equation along any path yields

$$p/\rho + \tfrac{1}{2} v^2 = \text{constant}, \tag{16.9}$$

the constant being determined by boundary conditions in the problem. If there were an external field derivable from a potential in addition to the forces due to the hydrostatic pressure, the potential per unit mass would need to be added to the left-hand side of Eq. 16.9. The term p/ρ can be thought of as a potential energy per unit mass in Eq. 16.9, but it is not strain energy because the liquid is incompressible and therefore does not show any strain (*cf.* Ex. 13.4).

The general equation of motion for compressible fluids, Eq. 16.4, can be solved only if we know the relation between density and pressure, *i.e.* the equation of state from a thermodynamic point of view. The motion depends then upon whether the system is isothermal, or whether we allow the temperature to be variable. A special case exists, namely the adiabatic case, where the equations can be solved fairly easily even when the temperature is a variable. Adiabatic conditions pertain if the motion is rapid enough to permit neglect of heat transfer by thermal conduction compared with transfer by convection.

Consider an ideal gas as the fluid under study, with an adiabatic equation of state:

$$\rho/\rho_0 = (p/p_0)^g, \tag{16.10}$$

where g is the ratio of the specific heat at constant volume to the specific heat at constant pressure. Consider irrotational steady flow, $\partial\mathbf{v}/\partial t = 0$, then Eq. 16.7 gives the equation of motion as:

$$(1/\rho)\nabla p + \nabla(\tfrac{1}{2}v^2) = 0. \tag{16.11}$$

Take the scalar product of this equation by the vector displacement $d\mathbf{r}$ and it becomes:

$$dp/\rho + \tfrac{1}{2}d(v^2) = 0, \tag{16.12}$$

the increments in the functions being those corresponding to the displacement $d\mathbf{r}$. This equation can be integrated along a path between any two positions in the fluid:

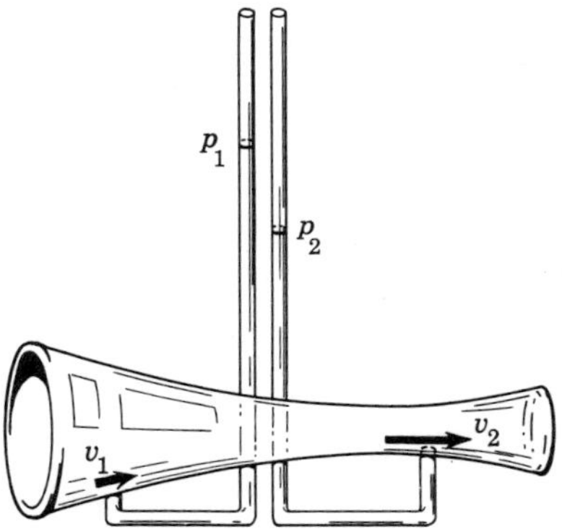

FIG. 16.1. Venturi's model to demonstrate Bernoulli's equation

$$\tfrac{1}{2}(v_b^2 - v_a^2) = -\int_a^b (1/\rho)dp. \tag{16.13}$$

To this we apply the equation of state, Eq. 16.10, to express ρ as a function of p in the integrand. The integral can then be evaluated easily:

$$\tfrac{1}{2}(v_b^2 - v_a^2) = [1/(1-g)](p_a/\rho_a)[1-(p_b/p_a)^{1-g}]. \tag{16.14}$$

This is the generalized Bernoulli equation for adiabatic flow of a compressible ideal gas. If the gas starts from rest at standard pressure, $v_a = 0$, and flows into an evacuated space maintained at almost zero pressure, $p_b = 0$, we have the maximum speed attainable given by v_b thus:

$$v_{\max}^2 = 2p_a/(1-g)\rho_a. \tag{16.15}$$

It is interesting to note that the existence of a finite maximum to the velocity of flow in the gas is essentially due to the adiabatic nature of the flow. For isothermal motion the equation of state would be simply

$$\rho/\rho_0 = p/p_0, \tag{16.16}$$

which would be equivalent to putting $g = 1$ in Eq. 16.10; thus Eq. 16.15 would in this case result in an infinite upper limit to the flow velocity.

The velocity of propagation of acoustic waves in a fluid is the velocity of propagation of compression waves derived in Eq. 14.26, but in a fluid we take the rigidity modulus zero, so the velocity of sound is

$$C = \sqrt{K/\rho}. \tag{16.17}$$

From Eq. 14.25, we may write

$$K = -Vdp/dV \tag{16.18}$$

and so the velocity of sound may be written in the form:

$$C = \sqrt{dp/d\rho}. \tag{16.19}$$

Sound waves are usually of sufficiently high frequency to permit us to assume the changes to be adiabatic. Using Eq. 16.10 in this last expression, we have

$$C = \sqrt{p/\rho g}. \tag{16.20}$$

Finally, we shall develop the equation for conservation of energy in fluid motion. Let Φ be the potential per unit mass due to any external field, and suppose this field to be time independent:

$$\partial\Phi/\partial t = 0. \tag{16.21}$$

In the presence of such a field, the equation of motion, Eq. 16.4, is modified to read

$$\rho d\mathbf{v}/dt = -\nabla p - \rho\nabla\Phi. \tag{16.22}$$

Take the scalar product of this equation with the velocity vector:

$$\rho\mathbf{v}\cdot d\mathbf{v}/dt = -\mathbf{v}\cdot\nabla p - \rho\mathbf{v}\cdot\nabla\Phi$$

and use the definition of the convective derivative to convert it into the form:

$$\tfrac{1}{2}\rho d(v^2)/dt = -\nabla\cdot(p\mathbf{v}) + p\nabla\cdot\mathbf{v} - \rho d\Phi/dt, \tag{16.23}$$

where the last term is the convective derivative of the potential (*cf.* Eq. 15.13). Next we note that

$$\nabla\cdot\mathbf{v} = \partial(\nabla\cdot\mathbf{u})/\partial t = -(1/K)(dp/dt). \tag{16.24}$$

See Eq. 14.25 for the last step. Three of the terms in Eq. 16.23 now have become time derivatives; we collect them on one side of the equation and multiply the equation by a volume element:

$$\tfrac{1}{2}\rho d\tau d(v^2)/dt + \rho d\tau d\Phi/dt + \tfrac{1}{2}d\tau d(p^2/K)/dt = -\nabla\cdot(p\mathbf{v})d\tau. \tag{16.25}$$

We can now integrate this equation term by term over some finite region in the fluid. In doing so we shall note that two time derivatives are convective, and we shall therefore allow the volume of integration to be carried along with the fluid motion—*i.e.* we integrate over a material region that contains at all times the same material. The quantities $\rho d\tau$ are therefore constants and can go through the time derivatives in the first two terms on the left; these terms then become the time derivatives of the kinetic energy density and the potential energy density, respectively. The third term on the left represents the time rate of change of the strain energy per unit volume integrated over the volume of the region. Thus:

$$(d/dt)\iiint(\tfrac{1}{2}\rho v^2 + \rho\Phi + \tfrac{1}{2}p^2/K)d\tau = -\iint \mathbf{v}\cdot p d\boldsymbol{\sigma} \qquad (16.26)$$

where Gauss' theorem has been used to convert the right-hand side into a surface integral. Now in this term $pd\boldsymbol{\sigma}$ is the force on the surface element and $\mathbf{v}\cdot pd\boldsymbol{\sigma}$ is therefore the rate of work on the moving surface element, or the flow of energy across the surface out of the volume. Eq. 16.26 is then an expression of the conservation of energy in the fluid.

Exercise II.16

16.1. Incompressible fluid motion occurs in a uniform gravitational field. Modify the equation of motion, Eq. 16.1, to include the effects of weight, and derive the appropriate Bernoulli equation to replace Eq. 16.8.

16.2. A pipe of varying cross section carries water, assumed incompressible, up a height of 10 feet. If the cross section at the top is twice that at the bottom, find the excess pressure at the bottom required to maintain a flow velocity at the top of (a) 10 feet per second, and (b) 15 feet per second. Discuss the results.

16.3. Find the bulk modulus K in the two cases, (a) isothermal compression and (b) adiabatic compression, of an ideal gas.

16.4. Integrate Eq. 16.7 for isothermal expansion of an ideal gas and prove that $\rho = \rho_0 \exp(-\tfrac{1}{2}mv^2/kT)$, where T is the absolute temperature, k is Boltzmann's constant, and m is the mass of a gas atom. Hint: the ideal gas equation is $p/\rho = kT/m$.

16.5. Consider adiabatic expansion of an ideal gas and ignore effects of gravity; prove from Eq. 16.14 that

$$p = p_0[1-(v/v_m)^2]^{1/(1-g)},$$

where p_0 is the starting pressure and v_m the maximum possible flow speed. Find the analogous expression for the density of the gas.

16.6. Using the results of Ex. 16.5, prove that the velocity of propagation of sound, C, can be expressed as a function of the flow velocity:

$$C^2 = \tfrac{1}{2}(g^{-1}-1)(v_m{}^2 - v^2).$$

16.7. Show that the flow velocity exceeds the sound velocity when

$$v > v_m[(1-g)/(1+g)]^{1/2}.$$

II.17 Two-Dimensional Fluid Motion. Let the velocity of the fluid be restricted to lie in one plane, which is taken to be the x-y plane of coordinates. Consider a slab of fluid one unit in depth parallel to the x-y plane. Any curve OP drawn in the x-y plane from the origin to any point of interest P is the section by this plane of a cylindrical strip whose generating lines are parallel to the z-axis. Let $d\mathbf{s}$ be a vector representing an element of area on the strip; it is perpendicular to the element of the curve OP and equal to the length of this element of OP at the base of the element of area. The flux of the fluid through the element of area is

$$d\Psi = \mathbf{v} \cdot d\mathbf{s}. \tag{17.1}$$

The components of $d\mathbf{s}$ are $(-dy, dx)$, where dx, dy are the components of the element of OP. The total flux of fluid through the strip OP is

$$\Psi(x, y) = \int_0^P (-v_x dy + v_y dx), \tag{17.2}$$

the integral being taken along the curve OP to the point P at (x, y). This expresses Ψ as a function of the position of P with respect to the origin. If the point P is moved through a displacement δx, δy, the increment in Ψ is

$$\delta\Psi = -v_x \delta y + v_y \delta x \tag{17.3}$$

and we may then write

$$v_y = \partial\Psi/\partial x, \quad v_x = -\partial\Psi/\partial y. \tag{17.4}$$

The equation of continuity for an incompressible fluid in two dimensions, from Eq. 15.5 or Eq. 15.15, is

$$\partial v_x/\partial x + \partial v_y/\partial y = 0. \tag{17.5}$$

Comparing Eqs. 17.5 and 17.4, we see that $\Psi(x, y)$ is a single-valued function for incompressible fluid flow in two dimensions. This means that the same flux passes through any strip joining the origin O to the given point P. This function $\Psi(x, y)$ is called the *stream function* or the *current function* for the flow.

We note that the component of the curl of $\mathbf{v}$ normal to the plane of the motion is $(\nabla \times \mathbf{v})_k = \partial v_y/\partial x - \partial v_x/\partial y$, and, from Eq. 17.4, this gives the Laplacian of Ψ. Since $\frac{1}{2}(\nabla \times \mathbf{v})_k$ is the angular velocity of the fluid, (*cf.* Eq. 11.16), we then have

$$\nabla^2\Psi = 2\omega, \tag{17.6}$$

where ω is the angular velocity of the fluid. In irrotational motion the stream function is therefore a solution of Laplace's equation.

Comparing Eqs. 17.4 with 15.7 and 17.6 with 15.8, we see that two

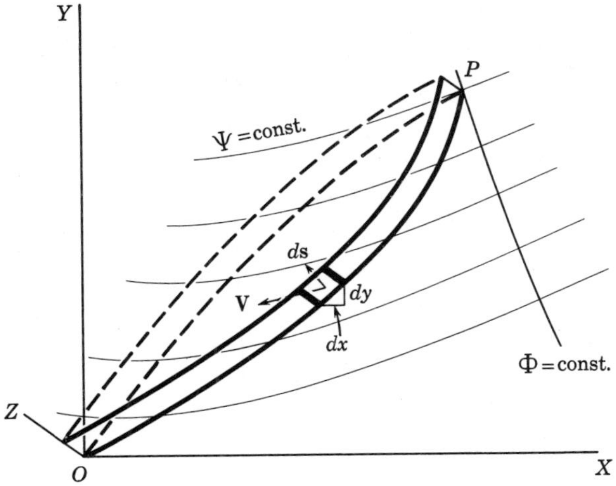

FIG. 17.1. The current function Ψ and the velocity potential Φ

functions exist, the potential Φ and the stream function Ψ, both of which are solutions of Laplace's equation in incompressible two-dimensional fluid motion. We can see that, from Eqs. 17.4 and 15.7, these two functions have gradients perpendicular to each other, which we can interpret as meaning that the lines of flow are perpendicular to the equipotential line. The two functions Φ and Ψ can be considered real and imaginary parts of one harmonic function in the complex plane. This is the basis of the well-known technique of solving two-dimensional problems in electrostatics, and the problem of incompressible fluid flow in two dimensions is mathematically equivalent to this.

When the fluid motion is not irrotational, the potential function ceases to exist as a single-valued function, but the stream function is still a good function, although no longer a solution of Laplace's equation; it is instead a solution of the Poisson equation, Eq. 17.6. As an example of rotational motion we consider a cylindrical vortex such that the angular velocity is zero outside of a radius a, and equal to a constant ω inside this radius. Expressing the equations in plane polar coordinates, we have

$$\left.\begin{aligned} r<a: &\quad (1/r)(d/dr)(r\,d\Psi/dr) = 2\omega \\ r>a: &\quad (1/r)(d/dr)(r\,d\Psi/dr) = 0 \end{aligned}\right\}. \tag{17.7}$$

We seek solutions of these equations that start out from zero at the origin, and join continuously with each other at the radius a. We find in fact that

$$\left.\begin{aligned} r>a: &\quad \Psi = \omega a^2 \ln(r/a) + \tfrac{1}{2}\omega a^2 \\ r<a: &\quad \Psi = \tfrac{1}{2}\omega r^2 \end{aligned}\right\}. \tag{17.8}$$

Cartesian coordinates can be rotated so that the x-direction is along the direction of increasing r, and the y-direction along increasing θ; then from Eq. 17.4 we would have

$$v_\theta = d\Psi/dr \tag{17.9}$$

and hence

$$\left.\begin{aligned} r<a: \qquad & v_\theta = mr/a^2 \\ r>a: \qquad & v_\theta = m/r, \quad m = \omega a^2 \end{aligned}\right\}. \tag{17.10}$$

Outside the vortex radius we may use the Bernoulli equation (16.9), because here the motion is irrotational. Therefore,

$$r>a: \qquad p/\rho = p_\infty/\rho - \tfrac{1}{2}m^2/r^2. \tag{17.11}$$

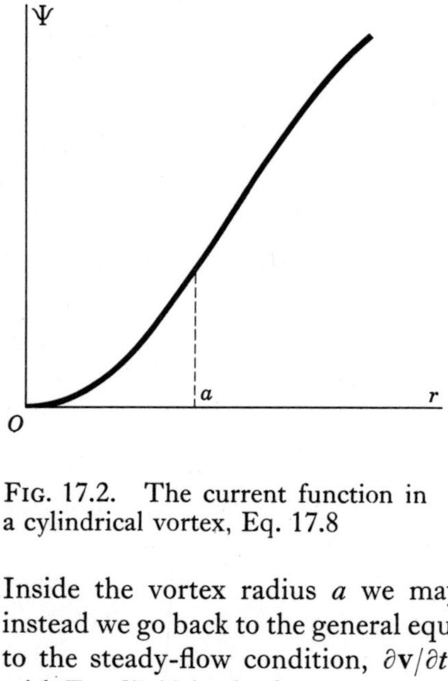

FIG. 17.2. The current function in a cylindrical vortex, Eq. 17.8

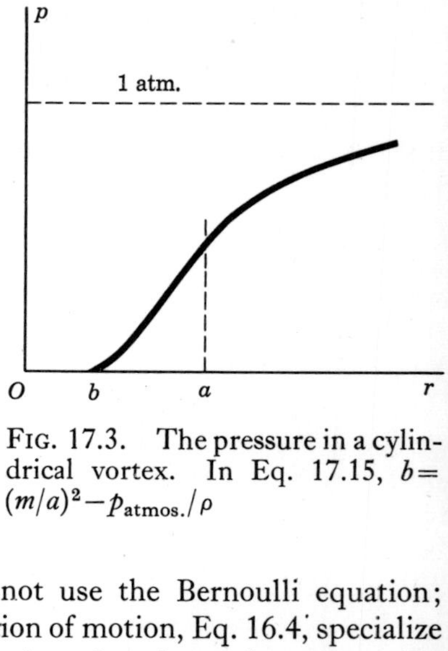

FIG. 17.3. The pressure in a cylindrical vortex. In Eq. 17.15, $b = (m/a)^2 - p_{\text{atmos.}}/\rho$

Inside the vortex radius a we may not use the Bernoulli equation; instead we go back to the general equation of motion, Eq. 16.4, specialize to the steady-flow condition, $\partial\mathbf{v}/\partial t = 0$, and evaluate the term $\mathbf{v}\cdot\nabla\mathbf{v}$ with Eq. 17.10 in the form

$$\mathbf{v} = \boldsymbol{\theta}_1(mr/a^2). \tag{17.12}$$

The operator $\mathbf{v}\cdot\nabla$ reduces to the single term $(mr/a^2)(\partial/r\partial\theta)$ so that

$$\mathbf{v}\cdot\nabla\mathbf{v} = (mr/a^2)^2(\partial/r\partial\theta)\boldsymbol{\theta}_1 = -(mr/a^2)^2(\mathbf{r}_1/r).$$

The equation of motion is then

$$-\partial p/\partial r = -(m^2r/a^4)\rho. \tag{17.13}$$

Integration yields

$$p/\rho = \tfrac{1}{2}(mr/a^2)^2 + \text{constant}, \tag{17.14}$$

where the constant is determined by matching pressures given at $r=a$ by the Eqs. 17. 11 and 17.14. Thus:

$$r < a: \qquad p/\rho = \tfrac{1}{2}(mr/a^2)^2 + p_\infty/\rho - m^2/a^2. \qquad (17.15)$$

It is interesting to notice that if the vortex is strong enough, *i.e.* if m or ω are great enough so that $(m/a)^2 \geqslant p_\infty/\rho$, then the pressure within the vortex falls to zero for some value of r at or greater than zero. A hole would then develop in the center of the vortex. The linear velocity at the rim of the vortex, $r=a$, is then

$$r = a: \qquad v_\theta = m/a = (p_\infty/\rho)^{1/2}. \qquad (17.16)$$

The value of this required to make the pressure vanish at the center is easily computed. Thus for water under one atmosphere pressure at remote points, $p=1$ atm., the value of v_θ is about $22\frac{1}{2}$ miles per hour at the rim of the vortex.

Exercise II.17

17.1. Prove that the gradients of the velocity potential and the stream function are perpendicular to each other.

17.2. Review the theory of analytic functions in the complex plane. The function $f=u+iv$ is said to be analytic if the *Cauchy-Riemann conditions* are satisfied: $\partial u/\partial x = \partial v/\partial y$, $\partial u/\partial y = -\partial v/\partial x$. Prove that both u and v are then solutions of Laplace's equation.

17.3. If the velocity potential and stream function are considered the real and imaginary parts of an analytic function, show that they must be solutions of Laplace's equation.

17.4. Show that the equation of continuity, Eq. 17.5, and the condition for irrotational motion in two dimensions are equivalent to the Cauchy-Riemann conditions on the complex function $v_x + iv_y$, which is therefore an analytic function in the complex plane.

17.5. Check that Eqs. 17.8 are solutions of Poisson's equation in the form of Eqs. 17.6 and 17.7.

17.6. Consider the vortex described by Eqs. 17.7 when the fluid is an ideal gas and the motion is adiabatic. Find the pressure as a function of radius, and find the velocity at the rim of the vortex just sufficient to produce zero pressure at the center. Hint: use Eq. 16.10 in Eq. 17.13 before integration, and use Eq. 16.14 for the region outside the vortex.

II.18 Propagation of Waves and Shocks in Fluids. The stress-strain relations obtained for elastic solids can be appropriately modified for fluids, and the wave propagation equations derived for solids can be translated into equivalent equations for fluids. The definition of strain in a fluid is somewhat arbitrary. In fact, there is no rigidity, at least for nonviscous fluids, so there is no natural shape from which to measure strain; also, specially for gases, there is no natural density at which we could say the compression strain is zero in any absolute sense. Instead, we arbitrarily count the volume of the fluid under one standard atmosphere and at normal temperature as the unstrained volume, and count both the dilation strain and stress as starting from that "zero." As the shear stress remains zero in a nonviscous fluid, the stress matrix in this case is simply

$$S = -\Delta p E, \tag{18.1}$$

where Δp is the excess pressure above one atmosphere. The strain matrix reduces also to the diagonal form:

$$\Phi = \tfrac{1}{3}(\Delta V / V_0)E, \tag{18.2}$$

where ΔV is the increase in volume above the standard volume V_0. The elasticity relation, Hooke's law, is then $S = K$ Trace $\Phi\, E$ or

$$-\Delta p = K \Delta V / V_0, \tag{18.3}$$

which is usually written in the alternative form:

$$K = -V \partial p / \partial V = \rho \partial p / \partial \rho. \tag{18.4}$$

Direct measurements of elastic coefficients are made under isothermal conditions, the stress being applied slowly enough for any heat generated to be uniformly distributed. Elastic vibrations, on the other hand, tend to occur under adiabatic conditions because, during the period of one vibration, there is usually insufficient time for heat to be dissipated. The ratio of the adiabatic to isothermal bulk modulus is easily found in the case of an ideal gas from

$$(\partial p / \partial \rho)_{\text{isothermal}} : (\partial p / \partial \rho)_{\text{adiabatic}} = g(\rho / \rho_0)^{1-1/g}$$

(*cf.* Eq. 16.10).

Reviewing the discussion of elastic waves in continuous media, we see that, for the simple elastic relations in Eq. 18.3, the equation for irrotational wave transmission, Eq. 14.7, becomes

$$\partial^2 \mathbf{u} / \partial t^2 = (K / \rho) \nabla^2 \mathbf{u}, \tag{18.5}$$

where, following the remarks of the last paragraph, the adiabatic coefficient has to be used:

$$K / \rho = (\partial p / \partial \rho)_{\text{adiabatic}}. \tag{18.6}$$

The velocity of propagation of irrotational or compression waves in the fluid is, therefore,

$$c = \sqrt{(\partial p / \partial \rho)_{\text{adiabatic}}}. \tag{18.7}$$

Transverse waves do not transmit because $\mu = 0$, and the equation for these waves reduces to $\partial^2 \mathbf{u} / \partial t^2 = 0$.

In deriving the wave equation originally we made an approximation, equating the material derivative of the velocity to the partial derivative. In solids for waves of not too large an amplitude, this neglect is permissible, but in fluids the neglected term turns out to be very significant; Eq. 15.12 gives the correction. Let us write down the more exact equation of motion and retain the previously neglected term and so derive a more precise wave equation. In one-dimensional motion, which suffices for the discussion of plane waves, Eq. 16.4 becomes

$$dv/dt = -(1/\rho)(\partial p/\partial x) \tag{18.8}$$

and if we write as above, $dp=(K/\rho)d\rho$, or $\partial p/\partial x = c^2\partial\rho/\partial x$, this equation yields

$$dv/dt = \partial v/\partial t + v\partial v/\partial x = -(c^2/\rho)(\partial\rho/\partial x). \tag{18.9}$$

The equation of continuity for one-dimensional motion is derivable from Eq. 15.3 in the form

$$\partial\rho/\partial t + v\partial\rho/\partial x = -\rho\partial v/\partial x. \tag{18.10}$$

We can eliminate either v or ρ between the last pair of equations, Eqs. 18.9 and 18.10. To eliminate ρ we proceed as follows. We differentiate Eq. 18.9 with respect to time, $\partial/\partial t$, and Eq. 18.10 with respect to x, $\partial/\partial x$, and eliminate the term $\partial^2\rho/\partial x\partial t$ between the results, and number this equation (C). We then use Eq. 18.9 to express the term $v\partial\rho/\partial x$ in Eq. 18.10 in terms of v, the result being another equation (D) for $\partial\rho/\partial t$ in terms of v. Eq. 18.9 for $\partial\rho/\partial x$ and Eq. (D) for $\partial\rho/\partial t$ used in Eq. (C) will then lead through a few simple steps to the equation

$$\partial^2 v/\partial t^2 + 2v\partial^2 v/\partial x\partial t + 2(\partial v/\partial x)(\partial v/\partial t + v\partial v/\partial x) = (c^2 - v^2)(\partial^2 v/\partial x^2). \tag{18.11}$$

This is a nonlinear equation for v as a function of x and t; if v is small enough to neglect the product of more than one factor containing v, it reduces to the simple wave equation. Otherwise we must proceed to solve the more exact equation by successive approximations. The zeroth order approximation is the solution of the simple wave equation, which we can write as

$$v = v_0 \exp(ikx - i\omega t) \tag{18.12}$$

where k and ω are constants related through $\omega/k = c$. Substituting this approximation into Eq. 18.11 and treating k and ω as constants, we come out with the equation

$$-\omega^2 v + 4k\omega v^2 - 3k^2v^3 + c^2k^2v = 0. \tag{18.13}$$

Canceling v and solving for ω/k, we find

$$\omega/k = (c^2 + v^2)^{1/2} + 2v. \tag{18.14}$$

This suggests that a better approximation than Eq. 18.12 would be

$$v = v_0 \exp i\omega(x/c' - t), \tag{18.15}$$

where ω is again constant, and $c' = \omega/k$, neither k nor c' being constant, but rather:

$$c' = (c^2 + v^2)^{1/2} + 2v. \tag{18.16}$$

So long as v is much smaller than c, c' is not much different from c, and the solution, Eq. 18.15, approaches Eq. 18.12. However, the speed of propagation c' is greater than c when v is positive. Each crest of the wave therefore travels faster than the troughs and eventually may catch up with the trough ahead of it, producing a saw-tooth wave form. When this happens, the crest and the next trough get so close together that they constitute a *shock front* where the gradient is so steep that the whole theory of elastic waves breaks down.

A simple and very useful theory of shock front propagation exists. Let U be the value of c', the velocity of propagation, in the limit when the wave front becomes a vertical shock front. Let this shock front be propagating to the right into initially undisturbed material having density ρ_0 at pressure P_0. Let the material immediately behind the shock front (*i.e.* in the crest of the wave) have density ρ and pressure P.

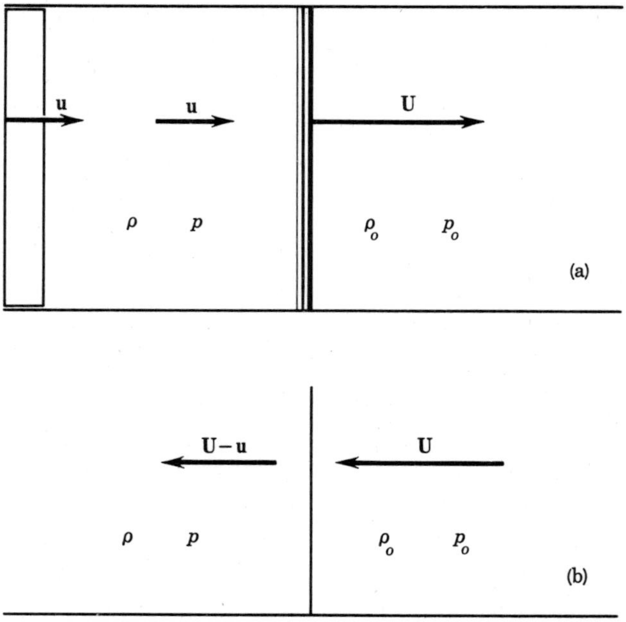

FIG. 18.1. Two different reference systems used to describe the propagation of a shock front

Both ρ and P are discontinuously greater than their initial values ρ_0 and P_0, respectively. The material just to the left of the shock front is moving to the right with a speed u, which is of course less than U and is called the particle velocity in the shock. The diagram, Fig. 18.1a, illustrates the notation just set up. We can visualize such a shock having been produced by *suddenly* starting to move a piston with the speed u into initially stationary material.

We shall now adopt a reference system of coordinates at rest relative to the propagating shock front, *i.e.* moving to the right with speed U relative to the undisturbed material (see Fig. 18.1b). On this system the undisturbed material is moving to the left with speed U, and the compressed material is moving to the left with speed $U-u$. We shall now apply the laws of conservation of mass, momentum, and energy to this situation.

The conservation of mass requires that the rate at which mass arrives per unit area per unit time from the right, in the undisturbed material, must equal the rate at which mass recedes from the front in the compressed material. This statement yields the relation

$$(U-u)\rho = U\rho_0. \tag{18.17}$$

The conservation of momentum at the shock front requires that the rate at which momentum arrives at the front from the right, minus the rate at which momentum leaves the front in the compressed material, must equal the net force per unit area on any element of material as it crosses the front. A column of material of length U enters the front in one second, and this column has momentum per unit volume equal to $\rho_0 U$; as it crosses the front this column becomes one of length $U-u$ and carries momentum $\rho(U-u)$ per unit volume. Hence, the statement yields the equation:

$$\rho_0 U^2 - \rho(U-u)^2 = P - P_0.$$

Using Eq. 18.17 in this, we reduce it to:

$$Uu\rho_0 = P - P_0. \tag{18.18}$$

It is a simple exercise to derive, from Eqs. 18.17 and 18.18, the two equivalent relations:

$$u^2 = (P-P_0)(1/\rho_0 - 1/\rho) \tag{18.19}$$

and

$$u = (1-\rho_0/\rho)U. \tag{18.20}$$

The conservation of energy at the shock front requires that the total energy arriving at the front per unit area per unit time from the right, minus the total energy leaving the front per unit area per unit time to the left, shall equal the rate at which work is done by the pressure difference across the front. The kinetic energy per unit volume on the right is $\frac{1}{2}\rho_0 U^2$, and the volume arriving per second is U; the kinetic energy per unit volume on the left is $\frac{1}{2}\rho(U-u)^2$, and the volume leaving the front per unit area per second to the left is $U-u$. The rate at which kinetic energy is being produced at the front is therefore $\frac{1}{2}\rho(U-u)^3 - \frac{1}{2}\rho_0 U^3$ per unit area per second. The rate at which work is being done by the pressure P_0 on the approaching material is $P_0 U$, and the work done by the receding material against the pressure P is $P(U-u)$ per second. If the internal energy (thermal and elastic) per unit mass in the undisturbed material is E_0, and in the compressed material is E, then the energy balance yields the equation:

$$\frac{1}{2}\rho_0 U^3 + \rho_0 E_0 U + P_0 U = \frac{1}{2}\rho(U-u)^3 + \rho E(U-u) + P(U-u). \quad (18.21)$$

We can use Eq. 18.17 to cancel the factor $\rho_0 U$ on the left with a factor $\rho(U-u)$ on the right of this equation, giving

$$\frac{1}{2}U^2 + P_0/\rho_0 + E_0 = \frac{1}{2}(U-u)^2 + P/\rho + E. \quad (18.22)$$

which may be compared with Bernoulli's equation, Eq. 16.9.

Rearranging terms in this equation yields

$$E - E_0 = P_0/\rho_0 - P/\rho + Uu - \frac{1}{2}u^2$$

and by using Eqs. 18.18 and 18.19 in the last two terms, respectively, we derive

$$E - E_0 = \frac{1}{2}(P+P_0)(1/\rho_0 - 1/\rho). \quad (18.23)$$

The three significant results of this theory comprise this last equation and the two following equations derivable from Eqs. 18.19 and 18.20—the velocity of shock front propagation in terms of the compression,

$$U = \sqrt{(\rho/\rho_0)(P-P_0)/(\rho-\rho_0)} \quad (18.24)$$

and the particle velocity, or piston speed, required to maintain this shock compression,

$$u = \sqrt{(P-P_0)(1/\rho_0 - 1/\rho)}. \quad (18.25)$$

In practice it is possible to measure both U and u. Therefore, by varying the piston speeds or particle velocities, u, and measuring the corresponding shock velocities, U, using Eqs. 18.24 and 18.25, a relation between P and ρ can be derived. Eq. 18.23 then gives the increase in internal energy caused in the shock. The relation between P and ρ derived in this way is called the *Rankine-Hugoniot equation of state*, after the two scientists who first developed the foregoing theory. It turns out that the energy difference given by Eq. 18.23 is not very different from what would occur in an adiabatic process, but the deviation is such that the rise in pressure associated with the shock compression is somewhat steeper than that which occurs in an adiabatic compression.

It is instructive to compare the shock velocity equation (18.24) with the propagation of sound waves, Eq. 18.7. It is clear that $U > c$ for two reasons; first, because of the factor ρ/ρ_0 in the expression for U, and, secondly, because the p–ρ curve is steeper in the shock than in the adiabat.

The Rankine-Hugoniot equations are just as valid for solid materials as for fluids, except that the pressure has to be interpreted as the longitudinal component of stress, and need not in principle be a hydrostatic pressure. The theory has been extensively used in understanding shock waves produced in solids by high explosive charges, and the Rankine-Hugoniot equations of state have been plotted for very many solids up to pressures as high as 10^5 atmospheres.

If one had a complete theory of the equation of state, one could use it to predict the shock velocity from Eq. 18.24. However, the usual laws of thermodynamics, from which such a theory might be derived, are useful only for slow processes. It is known that the rise-times for shock fronts may be as rapid as 10^{-11} second to pressures as high as several hundred kilobars, so it is certainly not safe to use ordinary thermodynamics to discuss shock compressions.

Exercise II.18

18.1. Calculate the ratio $K_{\text{adiabatic}}:K_{\text{isothermal}}$ for an ideal gas.

18.2. Derive Eq. 18.11 by the method outlined in the text.

18.3. Prove from Eq. 18.16 that c' is less than c when v is negative.

18.4. Show that spherical waves may develop shock fronts.

18.5. Use Eqs. 18.20 and 18.19 to express the pressure change $P-P_0$ and the density ratio ρ/ρ_0 as functions of U, u, and ρ_0.

18.6. If experiment shows that a particle velocity of 1 mm per microsecond produces a shock velocity of 6 mm per microsecond in a solid having a normal density 8 gm per cm^3, find the pressure increase accompanying the shock front.

II.19 Steady Motion of Viscous Fluids. So far we have considered only fluids in which no shear stresses exist, the so-called ideal fluids. In actual fluids viscosity is present and shear stresses exist, proportional to the space rate of change of velocity. In the simplest example of a fluid moving parallel to a plane surface, it seems to stick to the boundary, its velocity increasing with increasing distance from the boundary; a shear force is required across any plane parallel to the boundary in order to maintain the motion. The empirical situation is well described by the equation

$$X_y = \eta \partial v_x / \partial y, \tag{19.1}$$

where the plane boundary is normal to the y-axis, and the motion is along the x-direction; X_y is the shear stress per unit area of the boundary, and η is the *coefficient of viscosity*. We may compare this equation with the equation of shear-strain elasticity, *viz.* the last line in Eq. 13.4, which may be written in the form:

$$X_y = \mu(\partial u_x/\partial y - \partial u_y/\partial x). \tag{19.2}$$

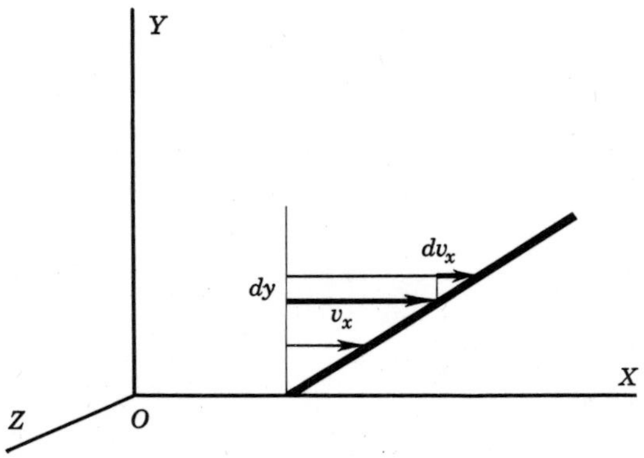

FIG. 19.1. The motion of a viscous fluid over the x–z plane, illustrating Eq. 19.1

In the viscous model we have $\partial v_y/\partial x = 0$, so Eq. 19.1 is the same as

$$X_y = \eta(\partial v_x/\partial y + \partial v_y/\partial x). \tag{19.3}$$

The analogy between Eqs. 19.2 and 19.3 is sufficiently close to suggest a more general statement of the law of viscosity, along the same lines as the general law of elasticity. Since the velocity **v** occurs in Eq. 19.3 where the displacement **u** occurs in Eq. 19.2, we may read Eq. 19.3 as a particular expression of a general linear relationship between the stress tensor and the rate of change of strain tensor in a viscous fluid. This analogy suggests that we could develop a viscosity law in an invariant form analogous to the invariant form of Hooke's law in isotropic solids, Eq. 13.15, with the rate of change of strain replacing the strain and the viscosity coefficient replacing the rigidity modulus, etc. Such an equation would give the stresses required to maintain any motion in the viscous fluid, not just the simple motion envisioned in connection with the empirical relation Eq. 19.1. These stresses required to maintain motion would have to be added to the stresses already required by the static law of elasticity that applies even for a nonmoving fluid, *viz.* the hydrostatic stress system given by Eqs. 18.1 and 18.3.

This analogy would give the following extra stress matrix required for the rate of strain:

$$S_{\text{viscous}} = B(\nabla \cdot \mathbf{v})E + \eta(\nabla\mathbf{v} + \widetilde{\nabla\mathbf{v}}), \tag{19.4}$$

where η and B are viscosity coefficients. If we consider motion in which there is no rate of change of the volume, $\nabla \cdot \mathbf{v} = 0$, we can identify this relation directly with Eq. 19.3. There is experimental evidence that rapid compression or rapid expansion of volume of a compressible viscous fluid involves irreversible work, so that volume viscosity does exist. This is taken care of in Eq. 19.4 by the first term on the right, with B as the coefficient of volume viscosity; the extra pressure is required to compress the volume at a finite rate, proportional to the rate of compression. Eq. 19.4 gives the stress required to change the strains at a finite rate; on top of these, as already stated, there is the hydrostatic stress persisting even when there is no motion. We therefore have to add these stresses to get the total. The generalized law of viscosity between stress and motion is, therefore,

$$S_{\text{total}} = -pE + B(\nabla \cdot \mathbf{v})E + \eta(\nabla\mathbf{v} + \widetilde{\nabla\mathbf{v}}). \tag{19.5}$$

This equation is essentially a guess, developed by analogy with the law of elasticity; to verify it we shall work out its consequences for the equation of motion of a viscous fluid when the stresses are not uniform. The divergence of the stress matrix is the body force per unit volume due to the stresses (*cf.* Eq. 12.13):

$$\nabla \cdot S_{\text{total}} = -\nabla p + \eta\nabla \cdot \nabla\mathbf{v} + (\eta + B)\nabla\nabla \cdot \mathbf{v}. \tag{19.6}$$

If, in addition to the stresses, there exists an external field of force $\mathbf{F}$ per unit mass of fluid—gravitation or weight, for example—the equation of motion, Eq. 12.15, becomes

$$\rho d\mathbf{v}/dt = \rho\mathbf{F} - \nabla p + \eta\nabla^2\mathbf{v} + (\eta + B)\nabla\nabla \cdot \mathbf{v}. \tag{19.7}$$

For small enough velocities, we may neglect the difference between the convective derivative $d\mathbf{v}/dt$ and the partial derivative $\partial\mathbf{v}/\partial t$ and write the approximate equation of motion instead:

$$\rho\partial\mathbf{v}/\partial t = \rho\mathbf{F} - \nabla p + \eta\nabla^2\mathbf{v} + (\eta + B)\nabla\nabla \cdot \mathbf{v}. \tag{19.8}$$

If we consider only incompressible fluids, we may take the divergence of the equation of motion and then equate $\nabla \cdot \mathbf{v}$ to zero everywhere. Thus Eq. 19.8 leads to the relation $\nabla^2 p = \rho\nabla \cdot \mathbf{F}$, and when the external field $\mathbf{F}$ is derivable from a potential, $\mathbf{F} = -\nabla\Phi$, we have simply:

$$\rho = \text{constant}: \qquad \nabla^2(p + \rho\Phi) = 0. \tag{19.9}$$

The sum of the hydrostatic pressure and the potential per unit volume of the external field forms a solution of Laplace's equation. These solutions are well known, and we shall discuss them in greater detail in a later section. As a simple example here we shall discuss only the application of Eq. 19.9 to *Poiseuille flow* of fluid through a tube.

Let the x-axis coincide with the axis of the tube, and assume the stream lines to be all parallel to the x-axis: $v_y = v_z = 0$. The fluid being incompressible, we also have $\nabla \cdot \mathbf{v} = 0$, which now becomes $\partial v_x/\partial x = 0$; therefore, from Eq. 15.11, we can show that in this special case $d\mathbf{v}/dt = \partial \mathbf{v}/\partial t$, even when the velocity is not small. There is then no approximation involved in Eq. 19.8. For steady-flow conditions, $\partial \mathbf{v}/\partial t = 0$, and we therefore now have an even stronger restriction, *viz.*:

$$dv/dt = dv_x/dt = 0. \tag{19.10}$$

This states that the flow down the tube is unaccelerated: v_x has the same value everywhere on a given stream line, *i.e.* for any given element of fluid, because d/dt is the convective derivative. The only velocity derivatives that do not need to vanish for the above reasoning are $\partial v_x/\partial y$ and $\partial v_x/\partial z$. Eq. 19.7 reduces, when $\mathbf{F}$ is zero, to

$$0 = -\nabla p + \eta \nabla^2 \mathbf{v},$$

and from this we can reason that, with $\mathbf{v}$ having only the one component v_x, the pressure gradient can also have only the one component $\partial p/\partial x$. Then in the absence of any external field Φ, Eq. 19.9 reduces to

$$\partial^2 p/\partial x^2 = 0, \tag{19.11}$$

which integrates at once to give

$$p = p_0 + Gx, \tag{19.12}$$

where p_0 and G are constants to be determined by boundary conditions. Note that the pressure gradient is G, a constant at all parts of the tube. Using this simplification in the equation of motion, Eq. 19.7, we find at once that

$$-G + \eta(\partial^2 v_x/\partial y^2 + \partial^2 v_x/\partial z^2) = 0, \tag{19.13}$$

recalling again the assumption that the fluid is incompressible, so that $\nabla \cdot \mathbf{v} = 0$ on the right-hand side of Eq. 19.6. Expressed in Cartesian coordinates, this equation may be solved to fit boundary conditions in a rectangular cross section, but a much simpler problem is presented by flow down a circular cross section. Changing to cylindrical polar coordinates, we have instead of Eq. 19.13:

$$(1/r)(d/dr)(r dv_x/dr) = G/\eta. \tag{19.14}$$

A particular integral of this is easily shown to be $v_x = \frac{1}{4}(G/\eta)r^2$. The complementary function is some solution of Laplace's equation in plane polar coordinates; since there is no dependence on the angle, this is necessarily of the form $A \ln r + C$, where A and C are constants. Because v_x must remain finite at $r=0$, we cannot allow the logarithm term; so we take the arbitrary constant $A=0$. This leaves the constant term only, and the complete solution has the form:

$$v_x = \tfrac{1}{4}(G/\eta)r^2 + C. \tag{19.15}$$

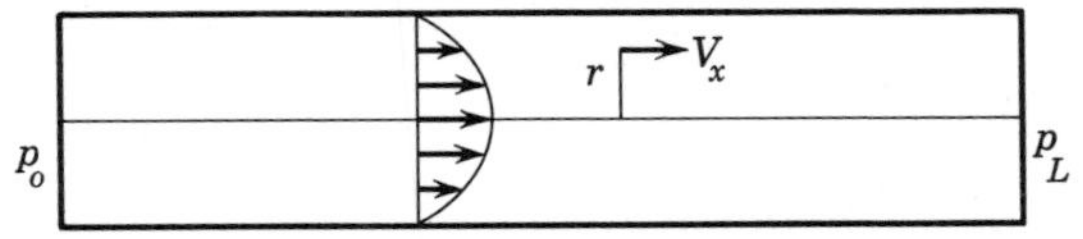

FIG. 19.2. Poiseuille flow through a tube, Eq. 19.12

If the tube has radius R, and the fluid is assumed to stick on the walls keeping $v_x=0$ at $r=R$, this fixes the constant C and we have

$$v_x = \tfrac{1}{4}(G/\eta)(r^2 - R^2) \tag{19.16}$$

for the velocity at radius r from the axis of the tube. It is proportional to the pressure gradient G, in the opposite direction, down the gradient. We can use this result to integrate the total volume flow of fluid through the tube per unit time, Q, finding in fact

$$Q = \pi R^4(p_2 - p_1)/8\eta L, \tag{19.17}$$

where $p_2 - p_1$ is the pressure drop down the length L. This result states that the volume flow is proportional to the fourth power of the radius of the tube; it is well verified experimentally, and Eq. 19.17 is the basis of a standard practical method of measuring viscosity.

Exercise II.19

19.1. The empirical relation between shear stress and velocity in a viscous fluid is quoted in Eq. 19.1 as $X_y = \eta \partial v_x/\partial y$. The analogous relation for flow in the y-direction is obviously $Y_x = \eta \partial v_y/\partial x$. If we were to subtract these two equations, we would have

$$X_y - Y_x = -\eta(\text{curl } \mathbf{v})_z.$$

However, we have seen that the stress matrix is necessarily symmetrical, $Y_x = X_y$; so this last equation would suggest that the motion of a viscous fluid is always irrotational. Can you detect the fallacy of this argument?

19.2. Derive Eq. 19.10 from Eq. 15.11 under the conditions of Poiseuille flow stated in the text.

19.3. Write down the complete set of individual equations contained in Eq. 19.4 for the case in which $\nabla \cdot \mathbf{v} = 0$ and $\partial v_x/\partial y$ is the only nonzero component derivative of the velocity.

19.4. Check that $A \ln r + C$ is a solution of Laplace's equation.

19.5. Verify Eq. 19.17 for the integrated flow.

19.6. Solve the problem of Poiseuille flow through the annular space between two coaxial cylinders of radii a and b, assuming the fluid sticks to both boundaries.

19.7. The annular space between two coaxial cylinders, radii a and b, is filled with incompressible viscous fluid. A constant relative rotational velocity is maintained about their common axis, taken as the z-axis. Assume that the fluid sticks to both cylinders, and that steady stream-line motion occurs in the fluid round the axis. The linear velocity at any point x, y in the fluid has components $v_x = -y\omega$, $v_y = x\omega$, where ω is the angular velocity, a function of r only, $r^2 = x^2 + y^2$. Use Eq. 19.5 to prove that the stress in the fluid is the negative hydrostatic pressure plus the following matrix:

$$\eta r d\omega/dr \begin{bmatrix} -\sin 2\theta & \cos 2\theta & 0 \\ \cos 2\theta & \sin 2\theta & 0 \\ 0 & 0 & 0 \end{bmatrix}.$$

19.8. Taking the normal unit vector $\mathbf{n} = \mathbf{i} \cos \theta + \mathbf{j} \sin \theta$, use Eq. 12.2 to prove that the force per unit area across any cylinder of radius r in the above exercise is a pure tangential shear equal to $S = \eta r d\omega/dr$.

19.9. The net torque on any annular element must vanish for steady motion. From this fact, set up the differential equation satisfied by ω in the last two exercises and show that, if $a > b$, $\omega = [a^2\Omega/(a^2 - b^2)](1 - b^2/r^2)$, where Ω is the relative angular velocity of the two cylinders.

Show also that the torque on the outer cylinder is $4\pi\eta h a^2 b^2 \Omega/(a^2 - b^2)$, where h is the height of the cylinder.

19.10. Show that $(\text{curl } \mathbf{v})_z = 2a^2\Omega/(a^2 - b^2)$, a constant. Find the linear velocity $\mathbf{v}$ and verify Stokes' theorem for any annular area bounded by the two cylinders.

II.20 Wave Motion in Viscous Fluids. In this section we shall consider the effect of viscosity on the transmission of small amplitude waves in a fluid. The final results are to produce damping of the compressional wave, and the possibility of a strongly damped transverse wave that is nonexistent in the ideal nonviscous case.

To discuss the compressional wave, we rewrite the approximate equation of motion, Eq. 19.8, in the form:

$$\rho \partial^2 \mathbf{u}/\partial t^2 = -\nabla p + \eta \nabla^2 \partial \mathbf{u}/\partial t + (\eta + B)\nabla\nabla \cdot \partial \mathbf{u}/\partial t. \tag{20.1}$$

Take the divergence of this equation and recall that all the partial derivative operators commute: then we get

$$\rho \partial^2(\nabla \cdot \mathbf{u})/\partial t^2 = -\nabla^2 p + (2\eta + B)\nabla^2 \partial(\nabla \cdot \mathbf{u})/\partial t. \tag{20.2}$$

Comparing Eq. 18.3 and Eq. 11.13 we write $\nabla \cdot \mathbf{u} = -(p - p_0)/K$ and get

$$(\rho/K)\partial^2 p/\partial t^2 = \nabla^2(p + D\partial p/\partial t), \tag{20.3}$$

where

$$D = (2\eta + B)/K.$$

This is the standard form of damped wave equation (*cf.* Eq. 9.2), the first-order derivative term being responsible for the damping. If, in Eq. 20.3, we substitute a solution of the form

$$p = p_0 \exp i(gx+\omega t), \tag{20.4}$$

we find

$$g^2(1+i\omega D) = \omega^2\rho/K. \tag{20.5}$$

If we require ω to be a real quantity, this demands that g be complex. Writing, therefore,

$$g = k+i\gamma, \tag{20.6}$$

we can solve Eq. 20.5 to find:

$$\left.\begin{aligned} 2k^2 &= (\rho\omega^2/K)[(1+\omega^2D^2)^{1/2}+1]/(1+\omega^2D^2) \\ 2\gamma^2 &= (\rho\omega^2/K)[(1+\omega^2D^2)^{1/2}-1]/(1+\omega^2D^2) \end{aligned}\right\}. \tag{20.7}$$

When the damping term ωD is small enough so that its square may be neglected, we can approximate to Eq. 20.7 and find

$$\left.\begin{aligned} k &= \omega(\rho/K)^{1/2} \\ \gamma &= -\tfrac{1}{2}k\omega D \end{aligned}\right\} \tag{20.8}$$

and the solution, Eq. 20.4, becomes

$$p = p_0e^{-\gamma x} \exp [i(\omega t+kx)]. \tag{20.9}$$

This represents a wave traveling in the negative x-direction, damping to an amplitude ratio $1/e$ in a distance $1/\gamma$ depending on the viscosity and the frequency of the wave; the velocity of propagation is $c=(K/\rho)^{1/2}$.

To discuss the possibility of a damped transverse wave in a viscous fluid, we return to the approximate equation of motion, Eq. 19.8, and write it in the form:

$$\rho\partial\mathbf{v}/\partial t = K\nabla\nabla\cdot\mathbf{u}+\eta\nabla^2\mathbf{v}+(\eta+B)\nabla\nabla\cdot\mathbf{v}, \tag{20.10}$$

where we have used $p=-K\nabla\cdot\mathbf{u}$. For solenoidal motion, both $\nabla\cdot\mathbf{v}$ and $\nabla\cdot\mathbf{u}$ may be taken zero, so that this equation reduces to

$$\rho\partial\mathbf{v}/\partial t = \eta\nabla^2\mathbf{v}. \tag{20.11}$$

This is the standard form of diffusion equation; in addition to the usual solutions to be discussed in the next section, it has a quasi-oscillatory solution that is strongly damped and highly dispersive. Consider plane waves only, propagated in the x-direction. Since $\nabla\cdot\mathbf{v}=0$, we have only transverse components (*cf.* Eq. 14.19), and in particular v_y satisfies the equation:

$$\rho\partial v_y/\partial t = \eta\partial^2v_y/\partial x^2. \tag{20.12}$$

A solution of this equation that is oscillatory at $x=0$ is easily found:

$$v_y = A \exp i(\omega t + kx)\, e^{kx}, \tag{20.13}$$

where

$$k = (\omega\rho/2\eta)^{1/2}.$$

The damping factor is so large that the amplitude drops to a factor $1/e$ in a distance equal to one wavelength, and the velocity of propagation is $(2\eta\omega/\rho)^{1/2}$, depending directly on the frequency ω.

In all the foregoing discussion of viscous motion we have assumed that the velocity is small enough so that $\partial\mathbf{v}/\partial t$ can be used in place of $d\mathbf{v}/dt = \partial\mathbf{v}/\partial t + \mathbf{v}\cdot\nabla\mathbf{v}$, on the left-hand side of Eq. 19.7. The nonlinear term $\mathbf{v}\cdot\nabla\mathbf{v}$ has been neglected throughout. If we consider a viscous fluid moving through a uniform tube, it is to be expected that when the velocity relative to the walls of the tube is allowed to become too great, the simple solutions we have found would break down. It is not easy to see theoretically exactly what the limiting speed would be for this breakdown to occur. We observe from Eq. 19.10, however, that so long as we can maintain the stream velocity exactly parallel to the walls of the tube, the nonlinear term remains exactly zero, and the simple solution remains true no matter how great the velocity. However, if a small disturbance were to introduce a component of velocity in any other direction, the nonlinear term would enter the equation and possibly produce a large result. The motion is therefore unstable and could break up into erratic eddy currents. This breakup by instability is observed to occur in flow through tubes when the dimensionless *Reynolds number* is greater than about 2000:

$$R = v\rho A/\eta > 2000, \tag{20.14}$$

where A is the radius of the tube. A similar relation appears to hold for boundaries of any shape, with appropriate definitions of the linear factor replacing A in each case. The theory of this effect is still under investigation. One may rationalize Eq. 20.14 somewhat as follows. We note that the nonlinear term $\rho\mathbf{v}\cdot\nabla\mathbf{v}$ must always remain small compared with the term $\eta\nabla^2\mathbf{v}$ for the simple solutions to be valid. This means that, for steady stream-line motion in the x-direction, we want

$$\rho v_y \partial v_x/\partial y < \eta\partial^2 v_x/\partial y^2$$

and hence

$$R = \rho v_x A/\eta < A(v_x/v_y)(\partial^2 v_x/\partial y^2)/(\partial v_x/\partial y). \tag{20.15}$$

But, in general, $A(\partial^2 v_x/\partial y^2)/(\partial v_x/\partial y)$ is of order unity, so that stability requires

$$R < v_x/v_y. \tag{20.16}$$

In practice it seems to be impossible to make sure that lateral deviations in speed remain less than one in a thousand of the forward flow speed, so that looking at Eq. 20.16 we see that there must be a maximum to the value of R beyond which the stability will break down; and from Eq. 20.14 this breakdown can be induced by increasing the forward speed v. The resulting turbulent motion is beyond the scope of the mathematical tools developed in the present course.

Exercise II.20

20.1. Find the solution analogous to Eq. 20.9 that represents propagation in the positive direction of x.

20.2. Find the solution of Eq. 20.3 in the same form as Eq. 9.9 and discuss its physical meaning. How would the motion be initiated?

20.3. Verify that Eq. 20.13 satisfies Eq. 20.12.

20.4. Eq. 20.11 states that in solenoidal motion the linear velocity diffuses through the fluid. Discuss this physically: What are the physical conditions required to ensure solenoidal motion; how would one go about checking the prediction experimentally?

20.5. Show that Eq. 20.11 also implies that the vorticity, measured by $\nabla \times \mathbf{v}$, diffuses through the fluid; discuss this in terms of experimental procedures.

20.6. Write the Laplacian operator in Eq. 20.11 in terms of cylindrical polar coordinates; assume that the velocity is directed always perpendicular to the axis of the cylinder and is independent of the axial coordinate. Consider the experimental arrangement of Exs. 19.7 to 19.10 and find the solution of the diffusion equation when the outer cylinder is forced to undergo oscillatory rotations about its axis.

II.21 Diffusion Equations. Consider first the diffusion of heat in a temperature distribution. Empirically we know that the flow of heat across unit area is at least approximately proportional to the temperature gradient:

$$\text{Heat flux} = -K\nabla T \tag{21.1}$$

and that the quantity of heat per unit volume at any point changes very nearly in proportion to the change in temperature:

$$\text{Increase in heat content} = \rho C dT. \tag{21.2}$$

In these equations K is the heat conductivity, C is the heat capacity per unit mass, and ρ is the density. The law of conservation of energy now states: The rate at which heat inside some volume increases is equal to the rate of generation of heat minus the rate of flux out through the surface. Mathematically this becomes

$$\iiint \rho C(\partial T/\partial t)dV = \iiint P dV + \iint K\nabla T \cdot d\boldsymbol{\sigma}, \tag{21.3}$$

where P is the rate of generation of heat per unit volume, and the integration is over a fixed volume. Provided all the integrands are well-behaved functions, we can take the volume small enough to treat all the functions as constants and can then cancel the integrations and derive the differential equation for diffusion:

$$\partial T/\partial t = Q + a^2\nabla^2 T, \tag{21.4}$$

where

$$Q = P/\rho C, \quad \text{and} \quad a^2 = K/\rho C.$$

Q is called the *temperature source* and a^2 is the *diffusion constant.*

Next consider the diffusion of chemical mixtures. Experimentally the flow rate is proportional to the gradient of the concentration:

$$\rho\mathbf{v} = -a^2\nabla\rho, \tag{21.5}$$

where ρ now stands for the concentration of the particular constituent, and a^2 is its diffusion constant. The equation of conservation of matter for this constituent reads

$$\iiint(\partial\rho/\partial t)dV = \iiint Q dV - \iint \rho\mathbf{v}\cdot d\boldsymbol{\sigma}, \tag{21.6}$$

where Q is the source rate per unit volume due, for example, to chemical reactions. We may again assume that the integrands are well-behaved and take the volume small enough to cancel the integrations:

$$\partial\rho/\partial t = Q + a^2\nabla^2\rho, \tag{21.7}$$

which is identical with Eq. 21.4.

Finally consider the diffusion of neutrons in an atomic reactor, where there exists the possibility of the capture of neutrons in the pile by a process completely different from the mechanism of production. The rate of disappearance by capture is proportional to the number present and can be written ρC, where C is called the cross section for capture, and ρ is the numerical density of neutrons. The equation for conservation is now modified to read

$$\partial\rho/\partial t = Q - \rho C + a^2\nabla^2\rho, \tag{21.8}$$

the capture being equivalent to a negative source function.

In the last section we discussed diffusion of velocity in viscous fluids and solved the diffusion equation for oscillatory boundary conditions. Here we introduce Green's method of solution for arbitrary boundary conditions. First we consider a simple problem. Let us limit the discussion to one space dimension and imagine a source that suddenly produces a mass m at a point x_0 at a definite time t_0, no other source existing either then or at any other time. What happens after this sudden introduction of the mass m? The hypothetical source function can be given an idealized mathematical expression:

$$Q = m\delta(t-t_0)\delta(x-x_0), \tag{21.9}$$

which is the product of two Dirac delta functions. Q is the rate of production of mass per unit length; the total mass produced is the integral of Q as a function of time and position:

$$\int\int Q dx dt = m\int\int \delta(t-t_0)\delta(x-x_0)dx dt = m. \tag{21.10}$$

Putting this source function into Eq. 21.7, it is clear that we want the solution, for times greater than t_0, of the equation

$$\partial\rho/\partial t = a^2\partial^2\rho/\partial x^2 + m\delta(t-t_0)\delta(x-x_0). \tag{21.11}$$

Now, for all times actually later than t_0, the source function is zero. What we really want is a solution of the diffusion equation with zero source, which, in the limit $t \to t_0$, approaches a point concentration of mass m at x_0. It turns out that such a solution can easily be written down:

$$\rho(x, t) = \frac{m}{2a\sqrt{\pi(t-t_0)}} \exp\left[-\frac{(x-x_0)^2}{4a^2(t-t_0)}\right]. \tag{21.12}$$

This can be verified by direct substitution to be a solution of the diffusion equation with zero source. But when we let t approach t_0, it becomes a concentrated mass at the point x_0, as can be seen at once if we sketch the shape of the function. We can also prove by integration that the total mass present is independent of time and equal always to m:

$$\int \rho(x, t)dx = m. \tag{21.13}$$

The solution equation (21.12) for a concentrated instantaneous source is called Green's function for the diffusion equation.

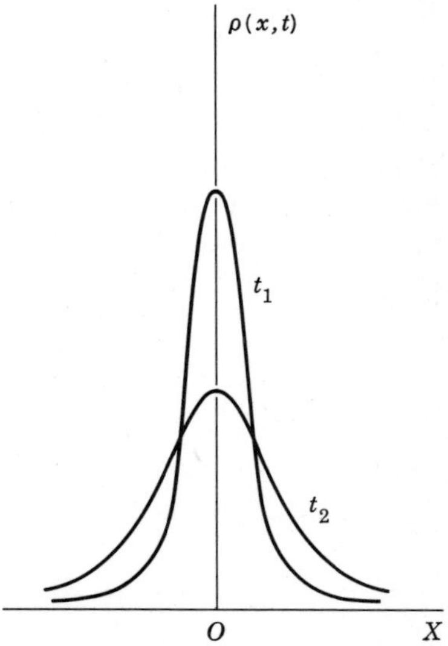

FIG. 21.1. Green's solution for the diffusion equation, drawn at two different times, $t_2 > t_1$. At zero time it is a Dirac delta function located on the origin

Consider next a general source function $Q(x, t)$. In any small element of position and small interval of time, dx_0 and dt_0, around the point instant x_0, t_0, the source produces a mass equal to

$$m = Q(x_0, t_0)dx_0 dt_0, \tag{21.14}$$

We can picture the whole distribution of source as a collection of sudden point sources, one to each small element of position and each small element of time. The problem is then to find the solution of the diffusion equation when we have to add the effects of all these small sources. We can easily see that the solution is just the sum of the solutions for the individual small sources. Thus, write Eq. 21.7 in the form:

$$\partial\rho/\partial t = \sum_n Q_n + a^2\nabla^2\rho, \tag{21.15}$$

with a sum over sources, $n=1, 2, \cdots$. If for any one source

$$\partial\rho_n/\partial t = Q_n + a^2\nabla^2\rho_n, \tag{21.16}$$

it is clear that $\sum_n \rho_n$ satisfies Eq. 21.15, while ρ_n is just Green's function for the source Q_n. Combining all the small point sources like those in Eq. 21.14 and summing the elements in the form of an integration, we have for the solution corresponding to the general source function $Q(x_0, t_0)$:

$$\rho(x, t) = \int_{-\infty}^{\infty} dx_0 \int_{-\infty}^{t} dt_0 \frac{Q(x_0, t_0)}{2a\sqrt{\pi(t-t_0)}} \exp\left[-\frac{(x-x_0)^2}{4a^2(t-t_0)}\right]. \tag{21.17}$$

Notice that the integration of source times t_0 goes only as far as the time t at which we estimate the function $\rho(x, t)$; later sources do not contribute to the result at this time.

We turn now to the following problem: given an initial distribution of density at time $t=0$, say $\rho(x, 0)$, a given function of x, and ask what happens in the future following this initial distribution of mass, with no sources present. We can solve this by using Green's functions through the following device. Suppose the initial density distribution is produced instantaneously from empty space by means of a suitably distributed collection of point-instant sources. To do this we would have to put Dirac delta function type sources everywhere, producing masses $\rho(x_0, 0)dx_0$ in the element of position dx_0 at the point x_0. This has to be written for the m in Eq. 21.12, and t_0 is to be set zero in that equation, and we then have for the combined effect of all these sources:

$$\rho(x, t) = \int dx_0 \frac{\rho(x_0, 0)}{2a\sqrt{\pi t}} \exp\left[-(x-x_0)^2/4a^2t\right]. \tag{21.18}$$

We can verify that when t approaches zero, this actually becomes the desired identity $\rho(x, t) \to \rho(x, 0)$, because the integral over x_0 acts like the Dirac delta function when t goes to zero and selects out the value of the integrand at which $x_0 = x$.

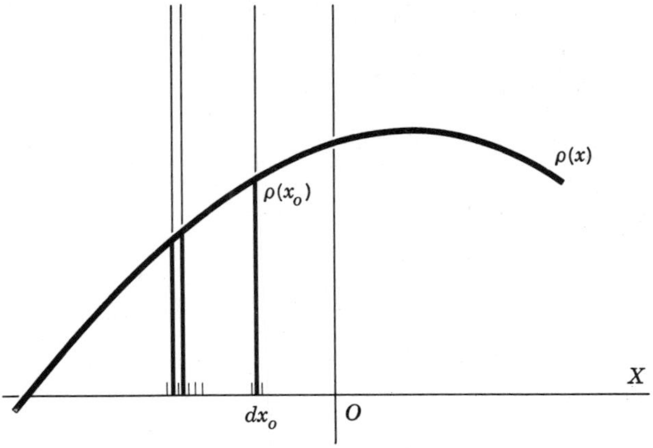

FIG. 21.2. Green's method for the initial value problem in diffusion: building up an arbitrary function by summing over a large number of Dirac delta functions

This discussion of diffusion is perhaps as far as one can go in discussing an essentially atomic process in terms of the continuum theory of matter. In fact, we have pushed the continuum theory a little too far already. Thus, consider the implications of the solution to the diffusion equation given in Eq. 21.12 for the effects to be expected immediately following the time t_0. The solution of course implies, as we wished it to imply, that at the time t_0 there is a concentration of material at x_0, and no matter elsewhere; then at a time infinitesimally close to t_0 but later than t_0, there is a finite though small density of matter to be found everywhere in space because the function drops only asymptotically to zero at great distances. There must then be an indefinitely great speed of propagation of a minute quantity of matter from the point of concentration out to great distances at the instant immediately following the initial concentration at t_0. Physically this is of course impossible; the maximum speed of propagation is obviously limited to the speed of the individual atoms, and this cannot be indefinitely great in any actual physical system.

This dilemma tells us that the diffusion equations (21.7 and 21.8, etc.) are erroneous at least for times too close to the occurrence of any very sudden change in the concentration. They can be accepted as reliable only when the concentration is not changing too rapidly, and to set up definite rules for deciding when the equation is acceptable we need to study the atomic processes in detail. It is found in fact from kinetic theory that the more exact equation replacing Eq. 21.7 is

$$(1/a^2)\partial\rho/\partial t = \nabla^2\rho - (1/c^2)\partial^2\rho/\partial t^2, \qquad (21.19)$$

where c^2 is related to the mean square speed of the atoms. The initial change is therefore more like a damped wave propagated with the speed c and, after the wave has damped out, the diffusion process follows in its wake.

Exercise II.21

21.1. Verify by substitution that Eq. 21.12 is a solution of Eq. 21.11.
21.2. Verify Eq. 21.13.
21.3. Verify by substitution that

$$\rho(x, y, t) = m(2a\sqrt{\pi t})^{-2} \exp(-r^2/4a^2t),$$

with $r^2 = x^2 + y^2$, is a solution of the diffusion equation in two dimensions, of Green's type.

21.4. Verify by substitution that

$$\rho(x, y, z, t) = m(2a\sqrt{\pi t})^{-3} \exp(-R^2/4a^2t),$$

with $R^2 = x^2 + y^2 + z^2$, is a solution of the diffusion equation in three dimensions, of Green's type.

21.5. Verify that in each of the last two exercises the total mass is m.
21.6. Consider a steady source with a wave form in space:

$$Q(x_0, t_0) = q \sin kx_0.$$

Use this source function in Eq. 21.17 and show that the resulting density is

$$\rho(x, t) = (q/k^2a^2) \sin kx.$$

Hint: write $y = x - x_0$ and integrate with respect to y, keeping x constant, using P.T. 508. Note that the source varies about zero, with both positive and negative phases. The factor $1/k^2a^2$ can be regarded as the admittance of the medium, the ratio of the steady response to the steady driving "force."

21.7. Repeat the last exercise with the source

$$Q(x_0, t_0) = q(1 + \sin kx_0).$$

21.8. Consider a source-free region in which we initially set up the density distribution:

$$t = 0: \qquad \rho(x, 0) = \rho_0(1 + \sin kx),$$

where ρ_0 is a constant. Show that the resulting density distribution tends to flatten out in a time proportional to the admittance of the medium.

References

The following texts contain useful material for collateral reading with the contents of this chapter. Material can also be found in the texts quoted at the end of Chapter I.

Mechanics, J. C. Slater and N. H. Frank, McGraw-Hill Book Co., Inc., New York, 1947.

Mathematical Physics, D. H. Menzel, Prentice-Hall, Inc., Englewood Cliffs, N.J., 1953.

Physical Mathematics, C. H. Page, D. Van Nostrand Co., Inc., Princeton, N.J., 1955.

Hydrodynamics, Horace Lamb, Dover Publications, Inc., New York, 1932.

A Treatise on the Mathematical Theory of Elasticity, A. E. H. Love, Dover Publications, Inc., New York, 1944.

Partial Differential Equations in Physics, A. Sommerfeld, Academic Press, New York, 1949.

Introduction to Solid State Physics, C. Kittel, John Wiley and Sons, Inc., New York, 1956.

Solid State Physics, A. J. Dekker, Prentice-Hall, Inc., Englewood Cliffs, N.J., 1957.

Supersonic Flow and Shock Waves, R. Courant and K. O. Friedrichs, Interscience Publishers, Inc., 1948.

Chapter III

THE MOLECULAR THEORY OF MATTER

Chapter III

THE MOLECULAR THEORY OF MATTER

III.22 General Dynamical Theorems. Up to this point we have been discussing the properties of continua, and we have found that many properties of actual materials do in fact coincide very closely with predictions based on the assumption of continuity within matter. However, to go further into detail requires that we take into account the molecular structure of matter, and for this purpose we must take up a study of the properties of very large numbers of particles interacting with each other in various ways.

Let $\mathbf{r}_k$ be written for the position vector of the kth particle relative to some arbitrary set of inertial coordinate axes. Let the mass of the kth particle be m_k. Let some field of force act in the region occupied by the system of particles, such that the force exerted on the kth particle be $\mathbf{F}_k$—which may be a vector function of both time and the position of the particle. Let the particles interact with each other, and write $\mathbf{F}_{jk}$ for the force acting on the kth particle due to the jth particle; we regard $\mathbf{F}_{jk}$ as a function of the positions of the two particles, but not explicitly a function of the time. According to Newton's third law of motion we have

$$\mathbf{F}_{jk} = -\mathbf{F}_{kj}. \tag{22.1}$$

The equation of motion for the kth particle in Newtonian mechanics is:

$$m_k d^2\mathbf{r}_k/dt^2 = \mathbf{F}_k + \sum_j \mathbf{F}_{jk}, \tag{22.2}$$

the sum being over all values of the subscript j. Note that $\mathbf{F}_{jj} = 0$ from Eq. 22.1—there is no force equal and opposite to itself. The position vector $\mathbf{r}$ of the center of mass of the system is defined by:

$$M\mathbf{r} = \sum_k m_k \mathbf{r}_k; \quad M = \sum_k m_k. \tag{22.3}$$

Summing up all the equations like Eq. 22.2 over the values of k, we can show that

$$Md^2\mathbf{r}/dt^2 = \sum_k \mathbf{F}_k. \tag{22.4}$$

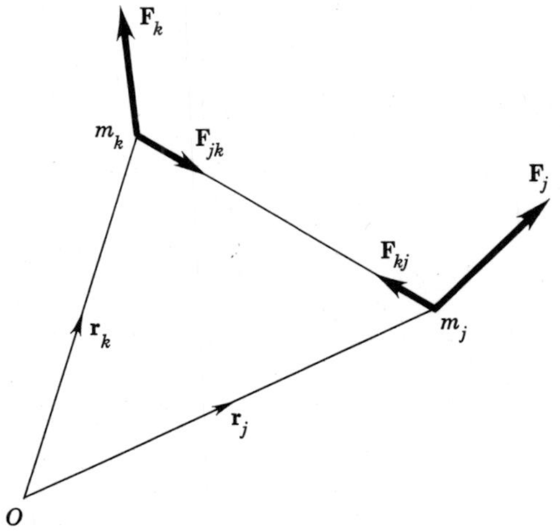

FIG. 22.1. The notation used to describe the forces on a pair of interacting particles

The interaction terms all cancel out and the result states that the position of the center of mass changes as if all the mass M were concentrated there and were acted on by the total resultant of all the external forces.

The total *angular momentum* of the system is defined as $\mathbf{L}$:

$$\mathbf{L} = \sum_k m_k \mathbf{r}_k \times d\mathbf{r}_k/dt. \tag{22.5}$$

If the interaction forces $\mathbf{F}_{jk}$ are parallel to the line joining the interacting particles, we call them *central forces*, and we can show from Eqs. 22.5 and 22.1 that the rate of change of the total angular momentum equals the total moment of the external forces only:

$$d\mathbf{L}/dt = \sum_k \mathbf{r}_k \times \mathbf{F}_k. \tag{22.6}$$

Write $\mathbf{r}_k'$ for the position vector of the kth particle relative to the center of mass:

$$\mathbf{r}_k' = \mathbf{r}_k - \mathbf{r}. \tag{22.7}$$

Then the kinetic energy can be written:

$$\text{K.E.} = \sum_k \tfrac{1}{2} m_k (d\mathbf{r}_k/dt)^2 = \tfrac{1}{2} M (d\mathbf{r}/dt)^2 + \tfrac{1}{2} \sum_k m_k (d\mathbf{r}_k'/dt)^2, \tag{22.8}$$

which means that the kinetic energy can be separated into two parts, one being that of the total mass M moving with the center of mass, the other being the kinetic energy of relative motion relative to the center of mass.

Take the scalar product of the equation of motion, Eq. 22.2, and the term $d\mathbf{r}_k/dt$, and sum over all values of k. The left side becomes

$$\tfrac{1}{2}(d/dt)\sum_k m_k(d\mathbf{r}_k/dt)^2,$$

so we may integrate the resulting equation with respect to time and find

$$(\text{K.E.})_B-(\text{K.E.})_A = \int_A^B \sum_k \mathbf{F}_k\cdot d\mathbf{r}_k+\int_A^B \sum_k\sum_j \mathbf{F}_{jk}\cdot d\mathbf{r}_k. \tag{22.9}$$

The first term on the right is clearly the decrease in the potential energy of the system due to its displacements in the external field of force. The second term on the right can, by Eq. 22.1 be reexpressed as

$$\int_A^B \tfrac{1}{2}\sum_j\sum_k \mathbf{F}_{jk}\cdot d(\mathbf{r}_k-\mathbf{r}_j)$$

and so is equal to the decrease in the potential of the interactions due to the changes in the relative positions of the particles. Thus the *conservation of energy*, (K.E.)+(P.E.)=constant, is a direct consequence of the Newtonian equations of motion.

The general statement of any dynamical problem involving several particles must specify the external field of force as a function of position and time, and the interaction between any pair of particles as a function of the positions of the two particles. The equations of motion, Eq. 22.2, are then simultaneous differential equations in the position variables as functions of time. If there are N particles, there are $3N$ variables and in effect $3N$ simultaneous equations. If the initial positions and the initial velocities of all the particles are given at some arbitrary initial time, the equations can in principle predict the positions and velocities of all the particles at any other time—past or future. The notation set up here is adequate when there are only two or three particles in the system. We are however interested here in problems where there are perhaps many millions of particles, and we need a more powerful notation, to which we turn in the next section.

Exercise III.22

22.1. From Eq. 22.2 write down the detailed equations of motion of a pair of particles in a uniform gravitational field and attracting each other with a force varying inversely as the square of the distance between them.

22.2. Derive Eq. 22.4.

22.3. Derive Eq. 22.6.

22.4. From Eq. 22.5 prove that

$$\mathbf{L}= M\mathbf{r}\times d\mathbf{r}/dt+\sum m_k\mathbf{r}_k'\times d\mathbf{r}_k'/dt,$$

where $\mathbf{r}_k'$ is the position of the kth particle relative to the center of mass, Eq. 22.7 ; interpret this equation.

22.5. Derive Eq. 22.8.

22.6. Set up the equation of motion for a single particle attracted to a straight line with a force acting in a direction perpendicular to the line and a function only of the radial distance from the particle to the line. Discuss the possible motions. Use cylindrical coordinates.

22.7. Set up the equation of motion for a particle attracted to a center by a force acting in the direction of the radius from the particle to the center and proportional to the radial distance. Do this both in spherical polar coordinates and in cartesian coordinates. Discuss the possible motions.

22.8. Two particles attract each other with a noncentral force defined as follows:

$$\mathbf{F}_{12} = k[\mathbf{r}_2 - \mathbf{r}_1 - (r/c)(d\mathbf{r}_2/dt - d\mathbf{r}_1/dt)],$$

where k is a constant, c a constant velocity, r the magnitude of the vector $\mathbf{r}_2 - \mathbf{r}_1$. Find the rate of change of the total angular momentum. Is Eq. 22.6 true in this case? Is the conservation of energy true here?

III.23 The Concept of Phase Space.

In discussing the motion of a single particle, *e.g.* the problem of a simple harmonic oscillator, we find that the equation of motion is a second-order differential equation that can in principle be integrated to give the position and velocity of the particle as a function of time if we are first given the position and velocity of the particle at one arbitrary initial time. Similarly, in the discussion of the motion of continuous media, the second-order equation of motion can in principle be integrated to give the configuration of the medium (string, membrane, elastic solid, etc.) as a function of the time if we are given the initial configuration and state of motion. In the last section we have shown that the same general theorem applies to systems of many interacting particles; given the initial positions and velocities of all the particles, we can in principle determine the future (and past) positions and velocities of all the particles, provided, of course, we know the law of force between them.

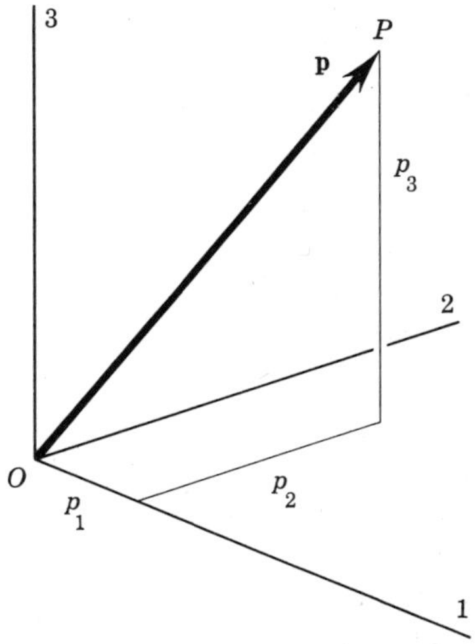

FIG. 23.1. The momentum space for a single particle

It is often convenient to substitute momentum for velocity in the foregoing theorem; given the position and momentum at time t_0, the equations of motion tell us the position and momentum at any other time t. The combined information involved in specifying all the positions and all the momenta of the system is called collectively the *phase* of the system. Given the phase at t_0, we can find the phase at t.

Consider any system with f degrees of freedom; for instance, we may have N particles, with $f=3N$ degrees of freedom; or we may have N spheres each of which has 6 degrees of freedom, so that $f=6N$. The number of degrees of freedom equals the least number of coordinates required to specify a configuration of the system. The number of momentum components required to fix the state of motion is also equal to f. In Cartesian coordinates there would be N vectors with $3N$ components for N particles, which is just the notation of §22. In general we shall now write

position coordinates:

$$q_j \quad (j = 1, 2, \cdots f) \tag{23.1}$$

and

momentum variables:

$$p_j \quad (j = 1, 2, \cdots f) \tag{23.2}$$

The ordered set of components

$$\mathbf{q} = (q_1, q_2, q_3, \cdots q_f) \tag{23.3}$$

may be regarded as an f-dimensional q-vector in the coordinate space of the system, and the ordered set

$$\mathbf{p} = (p_1, p_2, p_3, \cdots p_f) \tag{23.4}$$

as an f-dimensional p-vector in the *momentum space* of the system. *Phase space* is a combination of coordinate space and momentum space —a product space in a sense to be discussed later. Algebraic operations with the q-vectors are defined in coordinate space, and algebraic operations with the p-vectors are defined in momentum space. These geometrical terms have been discussed in Chapter I. A single point in coordinate space fixes the positions of all the parts of the system, and a single point in momentum space fixes the momenta of all the parts of the system. A line in coordinate space traces a history of the system's configuration, and a line in momentum space gives the *hodograph*, or history of momenta of all parts of the system.

As an almost trivial example we may discuss the phase space of a linear harmonic oscillator with one degree of freedom. Phase space is a product of a one-dimensional coordinate space and a one-dimensional momentum space. We may therefore conceive of it as being a two-dimensional space and represent it on a plane diagram. The diagram can picture all possible combinations of position x and momentum p. Possible paths in this space would be found from the equation of motion or, alternatively, from the conservation of energy:

$$E = p^2/2m + m\omega^2 x^2/2, \tag{23.5}$$

which may be written instead in the form

$$p^2/2mE + x^2/(2E/m\omega^2) = 1. \tag{23.6}$$

This form suggests the equation of an ellipse in the p-x diagram, with semi-axes $(2mE)^{1/2}$ and $(2E/m\omega^2)^{1/2}$, and total area $2\pi E/\omega$. These ellipses show all the possible paths of the system in phase space.

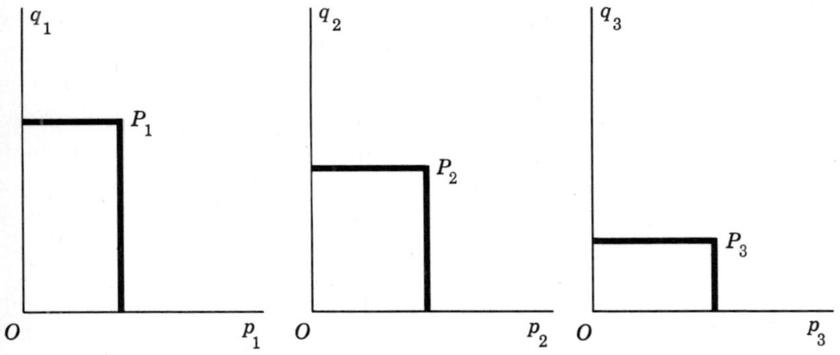

FIG. 23.2. Three sections through the phase space of one mass particle. The points P_1, P_2, and P_3 are the projections of a single representative point on the three planes. The six phase variables are determined by the coordinates of these three points

The three-dimensional oscillator requires a six-dimensional phase space, three dimensions for coordinates and three dimensions for momenta. It is impossible to visualize the product space in a simple picture; the best we can do is to draw three cross-section spaces, each section space being two-dimensional like the phase space for the linear oscillator. The possible paths in each section space are ellipses. Other problems can easily be presented in which Cartesian coordinates would be inconvenient; *e.g.* the problem of planetary motion of two gravitating masses. In general, we may start with the most obvious coordinates and define more generalized coordinates by setting up f independent functions of the original coordinates:

$$q_j' = q_j'(q_1, q_2, q_3, \cdots q_f). \tag{23.7}$$

We shall find it necessary to set up generalized momenta corresponding to these generalized coordinates:

$$p_j' = p_j'(p_1, p_2, p_3, \cdots p_f), \tag{23.8}$$

which are f more functions of the f original momenta. These functions must be correlated in some way with the functions employed in the generalized coordinates—in geometrical terms we say that a transformation of configuration space induces a transformation in momentum space.

Except in the very simplest cases, it is by no means obvious what transformation in momentum space is appropriate to go along with any given transformation in configuration space. For example, in going over from Cartesian coordinates to polar coordinates, one goes from linear momentum to angular momentum, and we are familiar with the procedure. With this as a guide, we may state the general principle of transformation in this way: the generalized momentum components are to be defined in such a way that Newton's laws of motion remain invariant—true with respect also to the generalized coordinates and momenta. In other words, the negative gradient of the potential energy in generalized coordinates equals the rate of change of the generalized momentum vector. Thus, in the original system, Newton's law states

$$-\partial V(q_1, q_2, q_3, \cdots q_f)\partial q_j = dp_j/dt, \tag{23.9}$$

with V standing for the scalar potential energy function. The new momenta are to be set up so that the equation is true in the new system:

$$-\partial V'(q_1', q_2', q_3', \cdots q_f')/\partial q_j' = dp_j'/dt, \tag{23.10}$$

where $V' = V$ is the potential energy expressed in terms of the new coordinates by means of Eq. 23.7. To follow this principle through in practice, we proceed as follows. In general, the kinetic energy is a quadratic function of the velocities, and in particular with Cartesian coordinates we have the kinetic energy in the form:

$$T(\dot{q}_1, \dot{q}_2, \dot{q}_3, \cdots \dot{q}_f) = \tfrac{1}{2} \sum_j m_j \dot{q}_j^2, \tag{23.11}$$

where m_j is the mass associated with the jth degree of freedom, and the dots denote differentiation with respect to time. The momentum $p_j = m_j \dot{q}_j$ is related to the kinetic energy by the equation

$$p_j = \partial T/\partial \dot{q}_j, \tag{23.12}$$

in Cartesian coordinates. In the next section we shall prove that

Eq. 23.10 is satisfied if we assume the new momenta are given by the equation obtained from Eq. 23.12 by assuming it invariant:

$$p_j' = \partial T'/\partial \dot{q}_j', \tag{23.13}$$

where $T' = T$ is the kinetic energy expressed in terms of the new coordinates by means of Eqs. 23.7. We may combine Eqs. 23.9 and 23.12 in one set of equations of motion in the form

$$(d/dt)(\partial T/\partial \dot{q}_j) = -\partial V/\partial q_j, \tag{23.14}$$

while Eqs. 23.10 and 23.13 are combined similarly to give:

$$(d/dt)(\partial T'/\partial \dot{q}_j') = -\partial V'/\partial q_j'. \tag{23.15}$$

The momenta have been eliminated here, so that if we can prove that Eqs. 23.14 are invariant for a coordinate transformation $q \rightarrow q'$, we have proved the required equivalence of the definition of p' in Eq. 23.13 with that in Eq. 23.10. That the invariance of Eq. 23.14 is plausible may be seen very easily in terms of the geometrical interpretation of phase space. Thus the right-hand side of Eq. 23.14 is the negative gradient in configuration space of the scalar V and therefore transforms like a vector. The left-hand side of Eq. 23.14 is the time derivative of a gradient in velocity space of the scalar T, and so also transforms like a vector: the velocity space is derivable directly from configuration space and the two transformations are unambiguously connected. The rigorous proof that Eq. 23.14 is invariant will be given in the next section.

Exercise III.23

23.1. Can a phase space have odd dimensionality?

23.2. The single equation, $E(\mathbf{p}, \mathbf{q}) = \text{constant}$, representing conservation of energy, selects a subspace having dimensionality $f-1$, where f is the dimensionality of the phase space. The phase orbit, or path in phase space, is a one-dimensional subspace. How many other equations are needed to determine the phase path of (a) a two-dimensional oscillator, (b) a sphere free to rotate about any axis, but having a fixed center?

23.3. A pair of particles free to move only in one plane attract each other with an inverse square law force. Sketch their phase space using plane polar coordinates, and indicate the subspaces of constant energy.

23.4. Set up the transformation equations from two-dimensional Cartesian coordinates to plane polar coordinates and use the principles involved in Eq. 23.14 to derive the momenta associated with the polar coordinates.

23.5. Do the same for three-dimensional spherical polar coordinates. Hint: show first that $T' = \frac{1}{2}m(\dot{r}^2 + r^2 \sin^2\theta\, \dot{\phi}^2 + r^2\dot{\theta}^2)$.

23.6. Do the same for three-dimensional cylindrical polar coordinates, with $T' = \frac{1}{2}m(\dot{r}^2 + r^2\dot{\theta}^2 + \dot{z}^2)$.

III.24 Hamilton's Variation Principle. Consider first a system with one degree of freedom represented by the coordinate q. Suppose we are given some function of position, velocity, and time, $L(t, q, \dot{q})$, and ask how a point should move in this one-dimensional space so that the time integral of L be a minimum or a maximum compared with the integral over other conceivable paths between the same two end points. In order to give a definite mathematical form to this question we can imagine the solution already known, *viz.*, the path is to be given by stating q as a function of time, $q = q(t)$. We can then form a family of slightly varied paths by taking functions $Q(t)$ to represent the altered motion, where

$$Q(t) = q(t) + \epsilon f(t), \tag{24.1}$$

ϵ being the family parameter, and $f(t)$ an arbitrary differentiable function defining the family. To compare all paths having the same end points

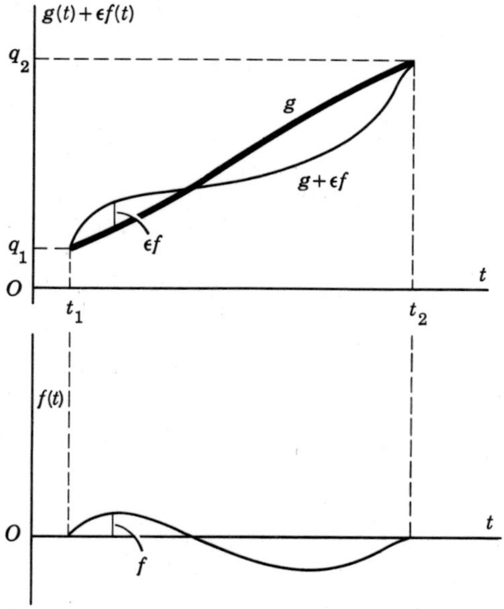

FIG. 24.1. The varied function $g(t) + \epsilon f(t)$ used in the derivations from Hamilton's variation principle, Eq. 24.1

(all alternatives start at a given instant and arrive together at another given instant), we make the functions $f(t)$ zero at both ends. The solution $q(t)$ coincides with the member of the family $\epsilon = 0$, whatever family $f(t)$ we may consider.

The minimum-maximum condition

$$\int L(t, q, \dot{q})dt = \text{minimum or maximum} \tag{24.2}$$

is then equivalent to

$$(\partial/\partial\epsilon)\int_{t_1}^{t_2} L(t, Q, \dot{Q})dt = 0, \quad \epsilon = 0, \tag{24.3}$$

and we want this to be true for all functions $f(t)$. Taking time derivatives of Eq. 24.1, we have

$$\dot{Q}(\epsilon, t) = \dot{q}(t) + \epsilon\dot{f}(t), \tag{24.4}$$

Since the limits t_1 and t_2 are independent of ϵ, the variations of L can be made without varying time, at each point on the path, and we can therefore differentiate the integral in Eq. 24.3:

$$\begin{aligned}(\partial/\partial\epsilon \int_{t_1}^{t_2} L(t, Q, \dot{Q})dt &= \int_{t_1}^{t_2} [(\partial L/\partial Q)(\partial Q/\partial\epsilon) + (\partial L/\partial\dot{Q})(\partial\dot{Q}/\partial\epsilon)]dt \\ &= \int_{t_1}^{t_2} [f(t)(\partial L/\partial Q) + \dot{f}(t)(\partial L/\partial\dot{Q})]dt. \end{aligned} \tag{24.5}$$

We have used Eq. 24.1 and Eq. 24.4 in the last step here. The second part of the last expression can be integrated by parts, and remembering that

$$f(t_1) = f(t_2) = 0, \tag{24.6}$$

we find that the integrated term vanishes and we have

$$\int_{t_1}^{t_2} \dot{f}(t)(\partial L/\partial\dot{Q})dt = -\int_{t_1}^{t_2} f(t)(d/dt)(\partial L/\partial\dot{Q})dt. \tag{24.7}$$

Eq. 24.5 now reduces to

$$\int_{t_1}^{t_2} f(t)[\partial L/\partial Q - (d/dt)(\partial L/\partial\dot{Q})]dt = 0. \tag{24.8}$$

According to Eq. 24.3, this equation has to be true when $\epsilon = 0$ no matter what the function $f(t)$ may be, consistent with Eq. 24.6. From Eq. 24.1, we therefore deduce that Eq. 24.8 is equivalent to

$$\int f(t)[\partial L/\partial q - (d/dt)(\partial L/\partial\dot{q})]dt = 0 \tag{24.9}$$

for arbitrary functions $f(t)$. We need a lemma, *viz.*, if

$$\int f(t)\, G(t)\, dt = 0$$

for arbitrary $f(t)$, then $G(t) = 0$ at all points in the range of integration. Proof: if $G(t)$ is not zero everywhere, there is some point at which it is positive (or negative). We can then find a small region around this point in which $G(t)$ has the same sign and is not zero, and an $f(t)$ that is zero everywhere except in this small region, where it is also of the same sign. Then the integral is positive, and not zero. Thus we have contradicted the initial hypothesis, which therefore requires $G(t)$ to be zero everywhere. Applying this lemma to Eq. 24.9 shows that everywhere

$$\partial L/\partial q - (d/dt)(\partial L/\partial \dot{q}) = 0, \tag{24.10}$$

which is known as *Lagrange's equation of motion*. This is a second-order differential equation relating the position q to the time t, and so determines the path of the system when the two end positions and times are given. If the function L were chosen

$$L = T(\dot{q}) - V(q), \tag{24.11}$$

Lagrange's equation (Eq. 24.10) would be identical with Eq. 23.14.

It is not difficult to generalize Lagrange's equation to apply to a system with any number of degrees of freedom. Thus we may consider a Lagrangian function of the form:

$$L(t, q_1, q_2, q_3, \cdots q_f; \dot{q}_1, \dot{q}_2, \dot{q}_3, \cdots \dot{q}_f), \tag{24.12}$$

with f degrees of freedom. We then replace Eq. 24.1 by f equations of the same type:

$$Q_j(\epsilon, t) = q_j(t) + \epsilon f_j(t), \tag{24.13}$$

and replace Eq. 24.3 by

$$(\partial/\partial\epsilon)\int_{t_1}^{t_2} L(t, Q_1 \cdots Q_f, \dot{Q}_1 \cdots \dot{Q}_f)dt = 0. \tag{24.14}$$

The first step in Eq. 24.5 must now contain in the integrand the sum over all the partial derivatives, each with respect to one of the f coordinates, or one of the f velocities. By choosing the arbitrary functions $f_j(t)$ all zero everywhere except one of them corresponding to one particular degree of freedom, we can remove all the other degrees of freedom from the problem. The selected degree of freedom then obeys equations identical with the one degree of freedom discussed already, namely Eq. 24.10. The final result is that every degree of freedom obeys a similar Lagrange equation:

$$\partial L/\partial q_k - (d/dt)(\partial L/\partial \dot{q}_k) = 0, \tag{24.15}$$

for all values of $k = 1, 2, 3 \cdots f$. These constitute f second-order differential equations and are sufficient in principle to determine the motion.

A generalization of Hamilton's principle is needed when the problem involves accessory conditions which restrict the possible variations that one is allowed to make in the paths. For example, we might be asked for the motion of a particle restricted to move on a given curved surface under the action of external forces. The functions $f_j(t)$ are then not all independent and arbitrary; they must be coupled in such a way that the particle remains on the given surface. The same mathematical techniques are applicable to many purely geometrical problems. For instance, one may ask for the surface of minimum area that can be constructed with a given periphery in space or for the curve of given length that encloses the maximum area on a given curved surface. In this second form, the area represents the Lagrangian function L, and the length of the curve is the accessory condition.

A formal statement of a general problem that includes most of these cases is the following: Find the path $q(t)$ between the times t_1 and t_2 for which the integral $\int L(t, q, \dot{q})dt$ is a minimum from among those paths for which the integral $\int G(t, q, \dot{q})dt$ has a given fixed value. G is some given function and represents the accessory condition. If we write

$$K(t, q, \dot{q}) = L(t, q, \dot{q}) + \lambda G(t, q, \dot{q}), \qquad (24.16)$$

we can prove that the problem is solved if $q(t)$ satisfies the equation:

$$\partial K/\partial q - (d/dt)(\partial K/\partial \dot{q}) = 0. \qquad (24.17)$$

Here λ is an arbitrary constant.

The proof of this theorem and its applications to interesting problems in geometry and dynamics will be found in more specialized texts on analytical dynamics or the calculus of variations.

The Lagrangian function of Eq. 24.12 is a scalar quantity under any coordinate transformation. The variation principle of Eq. 24.2 is therefore an invariant scalar equation for coordinate transformations. Hence the consequences of the variation principle must be true in any coordinate system and we are free to assume that any generalized coordinate system may be used in writing down the Lagrange equations in Eq. 24.15. These equations are, however, equivalent to the equations of motion, Eq. 23.14, if one uses the Lagrangian function, Eq. 24.11; thus, Eq. 23.14 is hereby proved to be invariant—this is what we needed in the discussion of the last section concerning the choice of generalized momenta to go along with any given generalized coordinate system.

A helpful way of looking at the Lagrange equations of motion, Eq. 24.15, is to think of $\partial L/\partial q_k$ as the generalized force component, and $\partial L/\partial \dot{q}_k$ as the generalized momentum component:

$$p_k = \partial L/\partial \dot{q}_k. \qquad (24.18)$$

With the particular Lagrangian given in Eq. 24.11, they are in fact the ordinary force and ordinary momentum, respectively. With Eq.

24.18, the Lagrange equations become obvious generalizations of Newton's law.

A convenient notation, often employed, sets up the Hamilton principle Eq. 24.2 in the form

$$\delta \int L(t, q, \dot{q})dt = 0 \tag{24.19}$$

The δ-symbol indicates that one varies the path of integration by varying q and $\dot{q}$ as functions of t. When the variation vanishes, the integral has its extreme value.

Exercise III.24

24.1. Given a Lagrangian function like Eq. 24.11 with

$$T = \sum_j \tfrac{1}{2} m_j \dot{q}_j^{\,2},$$

find the equations of motion from Eq. 23.14.

24.2. Write down in detail the mathematical steps between Eqs. 24.5 and 24.8.

24.3. Prove Eq. 24.15 as outlined in the text.

24.4. Set up the Lagrangian of Ex. 24.1 when spherical polar coordinates are appropriate and the potential V is a function only of the radius:

$$L = \tfrac{1}{2} m[\dot{r}^2 + r^2\dot{\theta}^2 + r^2 \sin^2\theta\, \dot{\phi}^2] - V(r).$$

Set up the equations of motion in these coordinates from Eqs. 24.15.

24.5. In Ex. 24.4, prove that the motion is necessarily in one plane through the center of the field. Hint: take the axis of coordinates perpendicular to the plane of motion at any arbitrary initial instant, $\theta = \pi/2$ and $\dot{\phi} = 0$, and prove that $\dot{\phi}$ is permanently zero.

24.6. Prove that in Ex. 24.4 the radius vector joining the particle to the origin sweeps out equal areas in equal times.

24.7. Develop the Lagrangian of Ex. 24.1 in cylindrical polar coordinates assuming the potential a function only of z. Set up the equations of motion from Eq. 24.15.

24.8. Consider a particle moving vertically in a constant gravitational field g where the path is given by

$$q(t) = u_0 t - \tfrac{1}{2} g t^2.$$

Set up the Lagrangian function $L(q, \dot{q}, t) = T - V$ for this motion and find the time integral $\int L dt$ between the time of projection vertically upward from some point to the time of return to the same point. Let the function defining the family of varied paths be

$$f(t) = \sin 3\pi t/T,$$

where T is the total time for the above path $q(t)$. Set up the Lagrangian for the varied paths $L(Q, \dot{Q}, t)$ with $Q(t)$ given by Eq. 24.1, and find the integral $\int L dt$ on the varied paths as a function of ϵ for the same T. Prove that the actual path coincides with the minimum of the integral.

III.25 The Principle of Least Action. Hamilton's variation principle was discovered in the nineteenth century by Sir William R.

Hamilton (1805–1865) and has become the basis of modern analytical dynamics; it has also come to be regarded as a model for universal physical laws. The search for a formulation of the laws of nature in terms of a basic variation principle was initiated more than 200 years ago by the metaphysical speculations of Count Pierre Louis Maupertuis (1698–1758). Maupertuis' *principle of least action* was given precise mathematical content, as far as the interpretation of dynamics was concerned, by Leonhard Euler (1707–1783) and Count Joseph Louis Lagrange (1736–1813). Here we shall give a deduction of the principle from (the historically much more recent) Hamilton's variation principle discussed in §24.

We note first that, in discussing Hamilton's principle, we compared all conceivable paths connecting two given points A and B at two given times t_1 and t_2; the system must travel from A to B in the same time for all the paths compared. The varied paths do not in general have to conform to any dynamical law—some may violate the conservation of energy in order to make the right total time of transit. We do not, in principle know any dynamical laws in setting up Hamilton's principle—we deduce them from the principle. On the other hand, in the principle of least action we shall restrict the comparison to all paths involving no violation of the conservation of energy, but relax the condition that all paths take the same length of time.

In Fig. 25.1 the two paths (1) and (2) connect the same two points A and B but in different times. The difference $b'-b$ on the diagram is $\epsilon f(t)_b$ and the time interval $t_{b2}-t_{b1}=\Delta t_b$. Evidently for small enough differences,

$$\epsilon f(t)_b = -\Delta t_b (dq/dt)_b. \tag{25.1}$$

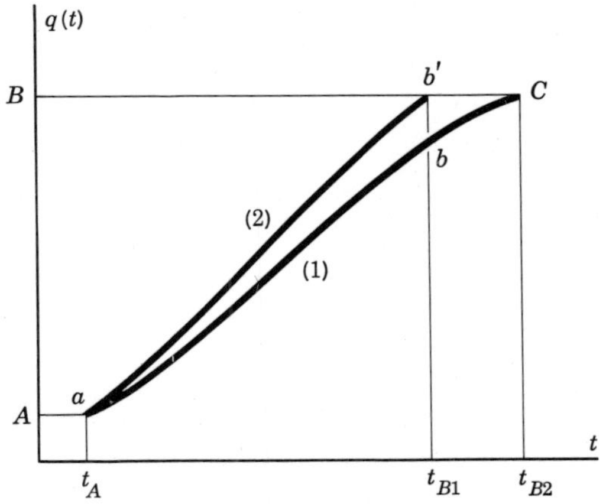

FIG. 25.1. The varied paths used in the principle of least action, Eq. 25.1

We may now evaluate the Lagrangian integral over the two paths *abc* and *ab′*. To do this, we can repeat the steps leading to Eq. 24.5, but Eq. 24.6 is not now correct. In the integration by parts preceding Eq. 24.7, the integrated term remains because Eq. 24.6 is not valid. The Lagrange equations, Eq. 24.10, can be assumed true for the actual paths, so that Eq. 24.5 now reduces to:

$$(\partial/\partial\epsilon)\int_a^b L dt = \left| f(t)(\partial L/\partial\dot{q})\right|_a^b, \tag{25.2}$$

and from Eq. 25.1 this gives:

$$\epsilon(\partial/\partial\epsilon)\int_a^b L dt = -\Delta t_b (dq/dt)_b (\partial L/\partial\dot{q})_b. \tag{25.3}$$

For infinitesimal values of ϵ this is the difference between the integrals from a to b' and from a to b. To get the difference between the integrals a to b' and a to c, we must subtract the integral from b to c which, because the path is infinitesimal, is simply $-L\Delta t_b$. Therefore, the complete difference is:

$$\Delta\int L dt = -\Delta t_b \dot{q}_b (\partial L/\partial\dot{q})_b + L\Delta t_b. \tag{25.4}$$

From Eq. 24.18, $\dot{q}(\partial L/\partial\dot{q}) = p\dot{q}$, and in Cartesian coordinates $p\dot{q} = 2T$, twice the kinetic energy. Therefore,

$$\Delta\int L dt = (L-2T)_b \Delta t_b = -H\Delta t_b, \tag{25.5}$$

where we have used Eq. 24.11 and defined

$$H = T+V. \tag{25.6}$$

H is called the *Hamiltonian function* and is equal, in this case, to the total energy. If we now restrict the variations to all paths for which H has the same constant value—obeying the conservation of energy—we have

$$H\Delta t_b = H\Delta \int_a^b dt = \Delta \int_a^b H dt,$$

noting that $\Delta t_a = 0$ by definition. Then Eq. 24.5 becomes

$$\Delta\int (L+H) dt = 0,$$

or

$$2\Delta\int T dt = 0. \tag{25.7}$$

This is the simplest form of the principle of minimum action. Although we have proved it only for Cartesian coordinates, it is an invariant form of equation and can therefore be accepted as true also in generalized coordinates. It is valid here only for one degree of freedom, but it is easy to prove that it is valid also for any number of degrees of freedom. In this general case, it is often stated in the equivalent form

$$\Delta\int\sum_j p_j\dot{q}_j dt = 0. \tag{25.8}$$

We emphasize again that Hamilton's principle, $\delta\int L dt = 0$, compares the integrated expression between two given points over all conceivable paths taking the same time, while Maupertuis' principle compares the integrated action over all paths having the same energy, not necessarily taking the same time. It is important to realize that from a logical point of view, one has to assume the conservation of energy to prove that Maupertuis' principle selects actual paths, while Hamilton's principle implies the conservation of energy without additional assumption.

Exercise III.25

25.1. Note that in the case of a free particle, $V=0$, the principle of least action compares paths of equal kinetic energy. Prove that this leads to two conclusions: (a) the path of a free particle between two fixed points is the line of minimum length, and (b) it is also the path of minimum time.

25.2. Derive Eq. 25.2 in detail.

25.3. Prove that in a constant gravitational field g, the path traced between two points $x=0$, $y=0$, and $x=L$, $y=0$, by a projectile of total energy E is one of two parabolas:

$$y = ax/b - \tfrac{1}{2}g(x/b)^2$$

or

$$y = bx/a - \tfrac{1}{2}g(x/a)^2,$$

where

$$a, b = \tfrac{1}{2}[(2E/m+gL)^{1/2} \pm (2E/m-gL)^{1/2}].$$

25.4. Find the kinetic energy as a function of time in the above example, and compute the action integral for the whole path.

25.5. Vary the path in the last two exercises as follows: let the particle follow the second parabola to a point nearly at the top; then let it change direction (but not its energy) to move horizontally at constant speed until it reaches the descending part of the original parabola, which it then follows to the final point. Compare the action integral in this case with that in Ex. 25.4.

III.26 Hamilton's Principle for Continuous Materials. Although we have presented Hamilton's variation principle as a general law of particle dynamics, it also is true for the dynamics of continuous materials, and it is worth digressing to see this in some detail. We consider, for example, the dynamics of an ideal fluid. The kinetic energy per unit volume of fluid is, in the notation of §11, $\frac{1}{2}\rho(\partial\mathbf{u}/\partial t)^2$, and the potential (elastic) energy per unit volume is given in Ex. 13.4 as $\frac{1}{2}K(\Delta V/V)^2 = \frac{1}{2}K(\nabla\cdot\mathbf{u})^2$. Therefore, the total kinetic energy of a given volume of material is

$$T = \tfrac{1}{2}\iiint\rho(\partial\mathbf{u}/\partial t)^2 dxdydz \tag{26.1}$$

and the total potential energy of the same volume is

$$W = \tfrac{1}{2}\iiint K(\nabla\cdot\mathbf{u})^2 dxdydz. \tag{26.2}$$

Hamilton's principle would then state, formally, that

$$\tfrac{1}{2}\delta\iiiint[\rho(\partial\mathbf{u}/\partial t)^2 - K(\nabla\cdot\mathbf{u})^2]dxdydzdt = 0, \tag{26.3}$$

assuming that $L = T - V$.

To give this formal statement a meaning, we note first that we cannot strictly speak of an integral along a path except in a very generalized sense. The actual problem would be to find the vector **u** as a function of position and time, and the variation would be considered as being caused by choosing varied conceivable functions for this in the integrand. The function $\mathbf{u}(x, y, z, t)$ that maximizes the integral would be the desired solution of the variation equation. When **u** is a vector we need three functions, one for each component, and we shall restrict the discussion to motion in one direction to simplify the analysis. A sufficiently general form of problem to cover the case of interest is, then, to find a scalar function $u(x, y, z, t)$ such that

$$\delta\iiiint\mathscr{L}(u, u_t, u_x, u_y, u_z, x, y, z, t)dxdydzdt = 0, \tag{26.4}$$

where $\mathscr{L}$ is some function of the indicated variables, u_t is $\partial u/\partial t$, u_x is $\partial u/\partial x$, etc. These last symbols are not to be confused with the components of a vector **u**. The function $\mathscr{L}$ is called a *Lagrangian density* because by integrating over space we get the Lagrangian function. The variation is effected by substituting $U(x, y, z, t)$ for $u(x, y, z, t)$ where, in analogy to Eq. 24.13,

$$U(x, y, z, t) = u(x, y, z, t) + \epsilon f(x, y, z, t). \tag{26.5}$$

Taking time derivatives, we have

$$U_t = u_t + \epsilon f_t, \tag{26.6}$$

and space derivatives

$$U_x = u_x + \epsilon f_x, \text{ etc.} \tag{26.7}$$

From these we can see that

$$\partial\mathscr{L}/\partial\epsilon = f\partial\mathscr{L}/\partial u + f_t\partial\mathscr{L}/\partial u_t + f_x\partial\mathscr{L}/\partial u_x + f_y\partial\mathscr{L}/\partial u_y + f_z\partial\mathscr{L}/\partial u_z, \tag{26.8}$$

when $\epsilon = 0$. This is the variation that is to be used in Eq. 26.4. Integrating this expression over space and time, as required in Eq. 26.4, one of the four integrations can be carried out partially in each of the last four terms. Thus we have

$$\int(\partial\mathscr{L}/\partial u_t)f_t dt = |(\partial\mathscr{L}/\partial u_t)f\,|_1^2 - \int f(\partial^2\mathscr{L}/\partial t\partial u_t)dt, \tag{26.9}$$

while

$$\int(\partial\mathscr{L}/\partial u_x)f_x dx = |(\partial\mathscr{L}/\partial u_x)f\,|_a^b - \int f(\partial^2\mathscr{L}/\partial x\partial u_x)dx, \tag{26.10}$$

and two similar expressions for the other two terms. If now we assume the arbitrary function $f(x, y, z, t)$ to be always chosen so as to vanish over all boundaries of the region—also at the initial time and the final time at all points in the region—the integrated parts of the last expressions are identically zero, and we end up with the following expansion for the variation integral:

$$\begin{aligned}\delta \iiiint \mathscr{L}\, dxdydzdt &= \iiiint f(x, y, z, t)[\partial\mathscr{L}/\partial u - \partial^2\mathscr{L}/\partial t\partial u_t \\ &\quad - \partial^2\mathscr{L}/\partial x\partial u_x - \partial^2\mathscr{L}/\partial y\partial u_y - \partial^2\mathscr{L}/\partial z\partial u_z]dxdydzdt \\ &= 0. \qquad (26.11)\end{aligned}$$

Since $f(x, y, z, t)$ is an arbitrary differentiable function, a simple generalization of the lemma proved in §24 suffices to deduce from Eq. 26.11 that

$$\partial\mathscr{L}/\partial u = \partial^2\mathscr{L}/\partial t\partial u_t + \partial^2\mathscr{L}/\partial x\partial u_x + \partial^2\mathscr{L}/\partial y\partial u_y + \partial^2\mathscr{L}/\partial z\partial u_z. \quad (26.12)$$

This is the Lagrange equation of motion for the continuous material, equivalent to the variation principle in Eq. 26.4. For the particular Lagrangian density,

$$\mathscr{L} = \tfrac{1}{2}\rho u_t^{\,2} - \tfrac{1}{2}K u_x^{\,2}, \quad (26.13)$$

it reduces simply to

$$0 = \rho\partial^2 u/\partial t^2 - K\partial^2 u/\partial x^2, \quad (26.14)$$

which we recognize to be the equation of wave propagation in the material, or the hydrodynamic equation of motion, Eq. 18.5, which we already know to be true from more elementary considerations.

The formalism illustrated in the foregoing simple example has been generalized and used extensively in many much more elaborate problems of field theory, some of which we shall touch upon superficially in a later chapter. The same formalism is applicable to the dynamical behavior of elastic solids. In this case the completely general expression for the potential energy and the elastic relations have to be used to find the integrated potential energy density in place of Eq. 26.2. The Lagrangian density having been determined in this way, the Lagrange equation, Eq. 26.12, leads to the true equations for propagation of elastic waves.

Exercise III.26

26.1. Write out the detailed steps in proving Eq. 26.8. Set up this equation in two different notations, (a) a numerical subscript notation, and (b) a vector notation, and carefully define the notations in each case.

26.2. Set up the analogy to this section applicable to a one-dimensional continuum, such as the vibrating string.

26.3. Consider a light elastic string embedded in a medium which possesses potential energy that is some function of the lateral displacement of the string.

Find the Lagrange equation of motion, assuming the Lagrangian function to be the difference between kinetic and potential energy.

26.4. Set up the analogy to this section applicable to a two-dimensional membrane using polar coordinates, and derive the equation of motion. Compare the result with that given in §10.

III.27 Canonical Transformations in Phase Space. In §23 we discussed transformations in configuration space and the corresponding transformations induced in momentum space. As stated in §24, the Lagrangian function is a scalar invariant for such transformations; the kinetic energy is a scalar defined in momentum space, and the potential energy is a scalar defined in configuration space. Now we shall broaden our point of view somewhat and consider the possibility that phase space as a whole undergoes transformations that cannot be separated into two parts, one in configuration space and one in momentum space. Analytically this situation occurs if the new coordinates are not defined as functions only of the original coordinates, but also depend on the original momentum components; and likewise the new momenta depend both on the original momenta and the original coordinates. Thus we shall write for the general transformation in phase space:

$$\left.\begin{aligned} Q_j &= Q_j(q_1, q_2, q_3, \cdots q_f, p_1, p_2, p_3, \cdots p_f) \\ P_j &= P_j(q_1, q_2, q_3, \cdots q_f, p_1, p_2, p_3, \cdots p_f) \end{aligned}\right\}, \qquad (27.1)$$

where $j = 1, 2, 3, \cdots f$. Two sets of functions of all the phase variables suffice to define the generalized coordinates Q_j and the generalized momenta P_j.

In such a general transformation it is not necessarily the case that either the kinetic or the potential energy would remain invariant. In fact, it would be quite surprising if they did, because, for example, the potential would depend on the new momenta as well as the new coordinates. In general, the Lagrangian function $L = T - V$ is not invariant either. However, there does exist a whole class of transformations of the type of Eq. 27.1 in phase space that do leave the Lagrangian equations invariant, and these are called *canonical transformations* or contact transformations.

Before considering the restrictions to be placed on Eq. 27.1 by the requirement that L be invariant, we consider the variation principle from the new point of view. In phase space there are $2f$ dimensions, with the phase components $q_1 \ldots q_f, p_1 \ldots p_f$, and there are $2f$ rates of change with respect to time, which we may write $\dot{q}_1 \ldots \dot{q}_f, \dot{p}_1 \ldots \dot{p}_f$. For the Lagrangian function, Eq. 24.12, we write instead some function of all the $2f$ phase components, their $2f$ rates of change, and time:

$$L = L(q_1, \cdots q_f, p_1, \cdots p_f, \dot{q}_1, \cdots \dot{q}_f, \dot{p}_1, \cdots \dot{p}_f, t). \qquad (27.2)$$

We may then go through all the same formal procedures in deriving the Lagrange equations of motion from Hamilton's variation principle, but in terms of this greater number of variables, ending up with $2f$ equations:

$$\left.\begin{aligned} \partial L/\partial q_j &= (d/dt)(\partial L/\partial \dot{q}_j) \\ \partial L/\partial p_j &= (d/dt)(\partial L/\partial \dot{p}_j) \end{aligned}\right\}. \tag{27.3}$$

To give these equations any meaning, we have to decide what to use for the Lagrangian function, and this is not a trivial question. We have here included both velocity and momentum components as independent variables, whereas in the old expression, Eq. 24.11, for L we have only the velocity appearing in the expression for the kinetic energy. Are we to retain the velocities in this way, or do we change some of them to momenta? Hamilton discovered that the following definition of L succeeds in yielding a physically useful meaning to the equations of motion, Eq. 27.3:

$$L = \sum_j p_j \dot{q}_j - H(q_1, \cdots q_f, p_1, \cdots p_f, t), \tag{27.4}$$

where H is the total energy, *i.e.* the sum of the kinetic and potential energies, expressed only in terms of the coordinates and momenta (not the velocities).

A word of reassurance may be in order here regarding the fact that we have used both momenta and velocities as independent variables; this does not really mean that physically we can treat a momentum and its corresponding velocity as independent of each other. What it means is that we agree mathematically to treat them as independent variables in the differentiation procedures used in arriving at the equations of motion from the variation principle. The logic of this procedure is perfectly valid regardless of the physical obscurity of the method.

It is easy to show that Eq. 27.4 coincides with Eq. 24.11 in the simple Cartesian case. Notice that the new Lagrangian does not contain the rates of change of the momenta, $\dot{p}_j$, explicitly. Putting Eq. 27.4 into Eq. 27.3, we find for the $2f$ equations of motion:

$$\left.\begin{aligned} -\partial H(\mathbf{p}, \mathbf{q})/\partial q_j &= dp_j/dt \\ dq_j/dt - \partial H(\mathbf{p}, \mathbf{q})/\partial p_j &= 0 \end{aligned}\right\}. \tag{27.5}$$

These are *Hamilton's equations of motion.* We note that there are twice as many of them as there are of the Lagrange equations, Eq. 24.15, and that they are first-order differential equations instead of second-order ones. In going over to the phase-space method we have succeeded automatically in performing one integration of the equations.

It is a simple exercise to show that Hamilton's equations, Eq. 27.5, are identical with Newton's equations of motion in Cartesian coordinates; in these coordinates the first set of Hamilton's equations states that "force equals rate of change of momentum," and the second set simply states that "velocity equals momentum divided by mass." In generalized coordinates there is no such simple physical or intuitive interpretation. However, from the way we have derived Hamilton's equations—from the variation principle—we know that they are true in any generalized coordinate system of the type defined in Eq. 27.1 provided the transformation is canonical in the sense defined previously. In general, therefore,

$$\left.\begin{aligned} \partial H'(\mathbf{P}, \mathbf{Q})/\partial Q_j &= -dP_j/dt \\ \partial H'(\mathbf{P}, \mathbf{Q})/\partial P_j &= dQ_j/dt \end{aligned}\right\}, \tag{27.6}$$

where H' is the Hamiltonian function expressed in terms of the Q's and P's. We return now to the question of selecting the canonical transformation.

We know from the way we have set out the variation procedure that we may accept only those transformations that leave unaltered the variation $\delta\int L dt$. Therefore, we require (see Eq. 27.4) that

$$\delta\int\{\sum_j P_j\dot{Q}_j - H'(\mathbf{P}, \mathbf{Q}, t)\}dt = \delta\int\{\sum_j p_j\dot{q}_j - H(\mathbf{p}, \mathbf{q}, t)\}dt. \tag{27.7}$$

If this is satisfied, then the same variation procedure will result in Eqs. 27.5 and Eqs. 27.6. But Eq. 27.7 will be satisfied identically if we arrange for

$$\sum_j P_j\dot{Q}_j - H'(\mathbf{P}, \mathbf{Q}, t) = \sum_j p_j\dot{q}_j - H(\mathbf{p}, \mathbf{q}, t) + dF/dt, \tag{27.8}$$

where F is any arbitrary function of all the variables p, q, P, Q. The difference between the two sides of Eq. 27.7 in this case is evidently

$$\delta\int_A^B dF = \delta(F_B - F_A) = 0, \tag{27.9}$$

because at A and B the varied paths coincide, so the δ operation vanishes.

Different types of function F lead to different types of transformations. As an example we may select some function of $\mathbf{q}$ and $\mathbf{Q}$ alone:

$$F = F(\mathbf{q}, \mathbf{Q}), \tag{27.10}$$

and proceed to deduce the transformation implied in Eq. 27.8. We have

$$dF/dt = \sum \dot{q}_j(\partial F/\partial q_j) + \sum \dot{Q}_j(\partial F/\partial Q_j), \tag{27.11}$$

so Eq. 27.8 becomes equivalent to:

$$\sum_j (p_j + \partial F/\partial q_j)dq_j - \sum_j (P_j - \partial F/\partial Q_j)dQ_j + (H' - H)dt = 0. \quad (27.12)$$

Since this result has to remain true for any arbitrary path in phase space, we may take dq_j and dQ_j as independent variations and so require that

$$p_j = -\partial F/\partial q_j, \quad P_j = \partial F/\partial Q_j, \quad H' = H, \quad (27.13)$$

for all j-values. The first of Eqs. 27.13 together with Eq. 27.10 can in principle be solved to give **Q** as a function of the q's and the p's, whereas the second of Eqs. 27.13 gives each P as a function of the q's and Q's, and so of the q's and p's, as we now have the Q's as a function of the q's and p's. Therefore, they define the transformation. To illustrate this procedure we shall consider the following function in a one-dimensional problem:

$$F(q, Q) = m\pi\nu q^2 \tan(2\pi Q), \quad (27.14)$$

with the idea of applying the results to a linear harmonic oscillator having mass m and natural frequency ν. We have from Eq. 27.13:

$$\left.\begin{aligned} p &= -2\pi m\nu q \tan(2\pi Q) \\ P &= 2\pi^2 m\nu q^2 \sec^2(2\pi Q) \end{aligned}\right\}. \quad (27.15)$$

We rearrange these to give the transformation equations in the form

$$\left.\begin{aligned} q &= (P/2\pi^2 m\nu)^{1/2} \cos(2\pi Q) \\ p &= -(2\nu m P)^{1/2} \sin(2\pi Q) \end{aligned}\right\}. \quad (27.16)$$

The Hamiltonian for a linear oscillator is

$$H(p, q) = p^2/2m + 2\pi^2 m\nu^2 q^2, \quad (27.17)$$

and putting Eq. 27.16 into this we have

$$H'(P, Q) = \nu P[\sin^2(2\pi Q) + \cos^2(2\pi Q)] \quad (27.18)$$

or simply

$$H'(P, Q) = \nu P. \quad (27.19)$$

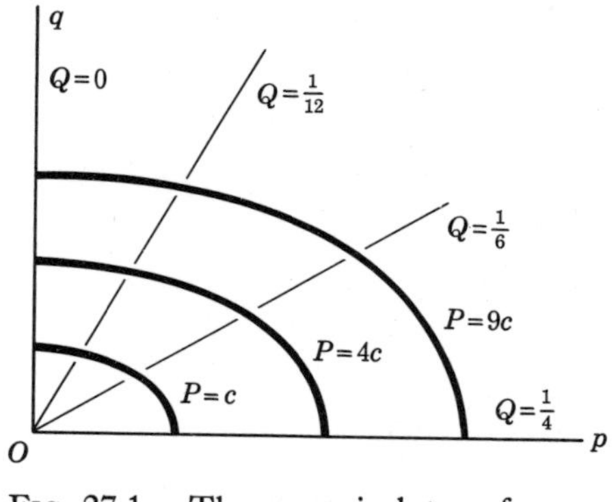

FIG. 27.1. The canonical transformation of Eq. 27.16. The lines P=constant are concentric ellipses whose diameters are proportional to $P^{-1/2}$

The generalized momentum P is called *action*; it has dimensionality energy multiplied by time. The new Hamiltonian is a function only of the generalized momentum and is independent of the generalized coordinate. The Hamiltonian equations of motion become specially simple in the generalized coordinate system; in fact, Eqs. 27.6 are now

$$dP/dt = 0, \quad dQ/dt = \nu, \tag{27.20}$$

which can be integrated at once:

$$P = J, \text{ a constant}; \quad Q = \nu t + \text{constant}. \tag{27.21}$$

Putting these integrals into Eq. 27.16 gives us the equations of motion in the original coordinate system in completely integrated form:

$$q = (J/2\pi^2 m\nu)^{1/2} \cos(2\pi\nu t + \text{const.}), \quad p = m\dot{q}. \tag{27.22}$$

The two arbitrary constants determine the amplitude and the phase of the vibrations.

The same kind of transformation can be generalized to suit any f-dimensional harmonic oscillator, by choosing

$$F(\mathbf{q}, \mathbf{Q}) = \sum_j m_j \pi \nu_j q_j^2 \tan(2\pi Q_j), \tag{27.23}$$

where j sums over all f degrees of freedom. Eqs. 27.13 then lead to the same kind of equations as Eq. 27.15 for each value of j separately. Also,

$$H(\mathbf{q}, \mathbf{p}) = \sum_j p_j^2/2m_j + \sum_j 2\pi^2 m_j \nu_j^2 q_j^2; \tag{27.24}$$

thus, analogously with Eq. 27.19, we can show that

$$H'(\mathbf{Q}, \mathbf{P}) = \sum_j \nu_j P_j. \tag{27.25}$$

Again each new momentum is a constant action, and each new coordinate Q is an angle variable:

$$P_j = J_j, \text{ a constant}; \quad Q_j = \nu_j t + \text{constant}, \tag{27.26}$$

and the familiar equations of motion like Eq. 27.22 follow for each degree of freedom separately.

If linear coupling exists between a set of oscillators, we represent the Hamiltonian in the form:

$$H(\mathbf{q}, \mathbf{p}) = \sum_j p_j^2/2m_j + \sum_j 2\pi^2 m_j \nu_j^2 q_j^2 + \sum_j \sum_i c_{ij} q_i q_j, \tag{27.27}$$

where c_{ij} are the coupling constants. Then, before effecting the canonical transformation, we make a rotation in configuration space so that the potential energy term is expressed in normal form. The rotation of course induces a rotation in momentum space, but, since the momenta occur symmetrically in $H(\mathbf{q}, \mathbf{p})$, it does not alter the form of this term:

$$\sum_j 2\pi^2 m_j \nu_j^2 q_j^2 + \sum_j \sum_i c_{ij} q_i q_j = \sum_j 2\pi^2 m_j \nu_j'^2 q_j'^2, \tag{27.28}$$

while

$$\sum_j p_j'^2/2m_j = \sum_j p_j^2/2m_j.$$

The canonical transformation is then carried out from q', p' to Q', P', and the same reasoning follows through exactly, arriving in place of Eq. 27.26 at the result:

$$P_j' = J_j', \text{ a constant}; \quad Q_j' = \nu_j' t + \text{constant}, \tag{27.29}$$

where ν_j' are the new frequencies of the normal modes. The constants J' are again action constants.

Any periodic system can be represented by a point in its phase space that goes around a closed loop in one period of the motion. Every normal coordinate, like q_j' in the last example, goes through a cycle of values, returning to its starting point in the normal mode's period $1/\nu_j'$. Now the function $F(\mathbf{q}', \mathbf{Q}')$, defining the canonical transformation, is a single-valued function in phase space, and so also returns to its initial value during any such cycle. Thus let us take one particular coordinate q_j' through one cycle, holding the other coordinates fixed—*i.e.* we excite only one normal mode of the system. We have then:

$$0 = \oint dF = \oint (\partial F/\partial q_j')dq_j' + \oint (\partial F/\partial Q_j')dQ_j'. \tag{27.30}$$

From Eq. 27.13 we therefore have:

$$\oint p_j' dq_j' - \oint P_j' dQ_j' = 0. \tag{27.31}$$

This proves that the area in phase space traced out by the representative point in one normal mode period is an invariant for the canonical transformation from $\mathbf{p}', \mathbf{q}'$ to $\mathbf{P}', \mathbf{Q}'$. The actual value of this area can be found by using Eq. 27.29 in the second part of Eq. 27.31:

$$\oint P_j' dQ_j' = J_j' \oint \nu_j' dt = J_j', \tag{27.32}$$

because the cycle goes through one period of time, $1/\nu_j'$, for the integral.

Eq. 27.31 shows explicitly that the area in phase space is an invariant scalar for canonical transformations; more generally it is true that the $2f$-dimensional extension-in-phase (generalized "area") is invariant for canonical transformations. The action constants are therefore scalar invariants for canonical transformations. They have played an important role in the historical development of quantum mechanics.

Exercise III.27

27.1. Check that Eq. 27.4 reduces to Eq. 24.11 in Cartesian coordinates.

27.2. Prove that Hamilton's equations of motion, Eq. 27.5, are identical with Newton's equations of motion in Cartesian coordinates.

27.3. Set up the kinetic energy of a particle in terms of spherical polar coordinates: $T(r, \theta, \phi)$. Define the momentum components in this system by means of Eq. 23.13 and then set up the Lagrangian function from Eq. 27.4, $L(r, \theta, \phi, \dot{r}, \dot{\theta}, \dot{\phi}, p_r, p_\theta, p_\phi)$, and find the equations of motion, Eq. 27.3.

27.4. A two-dimensional oscillator has a Hamiltonian:

$$H = \tfrac{1}{2}m\dot{x}^2 + \tfrac{1}{2}m\dot{y}^2 + 2m\pi^2(\nu_1{}^2x^2 + \nu_2{}^2y^2) + cxy.$$

Find the rotation in the $x-y$ plane that removes the coupling term, and find the frequencies of the two normal modes. Write down the solutions to the equations of motion like Eq. 27.22 in terms of these rotated coordinates, and then transform back to the original coordinates and discuss the motion.

III.28 Liouville's Theorem and Statistical Mechanics. The Hamilton-Lagrange method of analytical dynamics outlined in the last few sections has been applied to an extremely wide range of problems of practical interest, but we do not have space to discuss them in detail in this course. By way of illustration, we may mention the motion of rigid bodies of all shapes, the dynamics of crystal lattices, the motion of stellar masses and many related astrophysical problems, the behavior of electrical networks, the relativistic theory of high energy particles, *etc.* In general, the central problem is to choose a suitable set of generalized coordinates by means of which to describe the system and then set up a suitable Lagrangian function; once this has been done, the variation procedure for finding the equations of motion becomes a more or less routine matter following the steps outlined previously. The procedure can be followed in principle no matter how many particles there may be. The only trouble is that solutions of the equations of motion require a statement of the initial values of all the variables, and, when the number of particles is very great, this becomes impractical. It is impossible to predict the precise position and momentum of every atom in a gas, not because we do not know the equations of motion, but because we cannot record sufficient data on the initial positions and momenta of all the atoms. Fortunately in studying the properties of bulk matter, we are not really interested in just where each atom may be; much less precise information will suffice. We turn now to the techniques of statistics applied to such dynamical problems.

Consider at first a collection of noninteracting particles each of which has f degrees of freedom. Set up the phase space for a single such particle, having $2f$ dimensions: $q_1, q_2, q_3, \cdots q_f, p_1, p_2, p_3, \cdots p_f$. A point in this phase space tells the position and momentum of the particle or, more generally, the configuration and generalized momentum of the system when f is greater than 3. We do not know the positions or momenta of the N particles in the collection, but we can toss

N representative points at random into the phase space, to represent the collection in some fashion. N is so great a number that the representative points form a cloud or fluid in phase space, and we may discuss the average behavior of this fluid in accordance with the general hydrokinematic theorems proved in §15. We know the Hamiltonian function for any one of the particles, $H(\mathbf{p}, \mathbf{q})$, and we know that each particle—or each element of the fluid—obeys the Hamilton equations of motion, Eq. 27.5.

Let $\mathbf{v}$ be the $2f$-dimensional velocity of the representative points in phase space,

$$\mathbf{v} = (\dot{q}_1, \dot{q}_2, \dot{q}_3, \cdots \dot{q}_f, \dot{p}_1, \dot{p}_2, \dot{p}_3, \cdots \dot{p}_f). \tag{28.1}$$

The generalized divergence of this 2f-vector in phase space, in Cartesian coordinates, is:

$$\mathbf{div}\cdot\mathbf{v} = \sum_j \partial\dot{q}_j/\partial q_j + \sum_j \partial\dot{p}_j/\partial p_j. \tag{28.2}$$

Using Hamilton's equations of motion, Eq. 27.5, to change the time derivatives in this, we have

$$\mathbf{div}\cdot\mathbf{v} = \sum_{j=1}^{f}(\partial/\partial q_j)(\partial H/\partial p_j) - \sum_{j=1}^{f}(\partial/\partial p_j)(\partial H/\partial q_j). \tag{28.3}$$

Then supposing H to be a decent function of the q's and p's, the order of partial differentiation is commutable, and the right-hand side vanishes:

$$\mathbf{div}\cdot\mathbf{v} = 0. \tag{28.4}$$

Let ρ be the density of the cloud per unit extension-in-phase; then the equation of continuity, Eq. 15.3, can be understood in terms of this $2f$-dimensional problem:

$$\partial\rho/\partial t = -\mathbf{div}\cdot(\rho\mathbf{v}). \tag{28.5}$$

We now introduce the convective derivative in this $2f$-dimensional case:

$$D\rho/Dt = \partial\rho/\partial t + \mathbf{v}\cdot\mathbf{grad}\ \rho. \tag{28.6}$$

where **grad** is a $2f$-dimensional operator involving differentiation with respect to each of the $2f$ dimensions in the phase space. The equation of continuity can be transformed into

$$D\rho/Dt = -\rho\mathbf{div}\cdot\mathbf{v}. \tag{28.7}$$

Combining this with Eq. 28.4, we have

$$D\rho/Dt = 0. \tag{28.8}$$

This equation is *Liouville's theorem* which states that, if we consider a small part of the fluid occupying a certain extension-in-phase or "volume" $\Delta\tau$ at some time, and if we follow these particles in their motion, they will continue to occupy the same amount of extension-in-phase, because their density in phase is unchanged by the convection. Note that the "volume" in phase space is

$$\Delta\tau = dq_1 dq_2 dq_3 \cdots dq_f \, dp_1 dp_2 dp_3 \cdots dp_f \tag{28.9}$$

and that ρ is density in phase, or mass per unit τ, not by any means the same thing as ordinary density. We may now prove a *corollary* of Liouville's theorem:

If the cloud of representative points is given a uniform density in phase at any time, the density in phase will continue to be uniform for all subsequent time.

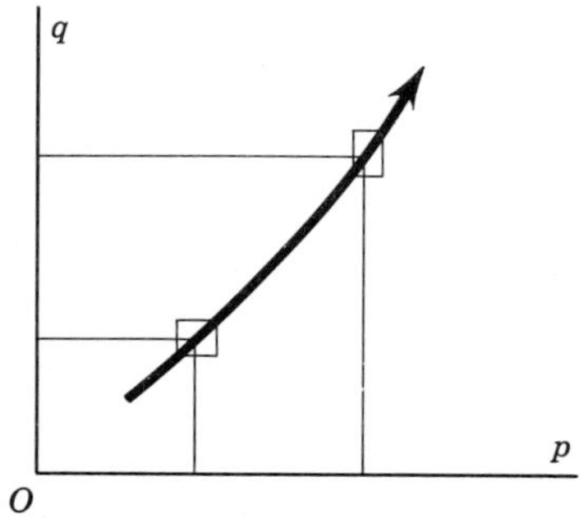

FIG. 28.1. An illustration of Eq. 28.8, an element of phase space in motion due to convection by the representative fluid

This follows from Eq. 28.5 because if ρ=constant at some time, we can take it out of the divergence operator and write $\partial\rho/\partial t = -\rho\,\mathbf{div}\cdot\mathbf{v}$; thus, with Eq. 28.4, we have at once $\partial\rho/\partial t = 0$, showing that ρ remains constant in time. Starting from its uniform value, it must therefore remain uniform.

We infer at once that a representative cloud having uniform density in phase actually represents a collection in equilibrium—it does not change its average characteristics.

We may now prove a second *corollary* to Liouville's theorem:

If the cloud of representative points is set up with an initial density in phase as some function only of any constant of the motion—e.g. *the energy of the system—then it remains constant in time, the same function of that constant of the motion.*

To see this, we may imagine surfaces of constant E drawn in the phase space: these are really hypersurfaces with dimensionality $2f-1$ in the phase space of dimensionality $2f$. We then set up a cloud of representative points having a density in phase a function of E. This means that two surfaces corresponding to energies E and $E+dE$ enclose a region in phase space having almost uniform density if dE is small enough. During the subsequent motion, no point originally within this region can get outside it, because to do so it must change its energy, which is contradictory to the Hamilton equations governing its motion. The total mass in the region is therefore a constant, and from Liouville's theorem its density must remain uniform since it starts out that way. Each small step in E can be discussed in the same way, so that the density remains the same function of E throughout the subsequent motion.

The corollaries to Liouville's theorem therefore allow us to infer that any cloud whose density in phase is any function of E represents a collection in equilibrium. In particular, if we know that the energy of each particle is very close to a certain energy value E, we can set up a cloud to represent this knowledge, filling one of the small regions, say between $E-\frac{1}{2}dE$ and $E+\frac{1}{2}dE$, with uniform density in phase. This cloud represents a collection in equilibrium, because its density is a function of the constant energy. The collection of particles or systems represented by such a cloud is called the *Gibbs microcanonical ensemble*. A more realistic situation would be one in which we know only the total energy of the collection of N particles, or, what is the same thing, we know the average energy per particle in the collection. The question then arises: What is the average distribution of the energy among the particles? We shall discuss this from the statistical point of view and find the correct kind of representative cloud in phase space in terms of which the average behavior of such a collection can be worked out. The collection of systems represented by this cloud is called the *Gibbs canonical ensemble*, and the distribution in energy is the Maxwell-Boltzmann distribution.*

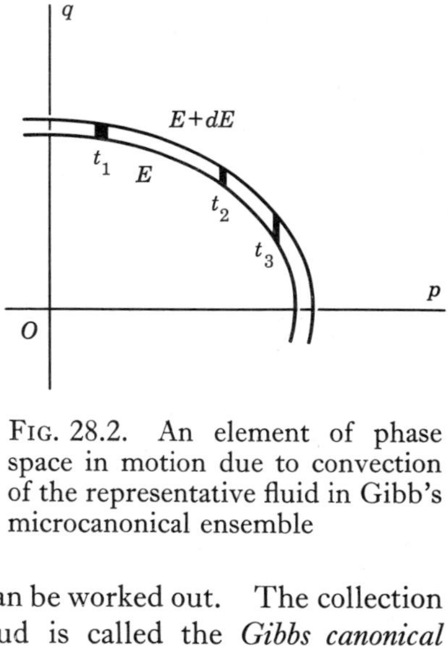

FIG. 28.2. An element of phase space in motion due to convection of the representative fluid in Gibb's microcanonical ensemble

Exercise III.28

28.1. Consider motion in one dimension with a potential field $V(x)$ such that $V(x)=0$, $x<a$; $V(x)=-V_0$, $a<x<b$; $V(x)=0$, $b<x$. Set up the phase space for the motion and draw lines of constant energy corresponding to $E=-\frac{1}{2}V_0$, 0, $\frac{1}{2}V_0$. Find the extension-in-phase between these lines and the lines $x=0$ and $x=L$. Find also the extension-in-phase corresponding to an infinitesimal increment in the energy, dE. Does this depend on V_0? Write out the generalized divergence of the velocity, in Eq. 28.3, for this case; also write out the generalized **grad** operator appearing in Eq. 28.6.

* Biographical note: James Clerk Maxwell, Scotland, 1831–1879; Josiah Willard Gibbs, U.S.A., 1839–1903; Ludwig Boltzmann, Austria, 1844–1906.

28.2. Consider the same kind of problem as above, but let the potential function form a barrier, $V(x) = +V_0$ between a and b, instead of a well. Draw the lines for $E = \frac{1}{2}V_0$, V_0, and $2V_0$ in this case. Also, answer all the other parts of Ex. 28.1 in this case.

28.3. Consider one-dimensional motion under constant gravitational force, the potential function being $V(x) = mgx$. Set up the phase space and draw lines of constant energy $E = \frac{1}{2}Km$, Km, $2Km$, where $K = 9.8$ meter^2 per sec^2. What is the extension-in-phase enclosed by any one of these lines and the momentum axis? Find the extension-in-phase between two lines of constant energy, E and $E + dE$, dE being infinitesimal. Write out **div·v** and the **grad** operator in this case.

28.4. Consider one-dimensional simple harmonic motion with a potential function $V(x) = \frac{1}{2}kx^2$. Set up the phase space and draw curves of constant energy $E = \frac{1}{2}Km$, Km, $2Km$; K being 1.0 meter^2 per sec^2. Find the extension-in-phase enclosed by each of these curves. Find the extension-in-phase between curves corresponding to E and to $E + dE$.

28.5. Consider a two-dimensional simple harmonic motion with a potential function $V(x, y) = \frac{1}{2}k(x^2 + y^2)$. Find the extension-in-phase enclosed by the hypersurface corresponding to total energy equal to $\frac{1}{2}m \times 1.0$ meter^2 per sec^2.

28.6. Consider a two-dimensional motion of a free particle restricted to a square area of side L in each dimension. What is the extension-in-phase for motion with constant unit energy or less?

III.29 The Maxwell-Boltzmann Distribution. Suppose we are given some specific dynamical system, *e.g.* a particle, and that we wish to discuss the statistical behavior of a large collection of N such systems or particles. We shall set up the phase space of the particle and mark out this phase space with a large number X of small equal extensions-in-phase $\Delta\tau$ (see Eq. 28.9). The shape of these small phase cells is to be such that they are bounded only by hypersurfaces of constant energy E and $E + dE$; thus the jth cell corresponds to all particle energies between E_j and $E_j + dE_j$. The intervals dE_j must be chosen so that all the cells are equal in extent. In Ex. 28.4, for instance, the cells would be the annular areas between elliptical curves representing constant energy. To describe the whole collection of particles —or the gas—we throw in the cloud of representative points. In doing this a certain number N_j fall into the jth cell; each of them therefore are ascribed energy E_j. The whole set of numbers N_j is called a *distribution in energy* of the collection. If the cloud of points were all tagged, so that we knew which points were in each cell, as well as their numbers, we would know what is called the *complexion* of the collection.

Having set up the cloud in this way, we know that the number of points is a function only of the particle energy, because each cell is uniquely associated with the energy value. Therefore, from Liouville's theorem and corollaries, we know that the cloud must represent an equilibrium collection. This will be so no matter what the result of our throw may be. Our physical intuition suggests that, if we restrict the throws to those that come out with some pre-assigned total energy,

$$E = \sum_{j=1}^{X} N_j E_j, \tag{29.1}$$

there must be some distribution more probable than the other possible ones. The more often one throws a cloud into the phase-space cell lattice, the more likely we shall get certain distributions and the less likely others, proportionately. Some distributions will obviously be ruled out as having the wrong total energy; others will be very improbable anyhow, even though they have the right total energy. For instance, we would hardly expect to throw all the points into one cell—*i.e.* the one cell that corresponds to the exact average energy per particle in the collection. We wish to find the most probable distribution consistent with the given total energy.

The fundamental assumption of our statistical method is this: When all the particles are alike, all complexions of the collection are equally probable. This assumption is a natural generalization from common-sense notions about the probabilities in games of chance. Consider, for example, a collection of four dice, each with six sides. Let the dice be labeled a, b, c, d. A throw resulting (A) in the dice showing scores $a=1$, $b=1$, $c=1$, $d=6$, and another throw (B) giving the dice scores $a=6$, $b=1$, $c=1$, $d=1$, would represent two distinct complexions but one and the same distribution; the distribution numbers are in each case $N_1=3$, $N_6=1$, all other $N_j=0$. In any long series of throws we would expect each of these complexions to occur with equal frequency—we say they are *a priori* equally probable. This common-sense notion is really the result of a long history of experience with true dice. Experiments that seemed to go against the rule would generally be interpreted as due to loaded dice. Similarly, thinking of the score one gets in dice throws as corresponding to the energy of particles in the cloud of representative points, we can assume that all complexions of the cloud are *a priori* equally probable, unless the particles are not similar for some reason. In a long series of throws, the distributions having the most complexions will occur most frequently.

The distribution $N_1=3$, $N_6=1$, found above in the example of four dice, corresponds to just four different complexions—according to which of the four dice happens to come up with the 6-face. A distribution of these same four dice in which all dice come up with the same score, say 1, would have only one complexion. A distribution in which all four dice come up with different scores would have 4! complexions. Throws that result in all four dice showing different scores are the most probable ones.

Looking a little more deeply into this question, we note that all the complexions are equally probable because each individual state of a die is equally probable. It is equally likely that a given die falls on any one of its six faces. In carrying this notion over to the cloud of representative points in the phase cell lattice, we have to be sure that each state of a point be equally likely—and this means that it must be equally likely to fall into any one of the cells. This was our reason for making all the phase cells equal in extent. Were we to make a completely

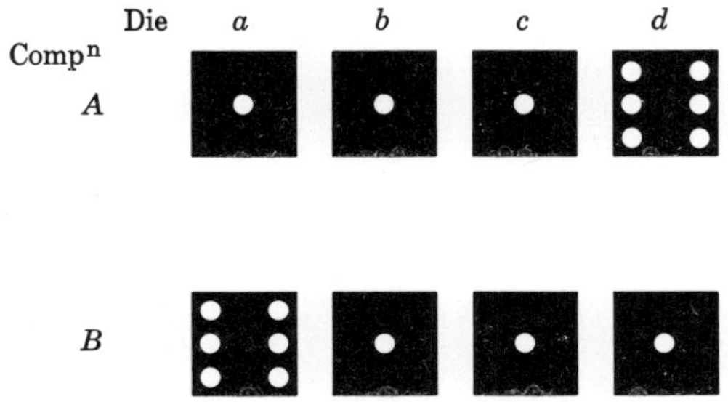

FIG. 29.1. Two complexions of four dice belonging to one distribution

random throw of points into the phase space, we would expect a uniform cloud; this would put equal numbers in equal extensions-in-phase —the probability is proportional to the extension-in-phase; so by making the small cells all equal in extent, we make their *a priori* probabilities equal. A great deal of philosophical discussion has occurred in attempts to give a deeper proof of these assumptions, but for our present purpose it is best to accept them because they actually work in practice.

Our program now calls for us to determine the number of complexions that correspond to every possible distribution in energy, and then to select that distribution that has the most complexions. This will be accepted as the most probable distribution, and we can examine the properties of a collection in this most probable distribution.

The number of ways in which N particles can be thrown into X cells when we know that N_1 must go into the first cell, $\cdots$ N_j into the jth cell, *etc.*, is

$$W = N!/(N_1!N_2! \cdots N_X!), \tag{29.2}$$

where of course

$$\sum_{j=1}^{X} N_j = N. \tag{29.3}$$

We are only interested in those distributions for which the total energy has the correct value—see Eq. 29.1. We wish to find those numbers N_j subject to the two restrictions, Eqs. 29.1 and 29.3, that maximize W.

From Eq. 29.2 we have

$$\ln W = \ln N! - \sum_j \ln N_j!. \tag{29.4}$$

When all the N's are large enough, we may use Stirling's approximation:

$$\ln N! \doteqdot N \ln N - N, \quad \ln N_j! \doteqdot N_j \ln N_j - N_j, \tag{29.5}$$

so that

$$P \equiv \ln W \doteqdot N \ln N - \sum N_j \ln N_j. \tag{29.6}$$

Now let us vary the N_j-values so that

$$\delta P = \sum_j (\delta N_j \ln N + \delta N_j - \delta N_j \ln N_j - \delta N_j) = \sum_j \delta N_j \ln(N/N_j) = 0. \tag{29.7}$$

Also

$$\delta N = \sum_j \delta N_j = 0 \tag{29.8}$$

and

$$\delta E = \sum_j E_j \delta N_j = 0. \tag{29.9}$$

All three variations, δP, δE, and δN are required to vanish: the last two because total number and total energy are known in advance, and the first because we want to maximize W, which is equivalent to maximizing P.

Multiplying Eq. 29.8 by $-\alpha$, and Eq. 29.9 by $-\beta$, two arbitrary parameters, and adding Eqs. 29.7, 29.8, 29.9, we find

$$\sum_j \delta N_j[\ln(N/N_j) - \alpha - \beta E_j] = 0. \tag{29.10}$$

Having introduced the two arbitrary parameters we are now free to treat all the variations δN_j as arbitrary, cancel them from Eq. 29.10 and assert that

$$N_j = N \exp(-\alpha - \beta E_j). \tag{29.11}$$

This set of numbers actually describes the cloud representing a Gibbs canonical ensemble. The parameters α and β are called *Lagrangian multipliers*. By comparing Eq. 29.11 for consistency with the total number N and the total energy E, we have two equations:

$$N = \sum_j N_j = Ne^{-\alpha} \sum_j e^{-\beta E_j} \tag{29.12}$$

and

$$E = \sum_j E_j N_j = Ne^{-\alpha} \sum_j E_j e^{-\beta E_j}. \tag{29.13}$$

These two equations can in principle be solved for the two parameters, α and β, which are therefore determined by the total number and total energy of the collection, provided we know all the energy values E_j.

The set of numbers given in Eq. 29.11 define the *Maxwell-Boltzmann distribution* in energy. The physical interpretation of this distribution depends of course on the physical systems we are discussing. A simple case will be examined later. Because the distribution is a function only of the particle energy, which is a constant of the motion, it represents an equilibrium situation, according to the corollary of Liouville's theorem.

Exercise III.29

29.1. Set up a table showing all the complexions of four dice for which the total score is 6 for a throw. Arrange the complexions in sets, each set corresponding to one distribution, and identify the distributions by listing the numbers N_j, $j=1, 2, \cdots 6$. Compute the relative probabilities of the several distributions, making use of Eq. 29.2.

29.2. One thousand coins are tossed simultaneously, heads or tails being equally probable. If heads scores as one unit of energy and tails scores as zero energy, write down Eqs. 29.12 and 29.13 for a total score 500 units for the throw. Solve them for the parameters α and β. What is the most probable distribution in this case?

29.3. In the experiment of Ex. 29.2, what are the parameters α and β when the total score is (a) 750 energy units, and (b) 250 energy units? What are the most probable distributions in these cases?

29.4. Use Eq. 29.6 and 29.11 to show that the parameter P of the most probable distribution for given E and N depends on E and N according to:

$$P = \alpha N + \beta E$$

and that hence

$$\beta = \partial P/\partial E.$$

Discuss the results of Ex. 29.2 and 29.3 in the light of this.

III.30 Spontaneous Approach to Equilibrium. We emphasize again that the Maxwell-Boltzmann distribution has been deduced for a collection of noninteracting particles. Essential to the discussion was the existence of the one-particle Hamiltonian, $H(\mathbf{p}, \mathbf{q})$ as a function of the phase of one particle, independent of all the other particles in the collection. We showed that any distribution set up as a function of the energy of a particle, as was done in §29, would be an equilibrium distribution according to the theorem of Liouville and its corollary, §28, provided always that the particles do not interact. The Maxwell-Boltzmann distribution is the most probable of these equilibrium distributions if we specify the total energy of the collection. In order to produce these distributions, we had to imagine some external agency capable of preparing the collection in any desired distribution—the experimenter must be able to "throw" the particles into their phase space much as a gambler is able to throw his dice on the table.

Now we consider a more realistic situation where the particles interact with each other according to some definite force law: they in fact will collide with each other after they have been "thrown" into their phase space. We still suppose that one-particle Hamiltonians can be set up and that the collisions are sufficiently sharp to cause almost instantaneous jumps among the particles from some initial phase to some final phase. The interaction between particles is negligible except at the moment of close approach, *i.e.* during the collision. The representative points of two colliding particles jump almost discontinuously to two new positions as a result of the collision. The whole collection is then undergoing jumps from one distribution to another. We assume that there is practically no predictable correlation between the particle motion before and after a collision and that the results of collisions can, to a good approximation, be regarded as causing random changes in the distribution of the collection. The distribution following collisions will then be governed by probability considerations alone; it will generally be one of the more probable distributions. If the collection—or gas—starts out in a very improbable distribution, the results of collisions will be to produce more and more probable distributions, and with collisions occurring as rapidly as they do in a gas, it takes only a few microseconds to reach the distribution of maximum probability, *viz*. the Maxwell-Boltzmann distribution. Thereafter only minute fluctuations will probably occur away from the most probable distribution, which we can therefore regard as the physical equilibrium distribution.

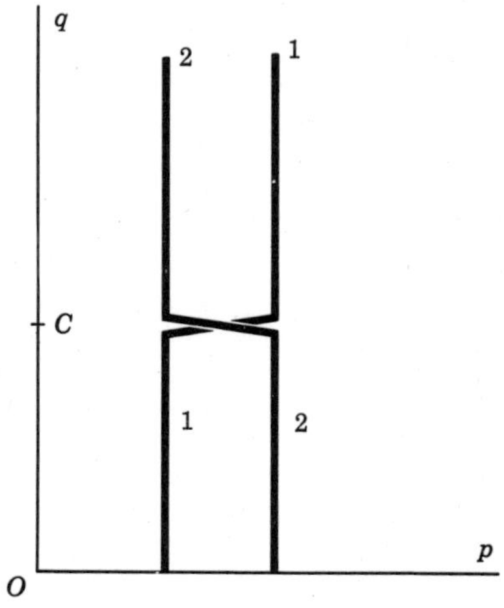

FIG. 30.1. A head-on collision between two equal elastic particles; particle 1 overtakes particle 2 at the point C

This entire argument is open to serious objections from the point of view of classical mechanics. When we assert that there can be practically no correlation between particle motion before and after a collision, we are merely admitting that we cannot in practice determine the initial precollision motion with sufficient precision to determine the motion after the collision. But in principle the motion is determined completely by the laws of dynamics applied to their interaction. Probability considerations can have nothing more than an heuristic value. Before the discovery of quantum theory and the uncertainty principle, this objection was met by Boltzmann in the following way. When in the steady or equilibrium condition represented by the Maxwell-Boltzmann distribution, all fluctuations away from this are actually more or less periodic; small fluctuations occur fairly frequently, large fluctuations very infrequently. This is a matter of exact calculation in terms of classical dynamics. It can be estimated that any deviation from the equilibrium condition in a gas of 10^{20} atoms, large enough to be easily observable, would occur roughly once in about a million years. If such a gas were prepared in such a violently nonequilibrium condition and released to its spontaneous behavior, it would not return to repeat that nonequilibrium condition within observable time. Microscopic fluctuations occur frequently enough to be observed, and they are well known as Brownian movements, which we shall discuss subsequently in this chapter.

As a slight digression, it is worth pointing out that the second law of thermodynamics, which states: *In any actual change of an isolated system the entropy always tends to increase*, is essentially equivalent to the statistical argument that actual changes tend always to go from less to more probable distributions. In fact, entropy can be interpreted as a measure of probability. In classical mechanics, therefore, the second law of thermodynamics is reduced to the status of an illusion arising from the purely accidental brevity of the human life span.

This conclusion satisfies neither the physicists nor the philosophers. Fortunately, in quantum mechanics the essential periodicity of large fluctuations no longer exists. This is because in quantum mechanics it is, in principle, impossible to determine the classical variables previous to a collision with sufficient precision to predict the exact result of the collision. The lack of correlation postulated in the statistical argument, which is indeed false in classical mechanics, is actually true in quantum mechanics. This point will become clear in a later chapter.

Exercise III.30

30.1. Two similar, hard, spherical, elastic particles each have a mass 1.00×10^{-23} gram and a radius of interaction 1.00×10^{-8} cm. They collide head on with speeds 1.00×10^{5}cm per sec in opposite directions. Find the motion after collision. If their paths may miss being head-on by plus or minus 1.00×10^{-8} cm, find the range of possible motions after the collision. Draw two two-dimensional sections through phase space, *viz.* the subspaces p_x, x and p_y, y to illustrate these collisions. Plot some of the collisions to illustrate the apparent lack of correlation between initial and final condition of the pair of particles.

30.2. Repeat the diagrams of the phase spaces in Ex. 28.1 to Ex. 28.4 and shade them to represent (a) microcanonical ensembles and (b) canonical ensembles, the density of shading to be proportional to the density of the representative cloud.

III.31 Statistical Theory of the Ideal Gas. An ideal gas is usually pictured as a collection of noninteracting mass points without internal structure—the phase space is six dimensional. Actually, if the mass points were strictly noninteracting, the collection would not behave like a gas: the particles must be allowed to collide with each other in order that the collection behave spontaneously in the way discussed in the previous section; collision is a form of interaction. However, we suppose that the collisions are so nearly instantaneous that they contribute nothing to the internal energy of the gas and that the one-particle Hamiltonian is an adequate basis for the dynamical theory of their motion.

The energy of a mass point is $(p_x{}^2+p_y{}^2+p_z{}^2)/2m$. The hypersurfaces of constant energy are, therefore, spherical surfaces in momentum space multiplied by the entire volume occupied by the gas in coordinate space. The small regions between two such hypersurfaces are spherical shells in momentum space multiplied by the entire volume in coordinate space. The extension-in-phase of the jth such region or cell in phase space is a five-dimensional product of three coordinate and two momentum components multiplied by dp:

$$\Delta\tau_j = 4\pi p_j{}^2 dp_j \cdot V \tag{31.1}$$

where p_j is the magnitude of the momentum—*i.e.* the radius of the sphere in momentum space—and V is the entire volume in coordinate space.

The cells are to be made all the same size, G, and therefore the increments dp_j must be adjusted to satisfy the relation

$$4\pi p_j{}^2 dp_j \cdot V = G, \text{ constant.} \tag{31.2}$$

The energy of a particle in this cell is $E_j = p_j{}^2/2m$, and the difference between the energies of adjacent cells is

$$dE_j = p_j dp_j/m. \tag{31.3}$$

The Maxwell-Boltzmann distribution assigns to the jth cell the number

$$N_j = N\exp(-\alpha-\beta E_j)$$

(see Eq. 29.11). Let us multiply this equation by unity on both sides, but on the right side we make the unity take the form $1=(1/G)4\pi p_j{}^2 dp_j \cdot V$ from Eq. 31.2. Thus we have

$$N_j = (N/G)4\pi V \exp(-\alpha - \beta p^2/2m) \cdot p^2 dp, \qquad (31.4)$$

where we have dropped the j-subscript from the momentum deliberately. If G is small enough to make all the dp's very small increments, we can now use an integration approximation to find the total number of particles:

$$N = \sum_j N_j \doteqdot 4\pi V(N/G)e^{-\alpha} \int_0^\infty e^{-\beta p^2/2m}\, p^2 dp. \qquad (31.5)$$

This corresponds to Eq. 29.12. From it we deduce that

$$e^\alpha = 4\pi(V/G) \int_0^\infty p^2\, e^{-\beta p^2/2m}\, dp = V(2\pi m/\beta)^{3/2}/G, \qquad (31.6)$$

and this is one equation connecting the two parameters α and β.

On the other hand, the total energy in the Maxwell-Boltzmann distribution is given by Eq. 29.13, which now becomes

$$E = \sum_j E_j N_j \doteqdot 4\pi V(N/G)e^{-\alpha} \int_0^\infty (p^2/2m)p^2 e^{-\beta p^2/2m}\, dp. \qquad (31.7)$$

This is the second relation between the two parameters α and β. Using the first such relation, Eq. 31.6, to eliminate $e^{-\alpha}$, we have

$$E = (NG/V)(2\pi m/\beta)^{-3/2}(4\pi V/G) \int_0^\infty (p^4/2m)e^{-\beta p^2/2m}\, dp.$$

Canceling various factors and performing the integration, we find

$$E = 3N/2\beta. \qquad (31.8)$$

Therefore, we conclude that the parameter $1/\beta$ is just two-thirds of the mean energy per particle in the Maxwell-Boltzmann gas. Eq. 31.6 then, of course, gives the other parameter in terms of known quantities:

$$e^\alpha = (V/G)(4\pi mE/3N)^{3/2}. \qquad (31.9)$$

From the point of view of elementary kinetic theory, we know that we can take $kT=2E/3N$ as the definition of absolute temperature, where k is Boltzmann's constant. Eq. 31.8 is therefore equivalent to:

$$\beta = 1/kT. \qquad (31.10)$$

We can now express the distribution Eq. 31.4 in the form:

$$N_j = 4\pi N(2\pi mkT)^{-3/2} \exp(-p^2/2mkT) \cdot p^2 dp, \qquad (31.11)$$

where N_j is the number of particles having momentum in the magnitude range p to $p+dp$. This is the fundamental formula of the kinetic theory of gases. The cells in phase space, used in setting up Eq. 31.11, were spherical shells in momentum space. The representative points were distributed uniformly throughout each such cell so as to be sure that the distribution was an equilibrium one according to the Liouville theorem. We may therefore divide each such cell again into smaller cells, and the number of points per cell will continue to be proportional to the size of the cell. Let us take Cartesian coordinates in momentum space so that the spherical shell is divided into a large number of rectangular cells:

$$4\pi p^2 dp = \sum dp_x dp_y dp_z. \tag{31.12}$$

We can then replace Eq. 31.11 by:

$$n(\mathbf{p})dp_x dp_y dp_z = (N/V)(2\pi mkT)^{-3/2} e^{-p^2/2mkT} dp_x dp_y dp_z, \tag{31.13}$$

where $n(\mathbf{p})$ now stands for the number of representative points, per unit volume, per extension in momentum space, at the momentum $\mathbf{p}$.

The fraction of the whole cloud that happens at any moment to have momenta within a particular cell $dp_x dp_y dp_z$ is the ratio $n(\mathbf{p})dp_x dp_y dp_z/N$ and this ratio is the probability that, if we select an atom at random in the gas, its representative point will be in this particular element in momentum space. If we make repeated observations of the momentum of atoms in the gas, the average result over a long series of observations will be the sum of all possible values multiplied by their respective probabilities:

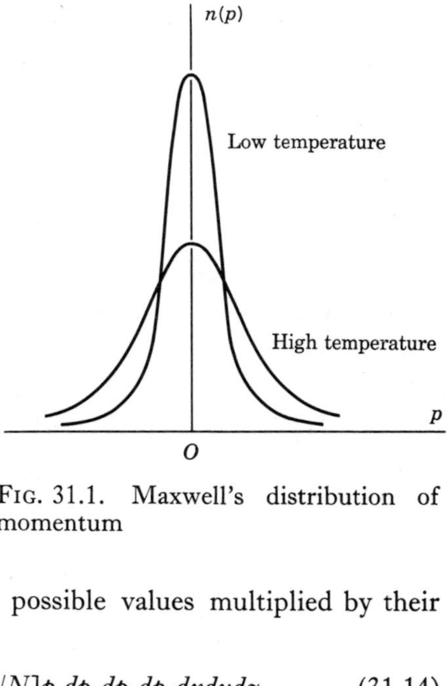

FIG. 31.1. Maxwell's distribution of momentum

$$\langle p \rangle_{av} = \iiiiiint [n(\mathbf{p})/N] p \, dp_x dp_y dp_z dx dy dz. \tag{31.14}$$

Using Eq. 31.13 in this, we have to go back to the polar coordinates in momentum space:

$$\begin{aligned} \langle p \rangle_{av} &= (2\pi mkT)^{-3/2} \int_0^\infty 4\pi p^3 e^{-p^2/2mkT} dp \\ &= (2/\pi)(2\pi mkT)^{1/2}. \end{aligned} \tag{31.15}$$

From this we see that the average speed of the atoms is

$$\langle v\rangle_{av} = 2(2kT/m\pi)^{1/2}. \tag{31.16}$$

The average of the square speed is obtained by multiplying each possible value of the square speed by its probability in the cloud of representative points and adding up, or integrating, over all phase space:

$$\langle v^2\rangle_{av} = \int\int\int\int\int\int[n(\mathbf{p})/N](p/m)^2\, dp_x dp_y dp_z\, dxdydz \tag{31.17}$$

$$= (2\pi mkT)^{-3/2}\int_0^\infty 4\pi(p^4/m^2)\, e^{-p^2/2mkT}\, dp$$

$$= 3kT/m. \tag{31.18}$$

The root mean square deviation of the speed from its average value can now be estimated. It is defined as the square root of the average of the square of the difference between speed and average speed, divided by the average speed. We have

$$\begin{aligned}\langle(\langle v\rangle_{av}-v)^2\rangle_{av} &= \langle(\langle v\rangle_{av})^2-2v\langle v\rangle_{av}+v^2\rangle_{av}\\ &= (\langle v\rangle_{av})^2-2(\langle v\rangle_{av})^2+\langle v^2\rangle_{av}\\ &= \langle v^2\rangle_{av}-(\langle v\rangle_{av})^2.\end{aligned} \tag{31.19}$$

Dividing the square root of this by $\langle v\rangle_{av}$ and using Eqs. 31.16 and 31.18, we have for relative r.m.s. deviation of the speed: $(3\pi/8-1)^{1/2}=0.42$. This is quite a wide variation of speed among the atoms of the gas and is independent of the temperature.

We shall calculate the number of atoms that hit unit area of a wall perpendicular to the x-axis in unit time. Consider first those atoms that happen to have a particular velocity, with x-component equal to p_x/m, toward the wall. The number of these hitting the wall in unit time is equal to the number of atoms in a cylindrical volume whose axis is equal to and parallel with the velocity, and whose base is the part of the wall we are interested in. This volume is just p_x/m. The number of such atoms in the gas is the density of the gas times the volume times the probability that an atom has the particular velocity. We use $n(\mathbf{p})dp_x dp_y dp_z/N$ for the probability, and Eq. 31.13, and sum over all velocities—*i.e.* all momenta—that are going toward the wall, in order to find the total current. The result is

$$J_x = (N/V)(2\pi mkT)^{-3/2}\int_0^\infty (p_x/m)\, e^{-p_x^2/2mkT}dp_x$$

$$\times \int\int_{-\infty}^{\infty} e^{-(p_y^2+p_z^2)/2mkT}dp_y dp_z \tag{31.20}$$

where we have separated out the p^2 in the exponent into its three parts so that the three integrations become standard forms. The integration over p_x is restricted to motion in one direction—toward the wall—while the other integrations extend over all possible momenta. Finally we find

$$J_x = (N/V)(kT/2m\pi)^{1/2}. \tag{31.21}$$

The sum of all the x-component momenta carried by these atoms is computed by multiplying the number of atoms by their x-component momentum before integrating; (p_x^2/m) then replaces the factor (p_x/m) in the integrand of the last equation. The integrals then give for this quantity

$$P_x = \tfrac{1}{2}NkT/V. \tag{31.22}$$

Now the atoms collide with the wall and are turned back into the gas; their net change in momentum is just double the quantity given in Eq. 31.22. This change in momentum occurs in unit time and is therefore effectively a force exerted by the gas on the unit area of wall. It is in fact the pressure of the gas, the apparently smoothed out result of an immense number of sharp impacts. The theory then predicts that the pressure of the gas depends on its temperature and its numerical density according to the relation

$$P = NkT/V. \tag{31.23}$$

This is the familiar *ideal gas law.*

The foregoing examples will here suffice to illustrate that all the important equilibrium properties of an ideal gas can be deduced from the Maxwell-Boltzmann distribution by statistical reasoning.

Exercise III.31

31.1. Use Eq. 31.11 to compute the number of He atoms having speeds between 10 and 11 cm per sec and between 10×10^2 and 11×10^2 cm per sec in one mole of helium gas at (a) 10°K and (b) at 1000°K.

31.2. Check the integrations involved in the steps from (a) Eq. 31.5 to Eq. 31.6 and (b) Eq. 31.7 to Eq. 31.8.

31.3. Check the integrations involved in deriving Eqs. 31.15 and 31.18.

31.4. Derive Eq. 31.22 in detail as outlined in the text.

31.5. Check the integrations in Eq. 31.20 and derive Eq. 31.21.

31.6. Consider mercury vapor at the boiling point 630°K to obey the ideal gas law approximately; find the approximate rate at which atoms are emitted from boiling liquid mercury both numerically and in terms of mass.

31.7. Consider an ideal gas confined to motion in a single plane, *viz.* a monolayer film. The energy of a mass point is $(p_x^2 + p_y^2)/2m$, and the extension in phase space for a cell is $\Delta\tau = A2\pi p dp$, replacing Eq. 31.1, where A is the area of the film. Develop the remaining calculations for this problem to derive the analog of Eq. 31.23 for the monolayer.

31.8. Develop the analog of Eq. 31.16 and Eq. 31.18 for the monolayer and find the r.m.s. deviation of speed for the monolayer gas.

31.9. What is the analog of Eq. 31.9 in a monolayer gas?

III.32 Perturbation Theory of Transport Processes. The Maxwell-Boltzmann distribution is in fact capable of explaining all the equilibrium thermodynamic properties of ideal gases. When, however, we seek to understand nonequilibrium effects like the conduction of heat down a temperature gradient or the flow of mass down a pressure gradient against the viscous effects at the walls, the Maxwell-Boltzmann distribution is no longer adequate. We can see this immediately if we try to set up a net flow of mass with the Maxwell-Boltzmann distribution. The number of particles moving to the right is given by Eq. 31.20, and the number moving to the left is given by the same expression, except that the limits of integration take p_x from minus infinity to zero. This exactly balances the flow to the right. In general, if the quantity being transported by the atoms, *e.g.* kinetic energy for heat, is a function of the speed only, $f(v)$, the net transport in any one direction due to the Maxwell-Boltzmann distribution is the complete integral appearing in Eq. 31.20 with the function $f(v)$ as an extra factor in the integrand, and the limits must be taken from minus to plus infinity: the result is necessarily zero by the symmetry of the formula.

The presence of a net flow of such quantities as heat or mass in a gas implies therefore that it is not in a Maxwell-Boltzmann distribution. A small current would imply a small deviation, and so long as we restrict the discussion to small effects, certain approximate devices are available. We consider here the generalization of kinetic theory developed by Boltzmann to account approximately for small, steady-transport phenomena.

The distribution given by Eq. 31.13 was developed specifically for equilibrium conditions. We shall write for the distribution function in general the symbol:

$$f(p_1, p_2, p_3, x_1, x_2, x_3, t)dp_1dp_2dp_3dx_1dx_2dx_3, \tag{32.1}$$

being the number of representative points of the cloud in the phase cell $dp_1 \cdots dx_3$, having momentum components p_j and coordinates x_j ($j=1, 2, 3$), at time t. This number is not in general distributed uniformly in coordinate space. The function f is the number of representative points per unit extension-in-phase. If we can find this function we can describe the average time behavior of the collection of particles represented by the cloud.

Suppose the particles to be in some field of force $\mathbf{F}$ so that, between collisions, they are accelerated according to Newton's law:

$$\partial p_j/\partial t = F_j. \tag{32.2}$$

During a short interval of time δt, the particles whose representative points were in the cell $dp_1 \cdots dx_3$ of Eq. 32.1 will move, all very nearly with the same speed and acceleration, to a different cell in a new position in phase space given by:

$$x_j' = x_j + p_j \delta t/m; \quad p_j' = p_j + F_j \delta t. \tag{32.3}$$

The distribution function at this new cell must be given by the same function f of these new phase variables, because the function in Eq. 32.1 was set up to represent the situation everywhere in phase space. The general expression for the total variation of a function of a number of independent variables yields:

$$f(\mathbf{p}', \mathbf{x}', t') = f(\mathbf{p}, \mathbf{x}, t) + \delta t\, \partial f/\partial t + \sum_j \delta p_j \partial f/\partial p_j + \sum_j \delta x_j \partial f/\partial x_j. \tag{32.4}$$

Using Eq. 32.3 in this, we find

$$[f(\mathbf{p}', \mathbf{x}', t') - f(\mathbf{p}, \mathbf{x}, t)]/\delta t = \partial f/\partial t + \sum_j F_j \partial f/\partial p_j + \sum_j (p_j/m) \partial f/\partial x_j. \tag{32.5}$$

The change in number of representative points per unit extension-in-phase that this expression indicates can be due only to collisions: it is the net result of collisions between particles that send representative points into and out of the part of the cloud that has the prescribed motion. Therefore, we may write

$$[\delta f/\delta t]_{\text{coll}} = \partial f/\partial t + \sum_j F_j \partial f/\partial p_j + \sum_j (p_j/m) \partial f/\partial x_j. \tag{32.6}$$

This is called *Boltzmann's equation*, and the problem remains to find the function f as a solution of this equation. If we are interested, as in this chapter, only in steady-flow conditions, the partial derivative $\partial f/\partial t$ must vanish and can be omitted from the equation as a special case.

Let us consider a system that has steady motion of some property in the x-direction. The perturbed distribution function f may then

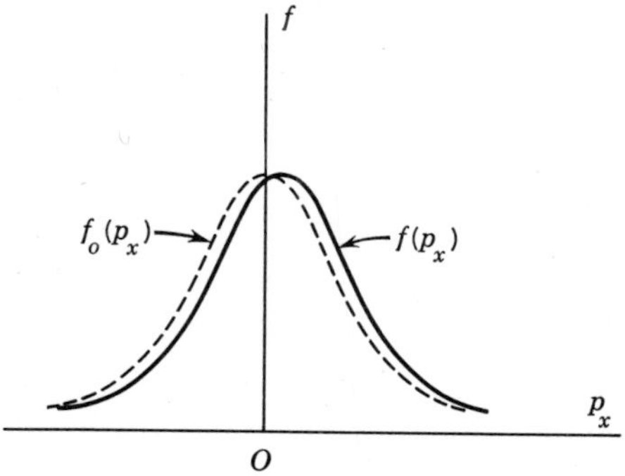

FIG. 32.1. The perturbation of Maxwell's distribution corresponding to a thermal gradient carrying heat current in the positive x-direction

depend on the x-component of velocity, p_x/m, in a unique fashion, while the unperturbed function (the equilibrium one) is of course symmetrical in all the velocity components. We assume then that

$$f = f_0 + p_x g, \tag{32.7}$$

where f_0 is the equilibrium distribution, and g is a small quantity depending symmetrically on all the components of momentum. This perturbation $p_x g$ is thus linear in the momentum in the direction of the transport. Now collisions always tend to change the collection toward the equilibrium distribution—this is the mechanism we have discussed in §30 for the attainment of equilibrium. We postulate here that for small perturbations like Eq. 32.7, the rate of return toward equilibrium is proportional to the magnitude of the perturbation and also proportional to the frequency of the collisions. We therefore write

$$[\delta f/\delta t]_{\text{coll}} = -p_x g/\tau, \tag{32.8}$$

where τ is called the relaxation time and is a measure of the time between successive collisions on any one particle.

For a simple model in which the particles are hard spheres, it can be shown that Eq. 32.8 is exact, with τ equal to the mean time between successive collisions and equal to l/v, where l is the mean free path. In general, it is a complicated problem to compute the exact value of the rate of change of f due to collisions, and the form Eq. 32.8 is a convenient parametric statement that permits us to postpone the awkward calculation. Both the mean free path and the mean collision time may be expected to depend on the velocity of the particle. In some discussions it is assumed that the free path is independent of the velocity, and an alternative way of simplifying the calculations is to assume that the time τ is independent of velocity; neither assumption can of course be exact. Here we shall take τ to be a constant.

By restricting Eq. 32.6 to one-dimensional flow, the distribution function is made to depend only on x_1, and the force $\mathbf{F}$ has only the one component F_1. We can drop the subscripts for these variables and, using Eq. 32.8, obtain

$$F\,\partial f/\partial p_1 + (p_1/m)\partial f/\partial x = -p_1 g/\tau. \tag{32.9}$$

Since the perturbation has been assumed small, we may neglect the derivatives of g contained in the first term of this equation and write to a good approximation

$$F\,\partial f_0/\partial p_1 + (p_1/m)\partial f_0/\partial x = -p_1 g/\tau, \tag{32.10}$$

where f_0 is the unperturbed distribution function. This equation is sufficient to determine the perturbation function g.

Comparing Eqs. 31.13 and 32.1, we can see that

$$f_0 = (N/V)(2\pi mkT)^{-3/2}\, e^{-p^2/2mkT}. \tag{32.11}$$

If now we allow both the density $\rho = N/V$ and the temperature T to be functions of position x, Eq. 32.10 gives the perturbation function:

$$g(p) = -(\rho\tau/m)(2\pi mkT)^{-3/2}\, e^{-p^2/2mkT} \times[-F/kT + \partial \ln \rho/\partial x + (\partial \ln T/\partial x)(p^2/2mkT - 3/2)]. \tag{32.12}$$

From this we can deduce all the transports of interest.

In general, we know from the discussion preceding Eq. 31.20 that the net current of particles in the x-direction is given by the integral of the distribution function f times the x-component of velocity. If the current per unit area is required, we integrate only over the momenta:

$$J = \iiint_{-\infty}^{\infty} (p_1/m) f\, dp_1 dp_2 dp_3. \tag{32.13}$$

The unperturbed part (f_0) of f contributes zero to this integral because it is symmetrical in the momenta. We then have

$$J = \iiint_{-\infty}^{\infty} (p_1{}^2/m) g(p)\, dp_1 dp_2 dp_3. \tag{32.14}$$

Since $g(p)$ depends only on the magnitude p, and the integral of $p_1{}^2$ is the same as with $p_2{}^2$ or with $p_3{}^2$, we may write Eq. 32.14 as

$$J = (4\pi/3m) \int_0^{\infty} p^4 g(p)\, dp. \tag{32.15}$$

by going over to spherical polars in momentum space.

The transport of heat in the x-direction is

$$W = \iiint_{-\infty}^{\infty} (p^2/2m)(p_1/m) f\, dp_1 dp_2 dp_3 \tag{32.16}$$

and again the unperturbed part of f cancels from the integrations leaving

$$W = \iiint_{-\infty}^{\infty} (p^2/2m)(p_1{}^2/m) g\, dp_1 dp_2 dp_3 = (4\pi/6m^2) \int_0^{\infty} p^6 g(p)\, dp. \tag{32.17}$$

We shall consider several particular cases of interest. First, suppose the temperature to be uniform and no external field to be present, so that the only term present on the right side of Eq. 32.12 for $g(p)$ is the one in the gradient of the numerical density ρ. If this expression is now put into Eq. 32.15, we find the current density to be

$$J = -(\tau kT/m)\, \partial\rho/\partial x, \tag{32.18}$$

showing that the diffusion constant, *cf.* Eq. 21.5, is $\tau kT/m$. This is proportional to the relaxation time and to the temperature; but we noted in Eq. 31.18 that the mean square speed is proportional to the absolute temperature, so that if the relaxation time depends on mean speed in some way, the dependence of the diffusion constant on temperature will be somewhat more complicated.

Secondly, suppose that the only dependence of $g(p)$ on position is through the temperature gradient, the other terms in Eq. 32.12 being zero. Inserting this into Eq. 32.17 gives the heat current:

$$W = -(5k^2 T\rho\tau/m)\,\partial T/\partial x, \tag{32.19}$$

so the thermal conductivity is $5k^2T\rho\tau/m$.

Thirdly, consider the presence of nothing more than an external field F with no temperature or density gradients. Taking only the term in F/kT on the right of Eq. 32.12, one obtains for the current density:

$$J = (\rho\tau/m)F \tag{32.20}$$

and the conductivity of the flow is $\rho\tau/m$.

A specially interesting application of these results has been made to the theory of conduction in metals. On the Drude-Lorentz theory of conductivity in metals, the free electrons are treated as an ideal gas moving through the atomic lattice and making collisions with the atoms. The time τ of the above theory is interpreted as the mean time for collisions, and the force F is equal to eE, where e is the charge on the electron, and E is the electrical field intensity. The electric current is eJ, which is given by Eq. 32.20 as:

$$eJ = (e^2\rho\tau/m)\,E. \tag{32.21}$$

On the same theory, the major contribution to the heat current in a metal is that due to the transport of kinetic energy by the free electrons. This is given by Eq. 32.19, m being the electronic mass. The ratio of thermal to electrical conductivity of the metal on this theory is a universal constant, thus:

$$\frac{\text{Thermal conductivity}}{T\times\text{electric conductivity}} = 5(k/e)^2 \tag{32.22}$$

which is named the *Wiedemann-Franz ratio*. If one works out the whole foregoing theory, using the free path as a constant instead of the collision time, one derives instead the value $2(k/e)^2$ for the Wiedemann-Franz ratio. Actual metals obey the law fairly well, the ratio being equal to about $3.4(k/e)^2$.

Finally let us derive an expression for the coefficient of viscosity.

To do this we picture the gas to be streaming with an average speed u in the y-direction, and allow u to be a function of x. The normal distribution function is written with respect to a coordinate system at rest with respect to the gas or moving with the speed u with respect to the laboratory; so we actually have in place of Eq. 32.11:

$$f_0 = (N/V)(2\pi mkT)^{-3/2} \exp[-(m/2kT)\{v_1^2+(v_2-u)^2+v_3^2\}] \quad (32.23)$$

and therefore

$$\partial f_0/\partial x = (m/kT)(v_2-u)f_0 du/dx. \quad (32.24)$$

We use this in Eq. 32.10 with F set zero, and the equation for the perturbation function is now

$$-p_1 g/\tau = (p_1/kT)(v_2-u)f_0 du/dx. \quad (32.25)$$

We are interested in conditions close to the boundary, so we may as well locate this boundary at the point where $u=0$, to correspond with the observed fact that the viscous gas sticks to the walls due to viscosity.

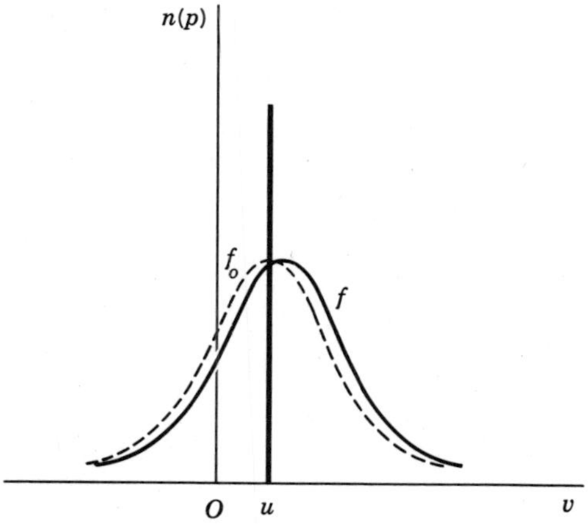

FIG. 32.2. Maxwell's distribution perturbed by viscous flow

Now the viscous drag on the wall is due to transfer of momentum. The momentum is in the y-direction, the direction of the velocity u, and the direction of transfer is in the x-direction across the wall. The net momentum carried per particle is mv_2, and the number of particles of any momentum transported in the x-direction across unit area in unit time is just $v_1 f$. Thus the net transfer of momentum per second per unit area is

$$m \iiint_{-\infty}^{\infty} v_2 v_1 f \, dp_1 dp_2 dp_3. \quad (32.26)$$

Write $f = f_0 + p_1 g$ in this, and observe that the integrals with f_0 in the integrand all vanish because of symmetry. Equating the momentum transfer rate to the viscous force, we then have the relation for the *coefficient of viscosity* η:

$$\eta\, du/dx = -m\iiint v_1 v_2 p_1 g(p)\, dp_1 dp_2 dp_3. \tag{32.27}$$

Putting Eq. 32.25 in this and setting $u = 0$ for the wall, we find

$$\eta = (N/V)\tau kT = kT\rho\tau. \tag{32.28}$$

Comparing this with Eq. 32.19 for the thermal conductivity, we have the relation involving only a universal constant:

$$\frac{\text{Thermal conductivity}}{\text{Coefficient of viscosity}} = 5k/m. \tag{32.29}$$

The observed ratio is considerably smaller than this value. The appropriate improvements in the theory required to give better agreement with the facts are elaborated in texts on kinetic theory. They involve a much more detailed discussion of the collision processes than has been given here.

One can derive many more interesting results from Eq. 32.12 by allowing simultaneous presence of two or more perturbing effects, such as the combined action of external field and temperature gradient. Or one can look for the heat current in the presence of an external field and the mass flow accompanying a temperature gradient. These effects are studied in detail in texts on kinetic theory. An application is possible again to the theory of electrons in metals, to study the electric currents due to temperature gradients and the heat currents accompanying electric fields: the theory of thermoelectric effects in general.

It turns out that the Drude–Lorentz theory of electrons fails in some very serious respects. In order to understand the observed, large, electrical conductivities, we have to postulate a large number of free electrons, about one free electron to every atom in the metal lattice; but then the theory predicts a very high heat-capacity contribution from the free electrons that would cause metals to have a higher heat capacity than insulators. This is just not the case. The Maxwell-Boltzmann distribution function is actually wrong for electrons; one has to use quantum statistics, a subject that will not be discussed in the present text.

Exercise III.32

32.1. Prove that Eq. 32.11 is consistent with Eq. 31.11.

32.2. Prove that the integral of f given by Eq. 32.7 over the entire phase space is equal to N for any function $g(p)$ of the magnitude of the momentum.

32.3. In place of Eq. 32.8, write

$$[\delta f/\delta t]_{\text{coll}} = -p_x p g/m\lambda,$$

where p is the magnitude of the momentum and λ is a constant free path. Use this to derive an alternative form of Eq. 32.12 in terms of λ instead of τ.

32.4. Using Ex. 32.3, derive an expression for thermal conductivity, for mass flow conductivity, and the Wiedemann-Franz ratio.

32.5. Derive Eq. 32.18 in detail.

32.6. Derive Eq. 32.19 in detail.

32.7. Derive Eq. 32.20 in detail.

32.8. Substitute f_0 from Eq. 32.23 into Eq. 32.13 in place of f and prove that the net current implied by Eq. 32.23 is Nu/V.

32.9. Carry out the integrations in Eq. 32.27 and derive Eq. 32.28.

III.33 The Theory of Random Motion. The perturbation theory developed in the last section is useful for discussing steady-flow phenomena, but fails in case the motion is irregular or transient. In particular the type of motion discussed in §21—dispersive motion—cannot be described accurately as due to small perturbations of the distribution function. In general, the methods of statistical mechanics outlined in the last few sections are best adapted to the study of equilibrium and steady-flow phenomena, and new or more powerful methods are required to describe the details of transient effects characteristic of matter on the atomic scale.

We shall introduce the study of transient effects in terms of the theory of random functions. The notion of a random function may best be approached in terms of simple experience with the throwing of dice. Thus we may throw a die any number of times and record the score (1 through 6) at the nth throw. Then $s(n)$, the score as a function of n, is called a random function because its value at any particular n is unrelated to its value at any other n. The random function is characterized only by the relative probabilities of finding its various possible scores; in the case of a true die there is a probability of 1/6 for each possible score.

Consider now a single particle P in a gas, subjected to collisions by the other particles of the gas; and suppose the collisions to occur at the mean collision rate, once every τ seconds, and that the particle P moves always at the mean speed v between collisions. We then postulate that the effect of each collision is random in the sense that the direction of motion after each collision is a random variable, or rather three random variables, *viz.* the three direction cosines of the velocity after each collision are completely unrelated to the direction of motion before the collision. This is a realistic assumption, essentially the same as was made in discussing the spontaneous approach to equilibrium in §30.

From this very general picture we can calculate the probability of finding P at any distance from its starting point as a function of time. The three-dimensional problem is rather complicated and we shall carry out the argument in detail only for a one-dimensional random walk.

In this problem a point P moves in one dimension in a series of jumps, each jump is of length $\ell = \tau v$, either forward or backward after each collision, independently of what happened at any previous collision. Starting at the point O, the particle P could, after N collisions, *i.e.* after time $N\tau$, be at any one of the following positions:

$$-N\ell,\ -(N-2)\ell,\ -(N-4)\ell, \cdots (N-4)\ell,\ (N-2)\ell,\ N\ell = n\ell, \tag{33.1}$$

where n is the net number of steps away from the origin at the end of the N collisions. Note especially that the point P can never be found at the position $(N-1)\ell$ after N steps of length ℓ. For a fixed N, the interval between possible positions is necessarily 2ℓ. We are interested in finding probabilities of the various possible final positions of P after N steps, as listed previously. We denote the probability of finding the particle at position $n\ell$ after N collisions by $P(N, n)$.

Since every jump is equally likely to go either left or right, each jump is equally probable, *viz.* $\frac{1}{2}$; and any sequence of N jumps is therefore equally probable, *viz.* $(\frac{1}{2})^N$. The question is, then: How many different, equally probable sequences are there that end up at any one final position $n\ell$? This number would tell us the probability of the final position. Clearly there is only one sequence that can end up at either extreme of the range of final positions, $N\ell$ or $-N\ell$: *viz.* the sequence of jumps all in the same direction, either right or left, respectively. But to reach the position $(N-2)\ell$, the particle must have made just one step to the left and $N-1$ steps to the right; it does not matter which of the N steps were taken to the left, and there are N different sequences by which it could arrive at the final position, corresponding to the N different places at which it could have taken its one step to the left. In general, to arrive at any given final position, let the particle make N_r steps to the right and N_l steps to the left; then we have necessarily $N_r + N_l = N$ and $N_r - N_l = n$. Solving these yields

$$N_r = \tfrac{1}{2}(N+n), \quad N_l = \tfrac{1}{2}(N-n), \tag{33.2}$$

giving the number of steps in either direction in terms of the number of steps and the final position.

Let us imagine the steps to have been given a serial number 1, 2, 3, $\cdots$ N; and let j or k represent this serial number. Consider some particular sequence of steps. Suppose that the jth step was to the left and the kth step was to the right; then if we make another sequence identical with this except that now the jth step is to the right and the kth step to the left, it will be counted a different sequence from the first one. If the jth and k^{th} steps were both in the same direction, the interchange would not have resulted in a different sequence. Starting from any given sequence, we can produce all the possible sequences ending up at the same position, in the same total number of steps, merely by making interchanges of pairs of steps in this way. Therefore the number of distinct sequences of N steps ending at position $n\ell$ is the number of ways we can interchange pairs of steps in opposite directions. Now there are N_r steps to the right and N_l to the left given by Eq. 33.2 for all equivalent sequences; in how many ways can pairs of steps in opposite directions be interchanged? The answer is the number of permutations of N steps, among which the permutations of N_r do not count, and permutations of N_l do not count. This number we know to be

$$X = N!/(N_r!N_l!). \tag{33.3}$$

Therefore, using Eq. 33.2 in the denominator and combining with the probability of each sequence, we have

$$P(N, n) = (\tfrac{1}{2})^N N!/\{[\tfrac{1}{2}(N+n)]![\tfrac{1}{2}(N-n)]!\}. \tag{33.4}$$

It is easy to verify that the sum over all the probabilities is unity:

$$\sum_{n=-N}^{+N} P(N, n) = 1, \tag{33.5}$$

in agreement with the fact that the particle necessarily ends up some place within the stated range of positions.

The case of particular interest is when N is very large compared with unity, and n is small compared with N; in other words, we ask for the probability that the particle be still rather near to its starting point, after a very large number of steps. When N is large we may use the Stirling approximation again, but we need a more exact form than we used in §29:

$$\ln N! = (N+\tfrac{1}{2}) \ln N - N + \tfrac{1}{2} \ln(2\pi) + \text{terms of order } 1/N. \tag{33.6}$$

This formula is to be used also for the logarithms of $\frac{1}{2}(N+n)!$ and $\frac{1}{2}(N-n)!$ with appropriate modifications. We take logarithms of Eq. 33.4:

$$\begin{aligned}\ln P(N, n) = {} & (N+\tfrac{1}{2}) \ln N - \tfrac{1}{2}(N+n+1) \ln \tfrac{1}{2}N(1+n/N) \\ & - \tfrac{1}{2}(N-n+1) \ln \tfrac{1}{2}N(1-n/N) - \tfrac{1}{2} \ln(2\pi) - N \ln 2. \end{aligned} \tag{33.7}$$

Using the further approximation

$$\ln(1 \pm n/N) = \pm n/N - n^2/2N^2 + \text{terms of order } (n/N)^3, \quad (33.8)$$

we can then show that

$$\ln P(N, n) = -\tfrac{1}{2} \ln N + \ln 2 - \tfrac{1}{2} \ln(2\pi) - n^2/2N + \text{smaller terms}, \quad (33.9)$$

or for sufficiently large values of N we have finally

$$P(N, n) = (2/\pi N)^{1/2} e^{-n^2/2N}. \quad (33.10)$$

This is a *Gaussian distribution*, with a maximum at $n=0$, dropping rapidly on both sides with a bell-shaped curve. It is obviously similar in form to the dispersive formula given by diffusion theory, Eq. 21.12, and we proceed to make this analogy exact.

Let x be the total distance traveled from the starting point to the position $n\ell$, $x=n\ell$, and let t be the total time taken for the journey, $t=N\tau$. Then if we write

$$a^2 = \ell^2/2\tau, \quad (33.11)$$

we can rewrite Eq. 33.10 in the form

$$P(N, n) = 2\ell(1/2a\sqrt{\pi t})e^{-x^2/4a^2t}. \quad (33.12)$$

Now 2ℓ is the distance between successive possible final positions, so in the infinitesimal notation we would substitute dx for 2ℓ, and if we write

$$P(N, n) = P(t, x)dx, \quad (33.13)$$

we have

$$P(t, x) = (1/2a\sqrt{\pi t})e^{-x^2/4a^2t}. \quad (33.14)$$

Consider that we start off a large number, m, of particles at time zero and position zero, and let them all be subject to the random walk process, each being entirely ignorant of the behavior of any of the others. Then the number of particles found within the interval dx—or at the final point $n\ell$—at time t must be equal to m times the probability of finding one particle there. The number of particles per unit range dx is therefore

$$\rho(t, x) = (m/2a\sqrt{\pi t})e^{-x^2/4a^2t}, \quad (33.15)$$

which is identical with Eq. 21.12 derived from diffusion theory. In fact, the diffusion constant given by Eq. 33.11 is the same as that given by the perturbation theory of Eq. 32.18. For we can write Eq. 33.11 in the form $a^2 = \frac{1}{2}\tau v^2$, and then, because the kinetic energy is $\frac{1}{2}Mv^2$, where M is the mass of the particle, and $kT = \frac{1}{2}Mv^2$, we derive

$$a^2 = \tau kT/M, \quad (33.16)$$

which is equivalent to Eq. 32.18.

When we discuss a three-dimensional random walk, the probabilities for steps in the three dimensions are regarded as independent of each other, so the probability of a path in any direction is the product of the probabilities of the three component paths required to produce it. In this way we can prove that the probability of finding the particle at position x, y, z after time t, per unit volume, is

$$P(t, x, y, z) = (4\pi a^2 t)^{-3/2} \exp[-(x^2+y^2+z^2)/4a^2 t]. \quad (33.17)$$

Given a large number of particles starting out simultaneously from the origin, the average displacement achieved in time t can be calculated from this formula. In fact,

$$\langle r \rangle_{\mathrm{av}} = (4\pi a^2 t)^{-3/2} \int_0^\infty r e^{-r^2/4a^2 t} 4\pi r^2 dr, \quad (33.18)$$

where $r^2 = x^2+y^2+z^2$. The integration is elementary:

$$\langle r \rangle_{\mathrm{av}} = 4a(t/\pi)^{1/2}. \quad (33.19)$$

The distance covered by random walk from any starting point is therefore proportional to the square root of the time, or the mean speed of random walk is inversely proportional to the square root of time.

The difficulty we had at the end of §21 with the infinite initial apparent speed of diffusion is here avoided very neatly. The form of Eq. 33.17 is of course identical with that given by diffusion theory, but in deriving it here we made the explicit assumption that the total number of steps N was very large compared with n, the number of steps removed from the starting point. If we look at the situation near the beginning where the diffusion equation gave trouble, we have to go back to the exact equation, *viz.* Eq. 33.4, which is valid for all values of N, particularly small ones. This gives the result that $P(N, n) = 0$ for all $n > N$. In other words, no effect can get away from the starting point faster than the speed of the individual steps themselves—the speed ℓ/τ.

Brownian movements of microscopic particles, suspended in a viscous fluid, liquid or gas, are immediately observable examples of random walk, but their detailed discussion is somewhat more complicated than the simple theory outlined above. The physical explanation is usually expressed as follows: the Brownian particles are large compared with the molecules of the suspending medium; actually they have dimensions of the order of 10^{-5} or 10^{-6} cm. They are, however, so small that minute pressure fluctuations in the medium can result in lack of balance between pressures on opposite sides of the Brownian particle. The latter is then subjected to a series of random impacts determined by the random pressure fluctuations of the medium.

Between impacts the Brownian particle loses the momentum it received at the last impact because of the viscosity of the medium through which it is moving, and this complicates the calculation of the free path ℓ_p between impacts. The mean time τ_p between impacts is also a difficult quantity to estimate theoretically. However, we can accept the result of the random walk calculation without committing ourselves to any particular value of ℓ_p or τ_p, and from Eq. 33.19 infer that the average distance traveled by a Brownian particle in time t is proportional to $t^{1/2}$:

$$\langle r \rangle_{\rm av} = 4\ell_p(t/2\pi\tau_p)^{1/2}. \tag{33.20}$$

By a detailed discussion that will not be repeated here, it can be shown that in fact

$$\langle r \rangle_{\rm av} = (kTt/\pi a\eta)^{1/2}, \tag{33.21}$$

where a is the radius of the Brownian particle and η is the viscosity of the medium. One can think qualitatively of Eq. 33.21 being derived from Eq. 33.20 by assuming that the time between impacts is proportional to the time between collisions between atoms in the medium and that, therefore, from Eq. 32.28,

$$\tau_p \propto \eta a^3/kT, \tag{33.22}$$

where we have also assumed that the mean step of the random walk is proportional to the size of the Brownian particle, $\ell_p \propto a$. Actually one would expect τ_p to be much longer than the mean time between atomic collisions, and the path length ℓ_p to be much greater than the linear dimensions of the Brownian particle. But neither theory nor experiment has yet succeeded in determining either τ_p or ℓ_p separately. However, the combination $\ell_p/\tau_p{}^{1/2}$, implied in Eq. 33.20 is well established both theoretically and by experimental verification of Eq. 33.19.

We have given here just a few illustrations of the usefulness of the concept of random functions. The technique has been very extensively developed in the past decade, both in fundamental theory and in engineering applications. Wider reading in this interesting field is suggested in the general references.

Exercise III.33

33.1. In the one-dimensional random walk, find the probability that the walker is just one step to the left of his starting point after he has taken N steps, with $N=1$, 3, 5, and 10^6, respectively. Use Stirling's approximation also for the cases where $N=3$ and 5, and compare with the exact results.

33.2. Using Eq. 33.15, show that the peak concentration of particles moves out from the origin at a speed $a/\sqrt{2t}$.

33.3. Show that at a fixed point x, the concentration ρ goes from zero through a peak value and returns to zero as time progresses, and that the peak value occurs at the time $t = x^2/2a^2$.

33.4. Show that the mean position of all the particles to the right of the origin at time t is $x = a\sqrt{t/\pi}$. Find the speed of the mean position of these particles.

33.5. Derive the formula quoted in Eq. 33.8.

References

The following texts contain useful material for collateral reading with the contents of this chapter.

Mathematical Methods for Scientists and Engineers, L. P. Smith, Prentice-Hall, Inc., Englewood Cliffs, N.J., 1953.

The Mathematics of Physics and Chemistry, H. Margenau and G. M. Murphy, D. Van Nostrand Co., Inc., Princeton, N.J., 1956.

Principles of Mathematical Physics, W. V. Houston, McGraw-Hill Book Co., Inc., New York, 1948.

Kinetic Theory of Gases, J. H. Jeans, The Macmillan Co., New York, 1940.

Dynamical Theory of Gases, J. H. Jeans, Dover Publications, Inc., New York, 1921.

Thermodynamics and Statistical Mechanics, Allis and Herlin, McGraw-Hill Book Co., Inc., New York, 1952.

An Introduction to Quantum Statistics, William Band, D. Van Nostrand Co., Inc., Princeton, N.J., 1955.

Theoretical Physics; Thermodynamics, F. W. Constant, Addison-Wesley Publishing Co., Inc., Reading, Mass., 1958.

Chapter IV

THE THEORY OF FIELDS

Chapter IV

THE THEORY OF FIELDS

IV.34 The Field Concept. In the last chapter we evaded, as far as possible, the question of interaction between particles. The collisions have been pictured as almost instantaneous, and the details of the motion during a collision ignored. Even where we have admitted the existence of fields, they have been thought of as "external," and the question of their sources has been removed from the problem discussed. This procedure is acceptable in first approximation in the study of gaseous materials, but becomes of doubtful value in liquids, whereas in the solid state the atoms are continually under the influence of each other's fields, and the concept of instantaneous collisions becomes completely fruitless. We cannot proceed further without actually coming to grips with the field concept in considerable detail.

Historically the field concept was not introduced to permit an understanding of the solid state; rather it had its origins in astronomy. The outstanding example of the field concept is Newton's law of universal gravitation. According to this law, a force of attraction acts between any two masses m and m', the magnitude of which is inversely proportional to the square of the distance r between them:

$$f = Gmm'/r^2, \tag{34.1}$$

where

$$G = 6.664 \times 10^{-11}\ \mathrm{M^3/KS^2}$$

on the meter-kilogram-second system of units. The force per unit mass on m' due to the mass m is Gm/r^2 in the direction of the line joining the two masses; the force per unit mass on m due to m' is Gm'/r^2 in the opposite direction. In the field concept, we picture the space in the neighborhood of a mass as a field in such a condition that, if a test particle were placed at some point, it would experience a force there due to the field. In vector notation, the field strength around a mass point m at the origin is

$$\mathbf{g} = -Gm\mathbf{r}_1/r^2, \tag{34.2}$$

where $\mathbf{r}_1$ is the unit vector drawn from the origin to the point of interest. The force experienced by a mass m' placed at this point would be $m'\mathbf{g}$, which is called the *weight* of m' in the field. Except when discussing immense masses like planetary and stellar bodies, the forces produced by gravitational fields are quite small. On the surface of the earth, g has a value about 9.81 M/S².

The dominant force between atomic particles is electrostatic. Here there is a force of repulsion between two charges q and q', proportional to the product of the charges and inversely proportional to the square of the distance between them. The magnitude of the force, when the charges are separated by a distance r in vacuo, is given by Coulomb's law:

$$f = qq'/4\pi\epsilon r^2 \tag{34.3}$$

and its direction is in the line joining the two particles. In rational M.K.S. units we have, in vacuo,

$$1/4\pi\epsilon = 9 \times 10^9 \text{ KM}^3/\text{Q}^2\text{S}^2, \tag{34.4}$$

where Q represents the coulomb unit of charge. In vector notation, the *electrostatic field strength* around the charge q is

$$\mathbf{E} = q\mathbf{r}_1/4\pi\epsilon r^2, \tag{34.5}$$

$\mathbf{r}_1$ being again the unit vector directed away from the charge q. This field strength is the force per unit charge experienced by a test particle placed at the point of interest: its dimensionality is KM/QS². The charge on an electron is 1.602×10^{-19} coulomb, and its mass is 9.106×10^{-31} kilogram. Using Eqs. 34.2 and 34.5, it is easy to compare the force of attraction due to gravitation and the force of repulsion due to electric charge on two electrons, the result being approximately

$$\frac{\text{Gravitational attraction}}{\text{Electrostatic repulsion}} = 2.42 \times 10^{-45}.$$

Coulomb enunciated a similar law for magnetostatic fields: the force of repulsion between two similar magnetic poles p and p' is

$$f = \mu pp'/4\pi r^2, \tag{34.6}$$

where, again in rational M.K.S. units,

$$\mu/4\pi = 10^{-7} \text{ KM/Q}^2. \tag{34.7}$$

The *magnetostatic field strength* around a pole p is

$$\mathbf{B} = \mu p\mathbf{r}_1/4\pi r^2; \tag{34.8}$$

its dimensionality is K/QS. The practical unit of magnetic field (M.K.S. system) is known as the *weber* per square meter:

$$1 \text{ weber/m}^2 = 1 \text{ K/QS}. \tag{34.9}$$

The dimensionality of the unit magnetic pole is correspondingly

$$[p] = \text{QM/S}. \tag{34.10}$$

From Eqs. 34.6 and 34.7, it is seen that two magnetic poles, each one unit on the rational system, repel each other with a force of 10^{-7} newton at a separation of one meter. Similarly from Eqs. 34.3 and 34.4 it is apparent that two electric charges, each one coulomb, repel each other with a force of 9×10^9 newtons at a separation of a meter. The rational system of units is not chosen for practical convenience in experimental measurements, but for the simplicity of the equations describing the laws of the electromagnetic field. This will become evident later. For more details concerning different systems of electromagnetic units, reference should be made to specialized texts on electromagnetic theory.

Only positive masses are recognized in the classical gravitation theory, although negative masses have appeared in recent speculations regarding nuclear forces. Positive and negative electric charges exist in all matter, normally in equal balancing amounts.

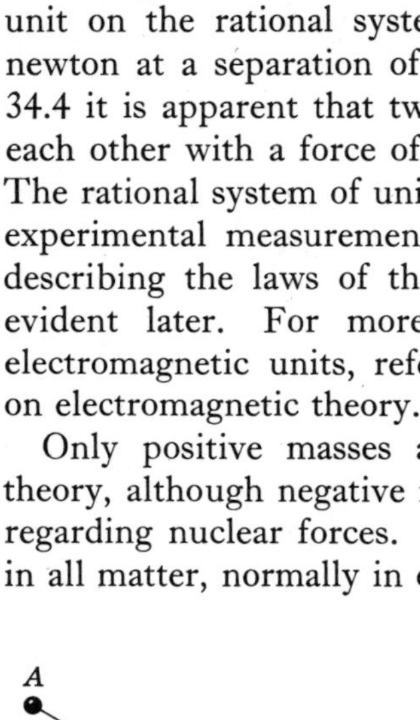

FIG. 34.1. Linear superposition of vector fields

The first basic rule of all field theories states that at any point the field strength due to a distribution of several sources (charges, masses, or poles, etc.) is the vector sum of all the field strengths due to all of the individual sources except any source that may be located at the point where the field is desired: this is the linearity rule. Without it, the mathematical handling of field theory would become impossibly cumbersome. The rule apparently fails in nuclear interactions, and this is one of the main reasons why nuclear theory is so much more troublesome than the theory of atomic interactions. In performing the vector sum over all individual fields at any point, we first remove, in thought, any charge (or other source) that may happen to be located at the point of interest, to leave a place for the test charge that we must in principle use to measure the field there. Incidentally, if we did not do this, there would be an infinite field strength at the point because of the singularity in the inverse square law at the origin, and the whole theory would become useless. If there is actually a charge at the point of interest we could, instead of removing it, use it as a test charge and this again in effect removes its contribution to the field there.

From Eq. 34.5, using the definition Eq. 6.18 it is easy to show that the divergence of the field strength due to a single point charge q is everywhere zero except possibly at the point occupied by the charge—*i.e.* at the singularity. Moreover, because the divergence is a linear operator, the divergence of the vector sum of a number of fields equals the sum of the divergences of the individual fields; since these are all zero, the divergence of the total field is also zero. Putting this result into Gauss' theorem, we have

$$0 = \iiint \nabla \cdot \mathbf{E} d\tau = \iint \mathbf{E} \cdot d\boldsymbol{\sigma}. \tag{34.11}$$

However, if we calculate the surface integral of **E** due to a single charge over any sphere with the charge at its center and use Eq. 34.5, we find at once

$$\iint \mathbf{E} \cdot d\boldsymbol{\sigma} = q/\epsilon. \tag{34.12}$$

The reason for this discrepancy between Eqs. 34.11 and 34.12 is of course that the Gauss theorem does not apply when singularities exist in the volume of integration. To overcome this inconvenience, it is customary to replace the discrete set of point charges by a "fictitious" continuous charge density. This density function is chosen so that in any volume $d\tau$ that is small compared with experimental measurements, but large on the atomic scale, the total charge $\rho d\tau$ due to the charge density ρ shall be equal to the sum of the discrete charges in that volume. Because ρ is a continuous, noninfinite function, we are justified in using the Gauss integral theorem, accepting Eq. 34.12 in the form

$$\iint \mathbf{E} \cdot d\boldsymbol{\sigma} = (1/\epsilon) \iiint \rho d\tau \tag{34.13}$$

and deducing that

$$\nabla \cdot \mathbf{E} = \rho/\epsilon. \tag{34.14}$$

The second basic rule of static field theory is that the fields shall be conservative, which means that the work done on a test particle taken around a closed loop shall vanish:

$$q \oint \mathbf{E} \cdot d\mathbf{r} = 0. \tag{34.15}$$

As we saw in Eq. 7.4, this implies the existence of a single-valued scalar potential function Φ such that

$$\mathbf{E} = -\nabla \Phi. \tag{34.16}$$

Combination of this equation with Eq. 34.14 yields *Poisson's equation* for the potential:

$$\nabla^2\Phi = -\rho/\epsilon. \tag{34.17}$$

In uncharged regions it reduces to Laplace's equation:

$$\nabla^2\Phi = 0. \tag{34.18}$$

A similar potential function, also satisfying Laplace's equation, exists for all conservative inverse square law fields. The many solutions of Laplace's equation are well known in pure mathematics, and the fitting of these solutions to practical situations comprised almost the entire subject of potential theory up until the last few decades.

We present here three basic theorems regarding the properties of the solutions of Laplace's equation.

THEOREM I. Any linear combination of solutions of Laplace's equation is also a solution of Laplace's equation.

As an immediate consequence of this theorem, if we have a number of differently located point charges at positions $\mathbf{r}_j$, the potential at any point $\mathbf{r}$ is

$$\Phi(\mathbf{r}) = (1/4\pi\epsilon)\sum_j q_j/|\mathbf{r}_j-\mathbf{r}|. \tag{34.19}$$

THEOREM II. Any partial derivative of a solution of Laplace's equation is also a solution of Laplace's equation.

This theorem provides a convenient device for seeking a variety of new solutions from any one that we may have found.

THEOREM III. If the values of Φ are given over the surface of any volume throughout which Φ satisfies Laplace's equation at all points, the solution within this volume is uniquely determined.

This theorem is called the *Uniqueness Theorem* and is the fundamental theorem of potential theory. For proof we may let Φ_1 and Φ_2 be, if possible, two different solutions having the same boundary values, and prove that they must be in fact identical throughout the volume. We define

$$\Phi = \Phi_1 - \Phi_2. \tag{34.20}$$

Now Φ is a simple linear combination of two solutions of Laplace's equation and is therefore, from Theorem I, a solution itself, thus:

$$\nabla^2\Phi = 0. \tag{34.21}$$

Consider the identity

$$(\nabla\Phi)\cdot(\nabla\Phi) = \nabla\cdot(\Phi\nabla\Phi) - \Phi\nabla^2\Phi = \nabla\cdot(\Phi\nabla\Phi), \tag{34.22}$$

by Eq. 34.21. Take the volume integral of this last equation and use Gauss' theorem:

$$\iiint(\nabla\Phi)\cdot(\nabla\Phi)d\tau = \iint\Phi\nabla\Phi\cdot d\boldsymbol{\sigma} = 0. \tag{34.23}$$

This is zero because over the surface Φ is the zero difference between the two functions Φ_1 and Φ_2, both of which have by hypothesis the same boundary values. But $(\nabla\Phi)\cdot(\nabla\Phi)$ is everywhere non-negative, because it is the square magnitude of the vector $\nabla\Phi$. Its volume integral can be zero, therefore, only if it is itself zero everywhere. Thus we have

$$\nabla\Phi = \nabla(\Phi_1 - \Phi_2) = 0 \tag{34.24}$$

throughout the volume. However, since $\Phi_1 - \Phi_2$ is zero on the boundaries, this last equation means that it must stay zero within the volume because its gradient is zero everywhere. This proves the theorem.

We have introduced the classical idea of a field here, in terms of the force exerted on a test particle. For this approach to be acceptable it is essential that the test particle have so small a charge that it causes no detectable disturbance in the sources of the field it is being used to measure. In macroscopic physics, astronomy, and classical mechanics, this condition is satisfied to a good approximation. Small test particles exist and we can handle them in the laboratory. In atomic physics, on the other hand, no suitable test particles exist. The electron is the smallest charge we know, and an extra electron placed in an atomic field produces a violent disturbance; even if the electron could be seen and handled—which it cannot—it would not do for a test particle. It becomes necessary, as it were, to disembody the field concept: we have to accept the existence of the field as a property of space even though its strength cannot be directly measured by the simple test-particle procedure.

The logic of modern field theory is the reverse of the classical procedure. The mathematical properties of the field are set up in a purely formal fashion and their consequences deduced for comparison with observation. In fact, the material particle of classical physics, instead of being the "building block" of the theory, becomes a derived concept, as a singularity of some kind in the field variables.

It is possible to understand the first half of this chapter in terms of the test-particle picture; it becomes difficult to do this later on, and it is practically impossible to understand the last chapters in that way.

Exercise IV.34

34.1. From Eqs. 34.2, 34.5, and 34.8, prove that the divergences of $\mathbf{g}$, $\mathbf{E}$ and $\mathbf{B}$ are all zero; but show that these results are not proved when r approaches zero.

34.2 Speculate on some possible effects on the law of gravitation of allowing the existence of both positive and negative mass. Contrast these with the effects of positive and negative charges in the electric case.

34.3. Find **E** from Eqs. 34.16 and 34.19. Prove that Eq. 34.19 satisfies Laplace's equation.

34.4. Prove Theorem I.

34.5. Prove Theorem II.

34.6. Enunciate and prove Theorem III for a two-dimensional space.

34.7. Find the field vector **E** at distance r from an infinitely extended straight line carrying a uniform linear charge density. To do this, use Eq. 34.12 on a cylinder of unit length and radius r, together with arguments from the symmetry of the arrangement.

34.8. Find the potential difference between two points at distances r_1 and r_2 from the charged line in Ex. 34.7 by integrating Eq. 34.16 in accordance with Eq. 6.11.

34.9. Consider the application of Eq. 34.19 to the problem in the last two exercises. Change the sum into an integral along the line of the charge (justify this step in detail) and evaluate the integral between infinite limits. Find the potential difference between the two points of Ex. 34.8, first by subtracting two complete integrals, and second by integrating the differences between the contributions at the two points from each element of the line. Discuss the infinities occurring in these calculations, and compare with the result of Ex. 34.8.

IV.35 Boundary Value Problems with Laplace's Equation. The simplest problems of field theory involve the determination of the scalar potential of the field, as a solution of Laplace's equation. As a simple illustrative example of methods that are also applicable in more complicated problems, we consider first the method of solution of Laplace's equation in two dimensions within a rectangular boundary. The value of Φ is to be prescribed as a function of position all around the boundary. We may give this problem a well-defined physical meaning by considering the mechanical equilibrium of a flexible membrane under tension in a wire rectangle lying at first in the x-y plane. The displacement of the membrane in the z-direction is $\Phi(x, y)$ and in static equilibrium satisfies the Laplace equation, from Eq. 10.3 with the time derivative set zero. The boundary values of Φ can be set to follow any desired function by distorting the wire frame in the z-direction as required. The solution of the two-dimensional Laplace's equation:

$$\partial^2\Phi/\partial x^2 + \partial^2\Phi/\partial y^2 = 0, \tag{35.1}$$

then gives the distorted shape of the membrane.

To tackle this problem it is strategic to begin with a still simpler one, *viz.* let the wire frame remain straight along three of the edges and bend to follow a function $f(x)$ on the fourth edge. Thus we let

$$\left.\begin{aligned} \Phi(x, y) &= 0 \text{ on the lines } x = 0,\ x = a,\ y = b \\ \Phi(x, y) &= f(x) \text{ on the line } y = 0. \end{aligned}\right\} \tag{35.2}$$

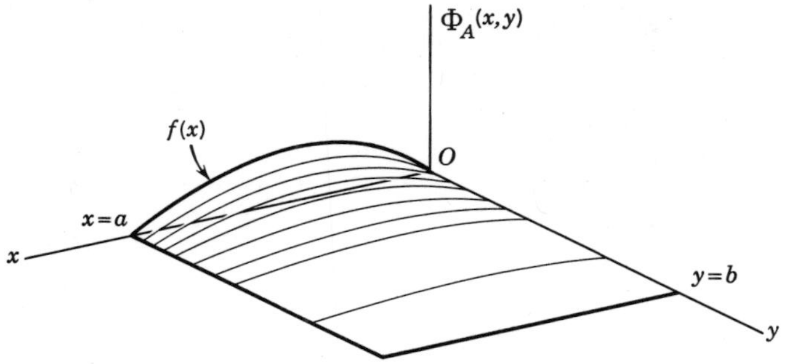

FIG. 35.1. A solution of Laplace's equation with boundary conditions described in Eq. 35.2

Theorem III of the last section shows that the solution is unique; we can therefore proceed by guessing a solution, and, if we find one, we have the only possible solution. We shall guess that the solution is separable in the Cartesian coordinates—because the boundary is rectangular:

$$\Phi(x, y) = X(x)\, Y(y). \tag{35.3}$$

Substituting this form into Eq. 35.1 and dividing by XY, we have

$$(1/X)d^2X/dx^2+(1/Y)d^2Y/dy^2 = 0,$$

which shows by the familiar reasoning that both

$$d^2X/dx^2 = -k^2X \tag{35.4}$$

and

$$d^2Y/dy^2 = k^2Y, \tag{35.5}$$

where k^2 is an arbitrary constant. The solution of Eq. 35.4 is $X=\sin kx$, and for this to vanish both at $x=0$ and $x=a$, to satisfy the stated boundary conditions, Eq. 35.2, we must restrict the arbitrary constant:

$$k = n\pi/a \tag{35.6}$$

and so

$$X = \sin(n\pi x/a), \tag{35.7}$$

with n an arbitrary integer. The solution of Eq. 35.5 that vanishes at $y=b$, to match the boundary conditions, is well known:

$$Y = \sinh[n\pi(b-y)/a], \tag{35.8}$$

and thus the complete solution in Eq. 35.3 is

$$\Phi_n = \sinh[n\pi(b-y)/a]\sin(n\pi x/a), \tag{35.9}$$

depending on the integer n. This satisfies automatically the boundary conditions on three of the edges of the rectangle, but it does not generally satisfy the boundary condition over the fourth side. To find a solution that satisfies this part of the boundary condition also, we make linear combinations of the solutions in Eq. 35.9 having different values of n. Thus by Theorem I of §34, the following function is also a solution

$$\Phi = \sum_n A_n \Phi_n, \tag{35.10}$$

where A_n are arbitrary constants. We then try to find constants A_n such that Eq. 35.10 satisfies all the boundary conditions. It is clear that it automatically satisfies the same boundary conditions as each Φ_n over the first three parts of the boundary: where each Φ_n vanishes, Φ also vanishes, as required. Over the fourth side, $y=0$, we want it to have a specified functional dependence on x:

$$f(x) = \Phi = \sum_n A_n \sinh(n\pi b/a) \sin(n\pi x/a), \tag{35.11}$$

from Eq. 35.9 at $y=0$.

Multiply the last equation by $\sin(m\pi x/a)$ where m is any integer, and integrate the result with respect to x over the length $0 \leqslant x \leqslant a$:

$$\int_0^a f(x) \sin(m\pi x/a) dx = \tfrac{1}{2} a A_m \sinh(m\pi b/a). \tag{35.12}$$

Since the function $f(x)$ is known, this equation determines each A_m, and substituting back into Eq. 35.10 we obtain the desired solution.

The procedure required to find Φ, when it is a given (in general nonzero) function on each of the four sides of the rectangle follows. Call the solution Eq. 35.12, $\Phi_A(x, y)$, and its value over the boundary $\Phi_A(x, 0) = f_A(x)$. Using similar methods we now find the solution $\Phi_B(x, y)$ that is zero over the three sides $y=0$, $y=b$, $x=a$ and has the value $\Phi_B(0, y) = f_B(y)$ over the side $x=0$. Again we find $\Phi_C(x, y)$ that is zero over the three sides $x=0$, $x=a$, $y=0$, while $\Phi_C(x, b) = f_C(x)$ on $y=b$. Finally find $\Phi_D(x, y)$, the solution that vanishes over the three sides $x=0$, $y=0$, $y=b$, while $\Phi_D(a, y) = f_D(y)$ on $x=a$. Then the sum of these four solutions is, by Theorem I of §34, also a solution, while its values on the four sides of the rectangle are the required functions:

$$\Phi = \Phi_A + \Phi_B + \Phi_C + \Phi_D \tag{35.13}$$

and

$$\left.\begin{aligned} \Phi(0, y) &= 0 + f_B(y) + 0 + 0 = f_B(y) \\ \Phi(x, b) &= 0 + 0 + f_C(x) + 0 = f_C(x) \\ \Phi(a, y) &= 0 + 0 + 0 + f_D(y) = f_D(y) \\ \Phi(x, 0) &= f_A(x) + 0 + 0 + 0 = f_A(x) \end{aligned}\right\}. \tag{35.14}$$

The solution Φ therefore satisfies the arbitrary boundary function all around the rectangle and is the uniquely determined solution.

As a second example we shall discuss the solutions of Laplace's equation in three dimensions, inside a spherical boundary. In this case it is convenient to express the Laplacian operator in spherical polar coordinates (*cf.* Ex. 8.11):

$$(1/r^2)(\partial/\partial r)(r^2\partial\Phi/\partial r)+(1/r^2 \sin\theta)(\partial/\partial\theta)(\sin\theta\ \partial\Phi/\partial\theta)+ (1/r^2 \sin^2\theta)\partial^2\Phi/\partial\phi^2 = 0. \quad (35.15)$$

We guess that the solution separates into three factors of the form

$$\Phi = R(r)T(\theta)F(\phi), \quad (35.16)$$

substitute this form into Eq. 35.15, divide by RFT and multiply by $r^2 \sin^2\theta$, and find

$$(1/R)(\sin^2\theta)(d/dr)(r^2dR/dr)+(1/T)(\sin\theta)(d/d\theta)(\sin\theta dT/d\theta)+ (1/F)d^2F/d\phi^2 = 0. \quad (35.17)$$

The last term on the left depends on the angle ϕ, the remainder of the expression on the left being independent of ϕ; each part is therefore equal to a constant so as to cancel permanently to zero as required by the equation. Thus we have

$$d^2F/d\phi^2 = -m^2F, \quad (35.18)$$

where m^2 is the constant. The solution of this is

$$F = \exp(\pm im\phi). \quad (35.19)$$

So far the constant m is arbitrary, but now we ask that the solution be single-valued in space, so that F must repeat itself when the angle ϕ goes through 2π. This is assured only if m is an integer. Eq. 35.19 then means that F is either $\sin m\phi$ or $\cos m\phi$ or some linear combination of these, with m an arbitrary integer. Putting Eq. 35.18 into 35.17 and multiplying through by $1/\sin^2\theta$ yields

$$(1/R)(d/dr)(r^2dR/dr)+(1/T\sin\theta)(d/d\theta)(\sin\theta\ dT/d\theta) = m^2/\sin^2\theta. \quad (35.20)$$

Here the r-dependent term separates from the rest, so each is constant again:

$$(1/R)(d/dr)(r^2dR/dr) = A. \quad (35.21)$$

To solve this we may guess that $R=r^n$ and find that this does indeed satisfy the equation with $A=n(n+1)$; or we might guess $R=1/r^{n+1}$ and find this also satisfies it, with A this time equal to $(n+1)n$, just as before. Since n is not specified, we may say that there are two sets of solutions of Eq. 35.21 provided we write $A = n(n+1)$, *viz.:*

$$R = r^n \quad \text{or} \quad R = 1/r^{n+1} \tag{35.22}$$

and no restriction is yet set on n.

We now make the following transformations:

$$u = \cos\theta, \quad T(\theta) = P(u). \tag{35.23}$$

Use of Eq. 35.21 in Eq. 35.20 now yields the result:

$$(1-u^2)d^2P/du^2 - 2udP/du + [n(n+1) - m^2/(1-u^2)]P = 0. \tag{35.24}$$

This is the *associated Legendre equation*, and it has been intensively studied in the theory of differential equations.

Consider first the special case when $m=0$, which, by Eq. 35.19, means that Φ is independent of the angle ϕ. Then Eq. 35.24 reduces to the *ordinary Legendre equation*:

$$(1-u^2)d^2P/du^2 - 2udP/du + n(n+1)P = 0. \tag{35.25}$$

This equation can be solved by expansion in a power series in u, but the series diverges when $u=1$, *i.e.* whenever the angle θ is either zero or π. The series is of little use therefore, unless it happens to terminate in a finite number of terms, the remaining terms being identically zero. This does happen if and only if n is an integer, which it therefore must be for the existence of well-behaved solutions of this type. The general solution is found to be:

$$T(\theta) = P_n(u) = \sum_{k=0}^{\leqslant \frac{1}{2}n} (-1)^k \frac{(2n-2k)!\, u^{n-2k}}{2^n(n-k)!(n-2k)!k!}. \tag{35.26}$$

Samples of this series are easily written down, and the fact that they are solutions of Laplace's equation may be verified by direct substitution:

$$\begin{aligned} &P_0(u) = 1; \quad P_1(u) = \cos\theta; \quad P_2(u) = (3/2)\cos^2\theta - \tfrac{1}{2}; \\ &P_3(u) = (5/2)\cos^3\theta - (3/2)\cos\theta; \\ &P_4(u) = (35/8)\cos^4\theta - (30/8)\cos^2\theta + 3/8; \quad \textit{etc.} \end{aligned} \tag{35.27}$$

Returning to the general Legendre equation, Eq. 35.24, with m not zero, we can prove that its solutions are derivable from the above Legendre polynomials by the formula:

$$P_{n,m}(u) = T_{n,m}(\theta) = (1-u^2)^{m/2}\, d^mP_n(u)/du^m. \tag{35.28}$$

We may note that $P_n(u)$ involves no powers of u higher than the nth; hence its mth derivative vanishes for all $m > n$ and so does $T_{n,m}(\theta)$. The fact that the only associated Legendre polynomials that exist are those with $m \leqslant n$ is important in many physical problems—it limits the number of possible solutions, as we shall see in §44 and again in §56. As examples we may list the following associated polynomials:

$$T_{n,0}(\theta) = P_n(u); \quad T_{n,1}(\theta) = \sin\theta\, dP_n/du;$$
$$T_{n,2}(\theta) = \sin^2\theta\, d^2P_n/du^2;\ etc.,$$

yielding

$$\begin{aligned} T_{1,1}(\theta) &= \sin\theta; \quad T_{2,1}(\theta) = \tfrac{3}{2}\sin 2\theta; \\ T_{3,1}(\theta) &= \tfrac{3}{2}\sin\theta\,(5\cos^2\theta - 1),\ etc., \\ T_{2,2}(\theta) &= 3\sin^2\theta; \quad T_{3,2}(\theta) = 15\sin^2\theta\cos\theta,\ etc. \end{aligned} \tag{35.29}$$

Combining all the results obtained for each factor in the expression of Eq. 35.16, *viz.* Eqs. 35.19, 35.22, and 35.28, we can set up the most general solution as a linear combination with all the permitted values of the two parameters m and n:

$$\Phi(r, \theta, \phi) = \sum_{n,m}\sum (A_{nm}r^n + B_{nm}r^{-n-1})e^{im\phi}T_{n,m}(\theta), \tag{35.30}$$

and we may hope to be able to match a wide variety of boundary values by means of functions like this, with proper choice of the arbitrary constants A_{nm} and B_{nm}.

If the region of interest includes the entire space inside some sphere, the terms in Eq. 35.30 that go to infinity at the origin must be removed; therefore, the arbitrary constants B_{nm} must all be taken zero. If the region of interest does not include the origin—*e.g.* it is space between two concentric spheres—then the B-series must in general be retained. If we are interested only in spherically symmetrical boundary conditions that are independent of the angle ϕ, only terms with $m=0$ are needed, and the solution must have the form:

$$\Phi(r, \theta) = \sum_n (A_n r^n + B_n r^{-n-1})\, P_n(\cos\theta). \tag{35.31}$$

We shall prove in this last case that it is possible to satisfy any symmetrical boundary conditions on the sphere, *e.g.*

$$\Phi(a, \theta) = f(\theta), \quad r = a, \tag{35.32}$$

where $f(\theta)$ is an arbitrarily chosen function of θ alone. To do this we must make use of a theorem that we shall quote here without proof, which states the following:

$$\int_{\theta=0}^{\pi} T_{n,m}(\theta)T_{p,m}(\theta)\sin\theta\, d\theta = \delta_{n,p}\frac{2}{2n+1}\frac{(n+m)!}{(n-m)!}. \tag{35.33}$$

Incidentally, the factor $\sin\theta\, d\theta$ is a zonal element of area on the unit sphere, and the integral here is essentially the scalar product of two functions in the function space over the domain of the spherical surface.

The δ-term on the right indicates that the functions are orthogonal when $n \neq p$.

The problem is this: Can we find appropriate constants A_n and B_n such that Eq. 35.31 actually matches the arbitrary surface values given by Eq. 35.32? To answer this we equate the two expressions:

$$f(\theta) = \sum_n A_n a^n P_n(\cos\theta), \tag{35.34}$$

having omitted the B-terms because the region includes the center of the sphere. Now multiply this relation on both sides by $P_p(\cos\theta) \times \sin\theta\, d\theta$ and integrate over the range $0 \leqslant \theta \leqslant \pi$. The right side reduces by Eq. 35.33, applied with $m=0$, to $\sum_n A_n a^n\, \delta_{n,p}\, 2/(2n+1)$, so that we have at once

$$A_n = (n+\tfrac{1}{2})a^{-n} \int_0^\pi P_n(\cos\theta) f(\theta) \sin\theta\, d\theta. \tag{35.35}$$

The coefficients A_n can therefore be found in general no matter what function $f(\theta)$ may be chosen.

Exercise IV.35

35.1. What is the solution of Eq. 35.1 that vanishes at all points of a rectangular boundary?

35.2. In Eq. 35.2, let $a=b=1$, and $f(x)=\sin(2\pi x)$. Find $\Phi(x, y)$.

35.3. Use Eq. 35.14 to find the solution $\Phi(x, y)$ for which $\Phi(0, y)=\Phi(x, b)=0$, and $\Phi(a, y)=\sin(\pi y/b)$, $\Phi(x, 0)=\sin(\pi x/a)$.

35.4. Find and discuss the solutions of Laplace's equation in the three-dimensional Cartesian coordinates that have the form $X(x)Y(y)Z(z)$. Consider only those solutions that vanish over all but one of the six faces of the rectangular parallelepiped bounding the region of interest. How would you set up the solution having arbitrary functional values over each of the faces?

35.5. Verify that Eq. 35.22 satisfies Eq. 35.21.

35.6. Verify by substitution that Eq. 35.26 satisfies Eq. 35.25.

35.7. Develop Eq. 35.27 from Eq. 35.26 and find several more polynomials.

35.8. Verify by substitution that Eq. 35.28 satisfies Eq. 35.24. Hint: note that $P_n(u)$ satisfies Eq. 35.25.

35.9. Develop Eq. 35.29 from Eq. 35.28 and find several more examples.

35.10. Find a solution to Laplace's equation like Eq. 35.30 such that $\Phi=f(\theta)$ over a sphere $r=a$, and $\Phi=g(\theta)$ over a sphere $r=b<a$, both functions $f(\theta)$ and $g(\theta)$ being arbitrary functions of the angle θ alone.

35.11. Find the solution of Laplace's equation in separable spherical polar coordinates when it is given an arbitrary functional dependence on both angles; $\Phi=f(\theta, \phi)$ over a sphere $r=a$, and remains finite at the origin, $r=0$.

35.12. Find the solution outside a sphere of radius a for which $(\partial\Phi/\partial r)_{r=a} = A\cos\theta$ and which approaches zero as r approaches infinity. Hint: use Theorem II of §34. In electrostatics this would be the field due to a spherical dielectric with a dipole moment.

35.13. Find the solution of Laplace's equation outside the sphere $r=a$ for which $(\partial\Phi/\partial r)_{r=a} = A\cos 2\theta$, and which goes to zero as $r \to \infty$. Hint: again use Theorem II of §34; note also $\cos 2\theta = \cos^2\theta - \sin^2\theta$, and use a linear combination of P_0, P_2 and $T_{2,2}$ in Eq. 35.30. In electrostatics this would be the field due to a spherical dielectric with a quadrupole moment.

35.14. Prove that the potential function inside a closed boundary may be determined except for an additive constant by giving the normal component of the potential as a function of position over the boundary. This is a modified form of the uniqueness Theorem III.

35.15. Consider the ammonia molecule NH_3 to be a dielectric sphere of radius 1.00 angstrom, having an electric dipole moment 1.47 Debye units. The Debye unit is 10^{-18} statcoulomb cm (3.335×10^{-30} coulomb meter). Find the interaction energy between two such molecules at a distance 4.00 angstroms between their centers. Compare this energy with the mean thermal kinetic energy at 300°K.

IV.36 The Dipole Field.

We state without proof the following theorem on Legendre polynomials:

$$\sum_n y^n P_n(\cos\theta) = (1 - 2y\cos\theta + y^2)^{-1/2} \quad (y < 1). \tag{36.1}$$

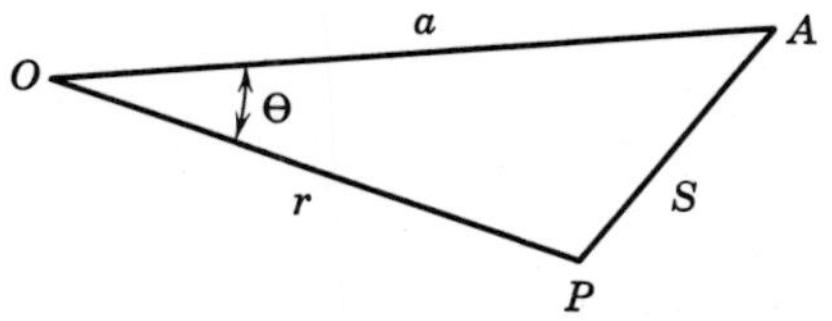

FIG. 36.1. Equation 36.2 illustrated

Consider the triangle shown in Fig. 36.1.

CASE I: $r < a$. We have

$$s^2 = a^2 + r^2 - 2ar\cos\theta = a^2(1 - 2y\cos\theta + y^2),$$

where $y = r/a < 1$. Then from Eq. 36.1 we have

$$1/s = (1/a)\sum_n (r/a)^n P_n(\cos\theta). \tag{36.2}$$

CASE II: $r > a$. We may now write $y = a/r < 1$, and find

$$1/s = (1/r)\sum_n (a/r)^n P_n(\cos\theta). \tag{36.3}$$

These results are useful in studying the potential field of a dipole. Consider the diagram of Fig. 36.2. The potential at P due to the charge $+q$ at the right and $-q$ at the left is Φ_P, where

$$4\pi\epsilon\Phi_P = q/r_1 - q/r_2. \tag{36.4}$$

We would prefer to express this as an explicit function of R and θ in plane polar coordinates with respect to the center 0.

CASE I: $R > d$. We have

$$\left.\begin{aligned} r_1^2 &= R^2+d^2-2Rd\cos\theta \\ r_2^2 &= R^2+d^2+2Rd\cos\theta \end{aligned}\right\} \tag{36.5}$$

and if we write $y = d/R < 1$, we can apply Eq. 36.3 to give

$$4\pi\epsilon\Phi_P = (q/R)[(1-2y\cos\theta+y^2)^{-1/2}-(1+2y\cos\theta+y^2)^{-1/2}],$$

so that

$$4\pi\epsilon\Phi_P = (q/R)[\sum_{n=0} y^n P_n(\cos\theta) - \sum_{n=0} (-y)^n P_n(\cos\theta)]. \tag{36.6}$$

The even terms all cancel and we are left with just double the odd terms:

$$4\pi\epsilon\Phi_P = (2q/R)\sum_{n=0} (d/R)^{2n+1} P_{2n+1}(\cos\theta). \tag{36.7}$$

When $R \gg d$ this reduces to the familiar expression for the potential of a dipole singularity at the origin:

$$4\pi\epsilon\Phi_P \doteqdot (2q/R)(d/R)P_1(\cos\theta) = 2qd\cos\theta/R^2. \tag{36.8}$$

These fields are due to an isolated *dipole* in infinite space. Now let us place the dipole inside some enclosure and demand certain boundary

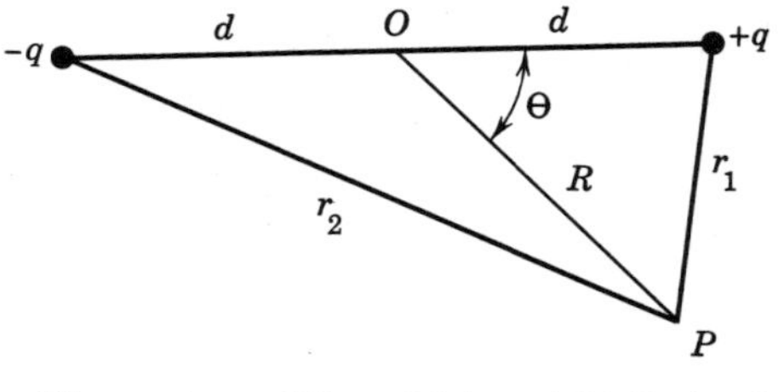

FIG. 36.2. An illustration of Eqs. 36.4 and 36.7, the field of a dipole

values of the potential on the walls of the enclosure. For example, we may surround the dipole of Fig. 36.2 by a sphere with center at O and radius R_0. Note first that the expression for Φ_P given by Eq. 36.7 is true for all points in space round the dipole—the field is symmetrical around the axis of the dipole and so is independent of the angle ϕ. We shall require that the potential at the sphere, $R = R_0$ shall have a prescribed dependence on the angles: $\Phi(R=R_0) = f(\theta, \phi)$, and ask for the field inside the sphere in the presence of the dipole near the origin.

The potential at the sphere $R = R_0$ due to the dipole itself can be written down at once from Eq. 36.7:

$$4\pi\epsilon\Phi_{\text{dip}}(R_0) = (2q/R_0)\sum_{n=0}(d/R_0)^{2n+1}P_{2n+1}(\cos\theta). \qquad (36.9)$$

The difference between this field and the required boundary values now forms a new boundary condition on a complementary solution of Laplace's equation, to be added to the dipole's field inside the sphere. Thus we seek a solution that (a) has no singularities inside the sphere, and (b) has as boundary values the difference:

$$R = R_0, \quad \Phi_{\text{comp}} = f(\theta, \phi) - \Phi_{\text{dip}}. \qquad (36.10)$$

Such a complementary solution can be formed by linear combinations of functions like those given in Eq. 35.30, in which we remove all the B-terms to avoid singularities inside the sphere. Thus

$$\Phi_{\text{comp}} = \sum_{n,m}\sum A_{nm}R^n e^{im\phi}T_{n,m}(\theta). \qquad (36.11)$$

The most simple and obvious example of this kind of problem occurs when the sphere is required to be an equipotential, $f(\theta, \phi) = 0$, as would be the case if the dipole were surrounded by a grounded conducting sphere. Under such circumstances, we require that

$$4\pi\epsilon\sum_{n,m}\sum A_{nm}R_0{}^n e^{im\phi}T_{n,m}(\theta) = -(2q/R_0)\sum_{n}(d/R_0)^{2n+1}P_{2n+1}(\cos\theta). \qquad (36.12)$$

Using the orthogonality theorem of Eq. 35.33, we can see that this requires all the coefficients A_{nm} to be zero when m is not zero and n is not odd. In fact, we find

$$4\pi\epsilon A_{2n+1}R_0{}^{2n+1} = -(2q/R_0)(d/R_0)^{2n+1}$$

or

$$4\pi\epsilon A_{2n+1} = -(2q/R_0)(d/R_0{}^2)^{2n+1}. \qquad (36.13)$$

The final solution is the sum of the dipole field and the complementary field required by the equipotential boundary condition, and, using Eqs. 36.11 and 36.7, we see now that

$$4\pi\epsilon\Phi = (2q/R_0)\sum_{n=0}[(R_0/R)(d/R)^{2n+1} - (Rd/R_0{}^2)^{2n+1}]P_{2n+1}(\cos\theta). \qquad (36.14)$$

If $R_0 \gg d$, this reduces to

$$4\pi\epsilon\Phi = (2qd\cos\theta/R^2)(1 - R^3/R_0{}^3). \qquad (36.15)$$

From this result we can calculate the field intensity in the neighborhood of the equipotential boundary and so derive the induced charge located there. This calculation is left as an exercise. It is generally true that many problems arise with particularly simple sources, such as dipoles, quadrupoles, or merely point charges, whose isolated fields are well known; if there were no boundaries other than infinity, there would be no problem; the presence of boundaries and the fitting of solutions to boundary conditions make the subject of potential theory much more interesting. We approach the general boundary problem more closely in the next section.

Exercise IV.36

36.1. Find the expression analogous to Eq. 36.7 that holds when $R < d$, and prove that this potential is zero at the center of the dipole.

36.2. A single point charge is placed a distance d on the x-axis from an infinite conducting plane at zero potential. Find the complementary solution of Laplace's equation needed to satisfy the boundary condition, and prove that it is equivalent to a field produced by an "image" charge of opposite sign placed a distance $-d$ on the x-axis from the plane.

36.3. Find the field intensity at the plane boundary in the last example, and determine the induced charge density. Verify that the total integrated charge is equal and opposite to the inducing point charge.

36.4. The dipole in Fig. 36.2 is placed in a half-space, the plane boundary being perpendicular to the dipole's axis at a distance D from its center. Find the complementary field required when the plane is a conductor at zero potential, and find the equivalent "image" distribution of charges.

36.5. Find the field at $R = R_0$ from Eq. 36.14, and find the induced charge density at the boundary. Find the total induced charge.

36.6. Repeat the last exercise, using Eq. 36.15 for the field.

36.7. Find the energy of an electron that results from the image effect when the electron is at a distance of 4.0 angstroms from the plane surface of a conductor.

IV.37 Boundary Value Problems with Poisson's Equation.

Poisson's equation, viz. Eq. 34.17, is an inhomogeneous equation of which Laplace's equation is the homogeneous form; the boundary value problem for Poisson's equation is therefore somewhat more difficult than that for Laplace's equation. Before embarking on its study we need first a theorem known as Green's equivalent stratum.

Green's equivalent stratum. Consider a finite region A bounded by a closed surface S. Inside the region, charge is distributed with a density ρ, and we are at first interested in the potential Φ_P produced at a point P outside the region A, by the charge inside A. We know of course that it is simply obtained by summing the potential contributions from all the charges in A, and, since none of them are too close to P, we do not need to worry about removing anything from the immediate vicinity of P; thus

$$\Phi_P = (1/4\pi\epsilon)\iiint(\rho/r)d\tau, \tag{37.1}$$

where r is the distance from the volume element $d\tau$ in A to the point P.

We now quote *Green's theorem* in integral transformations, which is derivable from Gauss' theorem, Eq. 7.8:

$$\iiint(\Phi\nabla^2\psi-\psi\nabla^2\Phi)d\tau = \iint(\Phi\nabla\psi-\psi\nabla\Phi)\cdot d\boldsymbol{\sigma}, \qquad (37.2)$$

where the volume integral is taken over a finite region bounded by the surface over which the surface integral is taken. The theorem is valid so long as no singularities occur in the integrands. In this theorem we let Φ be the potential function at the element of volume or surface,

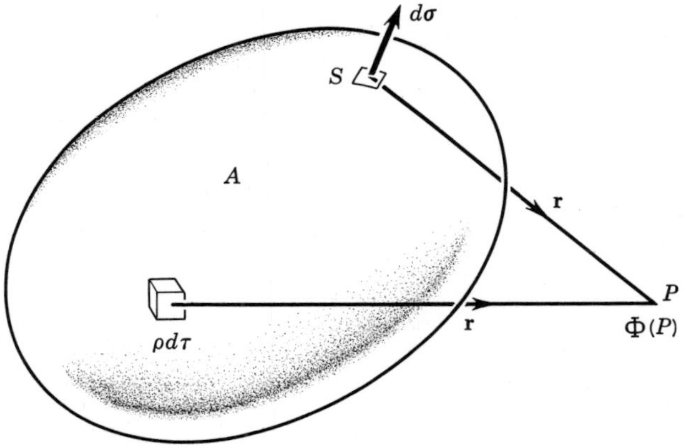

FIG. 37.1. An illustration of Eq. 37.1. The volume integration is carried out by taking the volume element all over the region A, and the surface integral by taking the element $d\boldsymbol{\sigma}$ all over the surface S. The point P remains fixed

respectively, and ψ be $1/r$, where r is the distance from the element (volume or surface) to the field point P. Since this point is outside the region, r is not zero in the integrands, so there is no trouble with singularities. In the volume integral we have always $\nabla^2\psi=\nabla^2(1/r)=0$ while $\nabla^2\Phi=-\rho/\epsilon$ from Poisson's equation, ρ being the charge density in the element $d\tau$. Using Eq. 37.1, we then see that the left-hand side of Green's equation, Eq. 37.2, reduces simply to $4\pi\Phi_P$. Thus,

$$4\pi\Phi_P = \iint[\Phi\nabla(1/r)-(1/r)\nabla\Phi]\cdot d\boldsymbol{\sigma}. \qquad (37.3)$$

This equation gives the field potential at P, outside the region A in terms of the condition at the surface of the region—the surface integral is called *Green's equivalent stratum*, and it has a convenient physical interpretation. The second part, *viz.*:

$$\Phi'' = -(1/4\pi)\iint(1/r)\nabla\Phi\cdot d\boldsymbol{\sigma}, \qquad (37.4)$$

is the potential of a surface distribution of charge $-\epsilon\nabla\Phi\cdot d\boldsymbol{\sigma}$ which has a density $-\epsilon\partial\Phi/\partial n$ per unit area. The first part, *viz.*:

$$\Phi' = (1/4\pi)\iint\Phi\nabla(1/r)\cdot d\boldsymbol{\sigma}, \tag{37.5}$$

is the potential at P of a surface dipole distribution, the surface element $d\boldsymbol{\sigma}$ having a dipole moment $\epsilon\Phi d\boldsymbol{\sigma}$; the distance between the two poles of this dipole layer is vanishingly small to make this statement true in the light of the exact field due to dipoles given in §36.

Evidently we could replace the actual charge density inside the region A by the appropriate Green's equivalent stratum of surface charges and dipole layer over the surface of the region, and produce exactly the same field Φ_P at all points outside the region, leaving a void inside A. A

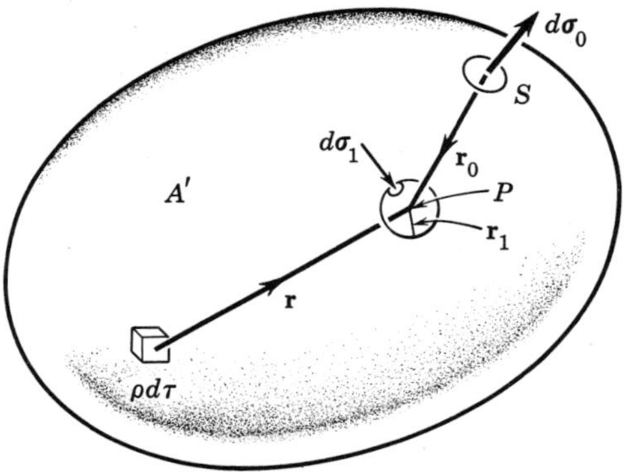

FIG. 37.2. An illustration of the discussion leading to Eq. 37.6. The volume integration is carried out by taking the volume element $d\tau$ over the entire region A', excluding the small sphere around the field point of interest, P. The surface integral consists of two parts; one part carries the surface element $d\boldsymbol{\sigma}_0$ over the outside surface of the region, the other carries the element $d\boldsymbol{\sigma}_1$ over the surface of the small sphere of radius r_i centered on P. The point P remains fixed throughout

discontinuity in potential and in potential gradient (field intensity) is then established at the surface of A; the charge on the surface establishes the field intensity and the dipole layer establishes the jump in potential. In going from the inside of a metal conductor through the metal surface to an electrical field in its surroundings, it is useful to picture the surface layer as containing this Green's equivalent stratum: the surface charge produces the field, and the dipole layer establishes the work function.

We shall now study the field inside the region A where the charge density exists; this is the essential problem of solving Poisson's equation with specified boundary conditions on the surface of the region A.

The method of solution begins again from Green's theorem, Eq. 37.2, with the same particular choice of functions Φ and ψ, except that this time the point P is inside the region A. This means that r can be zero in the integrand, so that special precautions have to be taken to remove the singularity from the integrands. This is done by first removing a small sphere having P as center from the region A, and adding the aurface of this small sphere to the domain of the surface integral on the other side of the equation. Figure 37.2 illustrates this. Because r can be zero in the integrand, we cannot assume the solution Eq. 37.1—it includes the contribution of charge indefinitely near to the field point. On the left-hand side of Green's theorem, we can assume that $\nabla^2(1/r)=0$ in the region A' however—because we have removed the small sphere around $r=0$ from A to form A'. We can also assume that Φ satisfies Poisson's equation—by hypothesis. Therefore, the left-hand side of the theorem reduces to $(1/\epsilon)\iiint(1/r)\rho d\tau$:

$$(1/\epsilon)\iiint(1/r)\rho d\tau = \iint_o [\Phi\nabla(1/r)-(1/r)\nabla\Phi]\cdot d\boldsymbol{\sigma}_o$$

$$+\iint_i [\Phi\nabla(1/r)-(1/r)\nabla\Phi]\cdot d\boldsymbol{\sigma}_i. \qquad (37.6)$$

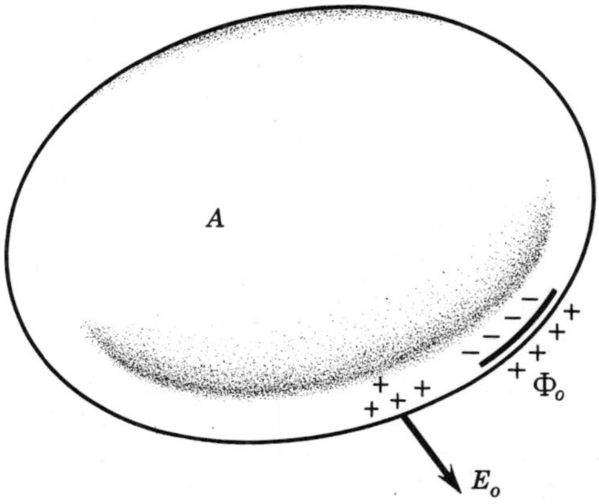

FIG. 37.3. Green's equivalent stratum of charges and dipoles required to produce an external field Φ_0 and $\mathbf{E}_0$ equivalent to that produced by the charges inside A and replacing these charges

In the first surface integral Φ is evaluated at the outer surface of A and r is the distance from the element of this surface to the point P; in the second surface integral the element $d\boldsymbol{\sigma}_i$ is directed into the little hole around P, r is the distance from this element to the center P, and Φ is the potential at the element $d\boldsymbol{\sigma}_i$ on the small sphere around P. We shall assume that Φ is a well-behaved function and that we can make the small sphere around P so small that Φ has practically no change over the sphere from its value at P itself. We can then remove Φ from the integrand, and for the same reason we can also remove $\nabla\Phi$ from its integrand; in each case they are written as their values at P. Thus the integral over the small spherical surface reduces to:

$$\Phi_P \iint_i \nabla(1/r)\cdot d\boldsymbol{\sigma}_i - (\nabla\Phi)_P \cdot \iint_i (1/r) d\boldsymbol{\sigma}_i.$$

Since r is a constant on the surface, each of the integrals is easily evaluated, and we have as the result: $4\pi\Phi_P + 4\pi r_i(\nabla\Phi)_P$. When we let the sphere approach zero, the second part of this vanishes, and we are left with $4\pi\Phi_P$ as the limiting value of the surface integral over the vanishingly small spherical surface—the last part of Eq. 37.6. Thus, we have, in place of Eq. 37.6,

$$4\pi\Phi_P = (1/\epsilon)\iiint (1/r)\rho d\tau - \iint_o [\Phi\nabla(1/r) - (1/r)\nabla\Phi]\cdot d\boldsymbol{\sigma}_o, \quad (37.7)$$

where the surface integral is now over only the outside boundary of the region A, and the volume integral is taken over the region A' that excludes the small sphere; but then the small sphere is allowed to shrink indefinitely and the limit of the integral is to be evaluated as this hole shrinks to zero. By this procedure we can generally define a precise value for the volume integral even though the domain of integration includes a singularity in the integrand.

We note first that the surface integral in Eq. 37.7 is just Green's equivalent stratum for the charges that may exist outside the region A; compared with Eq. 37.3 a negative sign appears in front of the expression in Eq. 37.7 because the points P are on opposite sides of the surface element. Secondly, we also note that if the region A contains all the charges, we can move the surface out far enough to reduce the potential to zero and so cause the surface integral to vanish. We then reduce Eq. 37.7 to the solution that one would write down intuitively:

$$4\pi\Phi_P = (1/\epsilon)\iiint(1/r)\rho d\tau \quad (37.8)$$

where the integral is over all space, and the singularity at $r=0$ is treated by the limiting procedure outlined previously.

The boundary value problem arises specifically when we keep the region A finite and do not restrict the charges to be inside. We can then think of Eq. 37.7 as the sum of two fields; one, the volume integral over A, is a particular solution of Poisson's equation; the other part, the surface integral, is a solution of Laplace's equation—it is essentially the complementary function added to the particular solution in order to satisfy the boundary conditions imposed by the field at the surface.

Now the solution of Laplace's equation for the complementary function is determined by the values of Φ on the surface or, alternatively, by the values of $\nabla\Phi$ on the surface; we cannot prescribe both function and its gradient in stating the boundary conditions. Looking at Eq. 37.7 as the solution of a boundary value problem, then, we may arbitrarily specify Φ as a function of position on the surface in the integral on the right side of Eq. 37.7, but then the $\nabla\Phi$ must be left to be determined by the equation itself—so the equation is essentially an integral equation for Φ. Numerical methods of computation can be set up for the detailed solution of such equations in any particular case, but it is difficult to write down a general solution in closed form.

Exercise IV.37

37.1. Prove Green's theorem, Eq. 37.2. Hint: apply Gauss' theorem to the vector $\Phi\nabla\Psi - \Psi\nabla\Phi$.

37.2. Prove that the surface integral expression on the right-hand side of Eq. 37.3 is a solution of Laplace's equation. Hint: note that the Laplacian operator differentiates with respect to the coordinates of P and that the terms $1/r$ are the only ones in the integral that depend on the position of P.

37.3. Prove that the integral over the small sphere, the last term in Eq. 37.6, reduces to $4\pi\Phi_P + 4\pi r_i(\nabla\Phi)_P$.

37.4. State, in greater detail than is given in the text, the conditions that must be imposed on Φ in order that the surface integral in Eq. 37.7 shall vanish when the surface is taken to infinity.

37.5. A point charge q and a point dipole of moment μ are placed at the center of a sphere of radius a. Find Green's equivalent stratum over the sphere, effective for points outside the sphere.

IV.38 Multipole Fields. We show in this section how the field due to a general charge distribution can be expressed as a sum over contributions from poles of different multiplicity: single poles, dipoles, quadrupoles, etc. To begin the discussion we need a generalization of Taylor's series theorem to three dimensions. We are familiar with the one-dimensional Taylor series:

$$f(x+h) = \sum_{n=0} (1/n!)h^n(d^n/dx^n)f(x). \tag{38.1}$$

If we think of $h\,d/dx$ as an operator acting on $f(x)$, it is convenient to write the operators appearing in the above expansion in the shorter form $(h\,d/dx)^n$. We state here, without attempting the proof, the corresponding theorem in three dimensions:

$$f(\mathbf{r}+\mathbf{s}) = \sum_{n=0} (1/n!)(\mathbf{s}\cdot\nabla)^n f(\mathbf{r}), \tag{38.2}$$

where the operator $\mathbf{s}\cdot\nabla$ replaces the operator $h\,d/dx$. If the components of $\mathbf{s}$ are $-x_0$, $-y_0$, $-z_0$, this operator is $-x_0\partial/\partial x - y_0\partial/\partial y - z_0\partial/\partial z$. Consider the situation in Fig. 38.1, where O is an origin of coordinates, Q a charge element, and P a field point. The vector $-\mathbf{r}_0$ takes the place of $\mathbf{s}$ in the Taylor expansion, and we can write Eq. 38.2 in the form

$$f(\mathbf{R}) = \sum_{n=0} (1/n!)(-\mathbf{r}_0\cdot\nabla)^n f(\mathbf{r}). \tag{38.3}$$

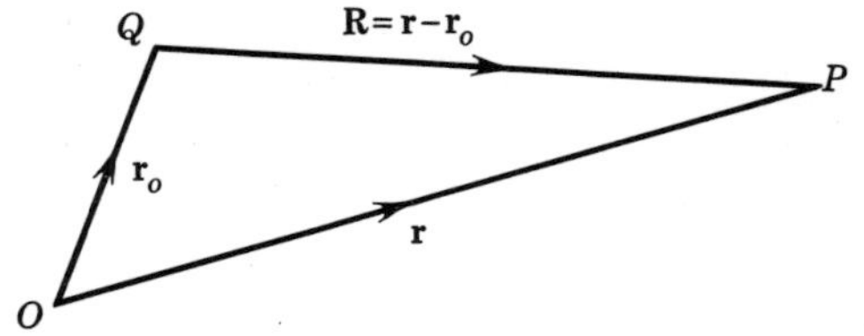

FIG. 38.1. An illustration of Eq. 38.3, concerning the Taylor expansion in three dimensions

In particular, we are interested in the potential of the charge Q, and so we take $f(\mathbf{R})=1/R$ and $f(\mathbf{r})=1/r$. Then we have the following series:

$$1/R = \sum_{n=0} (-1)^n \; (1/n!)(x_0\partial/\partial x + y_0\partial/\partial y + z_0\partial/\partial z)^n \; (1/r). \tag{38.4}$$

We can use the multinomial theorem here, *viz.*:

$$(A+B+C)^n = \sum_{j,k}\sum \frac{n!\, A^j\, B^k\, C^{n-j-k}}{(n-j-k)!\, j!\, k!} \quad (j+k \leqslant n). \tag{38.5}$$

If we let A be the operator $x_0\partial/\partial x$, B be the operator $y_0\partial/\partial y$, and C be $z_0\partial/\partial z$, the expression applies at once to the expansion in Eq. 38.4, which therefore becomes:

$$1/R = \sum_{n,j,k}\sum\sum \frac{(-1)^n\, x_0{}^j\, y_0{}^k\, z_0{}^{n-j-k}}{(n-j-k)!\, j!\, k!} \frac{\partial^n}{\partial x^j\, \partial y^k\, \partial z^{n-j-k}} (1/r). \tag{38.6}$$

This is the form of *Taylor's theorem* that we shall need directly.

Suppose that we are given some distribution of charge inside a finite region, and that we represent it by a density function $\rho(x_0, y_0, z_0)$ of position inside the region. We are to be interested in the potential outside the region due to this charge distribution. We can at once set down a general expression for this, from Eq. 37.1:

$$4\pi\epsilon\Phi_P = \iiint (1/R)\, \rho(x_0, y_0, z_0)\, dx_0 dy_0 dz_0. \tag{38.7}$$

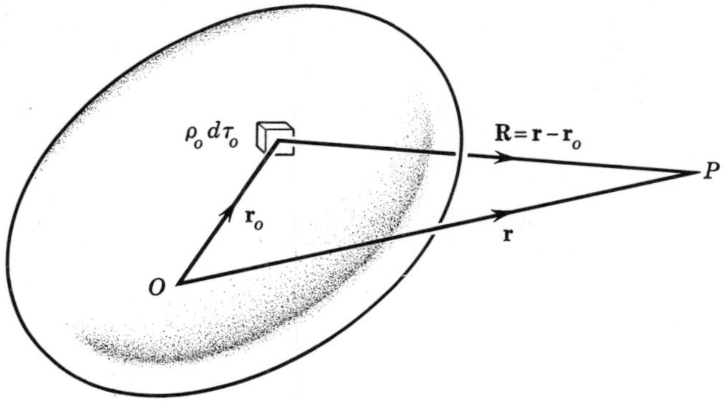

FIG. 38.2. An illustration of Eq. 38.7, concerning the multipole charge distributions

By means of the Taylor expansion of Eq. 38.6, it is possible to express the integrand here in terms of the distance r from some arbitrary origin in space, as illustrated in the diagram, Fig. 38.2. Thus, let us write:

$$D(j,\,k,\,n-j-k) = \frac{(-1)^n}{j!k!(n-j-k)!}\frac{\partial^n}{\partial x^j\,\partial y^k\,\partial z^{n-j-k}} \tag{38.8}$$

and

$$I(j,\,k,\,n) = \iiint x_0{}^j\,y_0{}^k\,z_0{}^{n-j-k}\,\rho(x_0,\,y_0,\,z_0)\,dx_0dy_0dz_0. \tag{38.9}$$

Then using Eq. 38.6 in the integrand of Eq. 38.7, we have

$$4\pi\epsilon\Phi_P = \sum_{n,j,k}\sum\sum I(j,\,k,\,n)\,D(j,\,k,\,n-j-k)(1/r). \tag{38.10}$$

The D's are of course differential operators. The integrals $I(j, k, n)$ are called the *moments* of the charge distribution. The first term in the expansion in Eq. 38.10 is obtained by setting $n=0$. The operators D in this term reduce to a single identity operator $D(0, 0, 0)$, and the moment $I(0, 0, 0)$ is nothing but the total integrated charge in the region:

$$I(0,\,0,\,0) = \iiint \rho(x_0,\,y_0,\,z_0)dx_0dy_0dz_0 = q \tag{38.11}$$

The first term in the expansion in Eq. 38.10 is therefore just q/r. We shall see soon that this is the only term we need if R is large enough compared with the dimensions of the region occupied by the charges; it is the Coulomb potential of the total charge considered located at the origin.

The next terms in the expansion are obtained by setting $n=1$. The operators are then $D(1, 0, 0)$, $D(0, 1, 0)$, and $D(0, 0, 1)$, which are simply $-\partial/\partial x_0$, $-\partial/\partial y_0$, and $-\partial/\partial z_0$. The moments are

$$\left.\begin{aligned} I(1, 0, 1) &= \iiint x_0\rho(x_0, y_0, z_0)dx_0dy_0dz_0 = M_x \\ I(0, 1, 1) &= \iiint y_0\rho(x_0, y_0, z_0)dx_0dy_0dz_0 = M_y \\ I(0, 0, 1) &= \iiint z_0\rho(x_0, y_0, z_0)dx_0dy_0dz_0 = M_z \end{aligned}\right\}. \tag{38.12}$$

These integrals are clearly analogous to the first moments in mechanics and are simply the generalized *dipole moments* of the charge distribution. We can in fact identify them as the three components of the vector dipole moment $\mathbf{M}$ as indicated in the preceding equations. Collecting the three terms obtained from setting $n=1$ in Eq. 38.10, it is easy to see that they become $-\mathbf{M}\cdot\nabla(1/r)$. Thus we now have

$$4\pi\epsilon\Phi_P = q/r - \mathbf{M}\cdot\nabla(1/r) + \text{additional terms}. \tag{38.13}$$

The next terms in the series are obtained by setting $n=2$ in the general expansion of Eq. 38.10. The differential operators in this case comprise all the second-order ones, $D(2, 0, 0) = \frac{1}{2}\partial^2/\partial x^2$, $D(1, 1, 0) = \partial^2/\partial x\partial y$, $D(0, 0, 2) = \frac{1}{2}\partial^2/\partial z^2$, etc., while the moment integrals are as listed below:

$$\left.\begin{aligned} I(2, 0, 2) &= \iiint x_0{}^2\rho_0 d\tau_0 = Q_{xx}; & I(0, 2, 2) &= \iiint y_0{}^2\rho_0 d\tau_0 = Q_{yy} \\ I(0, 0, 2) &= \iiint z_0{}^2\rho_0 d\tau_0 = Q_{zz}; & I(1, 1, 2) &= \iiint x_0y_0\rho_0 d\tau_0 = Q_{xy} \\ I(0, 1, 2) &= \iiint y_0z_0\rho_0 d\tau_0 = Q_{yz}; & I(1, 0, 2) &= \iiint z_0x_0\rho_0 d\tau_0 = Q_{zx} \end{aligned}\right\}. \tag{38.14}$$

These integrals are exactly analogous to the second moments that one uses to calculate the rotational inertias of a rigid body; they are called the six components of the *quadrupole moment tensor* of the electric charge distribution. Their contribution to the potential is of the form $\frac{1}{2}(Q_{xx}\partial^2/\partial x^2 + 2Q_{xy}\partial^2/\partial x\partial y + \cdots)(1/r)$; in subscript notation the whole contribution is simply: $\frac{1}{2}\sum\sum_{j,k} Q_{jk}(\partial^2/\partial x_j\partial x_k)(1/r)$. In dyadic notation it is, see Eq. 52.13: $\frac{1}{2}\mathbf{Q}:\nabla\nabla(1/r)$. The complete expression for the potential up to the quadrupole terms is, then,

$$4\pi\epsilon\Phi_P = q/r - \mathbf{M}\cdot\nabla(1/r) + \tfrac{1}{2}\mathbf{Q}:\nabla\nabla(1/r) - \cdots. \tag{38.15}$$

Evidently we could proceed indefinitely in this way to higher and higher order terms, the order being reflected in the number of differentiations contained in the operator. If we are far enough from the charged region, a few terms will suffice for an accurate computation; the nearer we are to the region, the more terms are needed.

A significant conclusion may be drawn if we put the argument the other way around. Suppose we are observing a charged region from outside, and we wish to determine, from measurements of its potential field outside, the distribution of charge inside the body. Clearly, if we are far from the region we can detect only the first term in the expansion of Eq. 38.15, which means we can determine only the total charge in the region. If we are nearer to the region, we may determine its dipole moment. We have to get rather close to the body to determine its quadrupole moment. If the charged body is small, like a molecule, we have to get observations within molecular distances in order to find its dipole and quadrupole moments. The quadrupole moment of an atomic nucleus is extremely difficult to determine, although it has been measured for a number of nuclei.

A certain type of quadrupole moment is in fact completely undetectable from outside the charge distribution. This occurs when the distribution has complete spherical symmetry—a point charge surrounded by an equal and opposite charge distributed uniformly on a sphere concentric with the first charge. This distribution has no net charge, and no dipole moment, and we shall see that, although its quadrupole moment is not zero, it produces no field outside—*i.e.* no contribution to Φ_P in Eq. 38.15. To show this, we note that the quadrupole tensor is a constant times the identity matrix because the second moment components are all the same about each of the three principal axes because of symmetry, and there are no cross terms. The field contributed by such a quadrupole moment is Φ', where

$$4\pi\epsilon\Phi' = \tfrac{1}{2}K\mathrm{E}:\nabla\nabla(1/r), \qquad (38.16)$$

K being a constant and E the identity matrix. But $\mathrm{E}:\nabla\nabla = \nabla\cdot\nabla$ and $\nabla\cdot\nabla(1/r) = 0$, so that $\Phi' = 0$ identically. From the point of view of measurements from outside the charge distribution, the measured quadrupole moment is therefore ambiguous to the extent that any such spherically symmetrical distribution could be added to it without making any difference.

If the charge distribution has an axis of symmetry, which we may select for the z-axis of coordinates, it is possible to represent the quadrupole tensor uniquely by a single number that is independent of the ambiguity due to the spherically symmetrical charge distributions. The quantity is called the *quadrupole strength*† of the distribution. It is defined as Q^* where

$$4Q^* = 2Q_{zz} - Q_{xx} - Q_{yy} = \iiint (3z_0^2 - r_0^2)\rho_0 d\tau_0, \qquad (38.17)$$

where $r_0^2 = x_0^2 + y_0^2 + z_0^2$. To show that this number is not sensitive

† Nuclear physicists have used the word "moment" here for the "strength" and much confusion has resulted with the tensor moment. We adopt the suggestion of Dr. Margenau in using the word "strength" (*Mathematics of Physics and Chemistry*, by H. Margenau and G. M. Murphy, 2nd ed., D. Van Nostrand Co., Inc., Princeton, N.J., 1956).

to the ambiguity in question, we may add a spherically symmetrical charge distribution to the original one, so that the new components of the quadrupole tensor become $Q_{xx}'=Q_{xx}+K$, $Q_{yy}'=Q_{yy}+K$, $Q_{zz}'=Q_{zz}+K$. From this we have

$$4Q^{*\prime} = 2Q_{zz}'-Q_{xx}'-Q_{yy}' = 4Q^*+2K-K-K = 4Q^*. \qquad (38.18)$$

It is this quadrupole strength Q^* that is usually referred to in the experimental literature, atomic and nuclear particles presumably having an axis of symmetry.

Exercise IV.38

38.1. Compare Eqs. 38.13 and 36.8.

38.2. A charge distribution consisting of two point charges on the x-axis $+q$ at $x_0=\frac{1}{2}d$ and $-q$ at $x_0=-\frac{1}{2}d$, may be thought of as an exact dipole in the sense of §36. Compute the integrals in Eq. 38.12 and Eq. 38.14 for the dipole and quadruple components.

38.3. Write down the general expression for the integrals $I(3, 0, 3)$ and $I(2, 1, 3)$ from Eq. 38.9. Compute these integrals for the charge distribution in Ex. 38.2—the "exact dipole."

38.4. Compare Eq. 38.13 with the exact dipole field in Eq. 36.7. Evidently the pure dipole field given by Eq. 38.13 is not an exact dipole in the sense of §36. Discuss this in the light of your result in Ex. 38.3. Prove that the exact dipole has non-zero multipole components of all odd orders in the general expansion of Eq. 38.10 where the parameter n defines the order.

38.5. Consider a charge distribution where point charges are placed as follows: $+q$ on the x-axis at $x_0=+\frac{1}{2}d$ and $-\frac{1}{2}d$; $-q$ on the x-axis at $x_0=+d$ and $-d$. Find the quadrupole components of this distribution from Eq. 38.14 and the quadrupole strength from Eq. 38.17. Prove that this distribution is not a pure quadrupole, but has higher even order terms.

38.6. Find the quadrupole tensor components from Eq. 38.14 and the quadrupole strength from Eq. 38.17 for the following charge distribution in the x-y plane:

$$+q \text{ at } x_0 = d,\ y_0 = 0, \text{ and at } x_0 = 0,\ y_0 = d$$

and

$$-q \text{ at } x_0 = 0,\ y_0 = 0, \text{ and at } x_0 = d,\ y_0 = d.$$

38.7. Would a shift of origin in the last example alter the tensor components? Would it alter the quadrupole strength?

38.8. A point quadrupole of strength Q is placed at the center of a sphere of radius a. Find Green's equivalent stratum over the surface of the sphere, effective for points outside it.

IV.39 The Energy of Charge Distributions. The potential energy of a point charge q at a point located by the vector $\mathbf{r}$ in a field whose potential is $\Phi(\mathbf{r})$ is simply $q\Phi(\mathbf{r})$. Extending this to the continuous charge distribution of density $\rho(\mathbf{r})$, the potential of a volume element $d\tau$ is $\rho(\mathbf{r})\Phi(\mathbf{r})d\tau$. The energy of a charge distribution is then the integral of this over the region of interest:

$$W = \iiint \rho(\mathbf{r})\Phi(\mathbf{r})d\tau. \qquad (39.1)$$

Consider a charge distribution over some region in which the external field does not vary too much, so that it can at any point in the region be expressed as a Taylor series with respect to some arbitrary point O' inside the region (see Fig. 39.1). Thus, the field at the point P in the region can be expressed in the form:

$$\Phi(\mathbf{r}) = \sum_{n=0} (1/n!)(\mathbf{r}\cdot\nabla_0)^n\, \Phi(\mathbf{r}_0) \qquad (39.2)$$

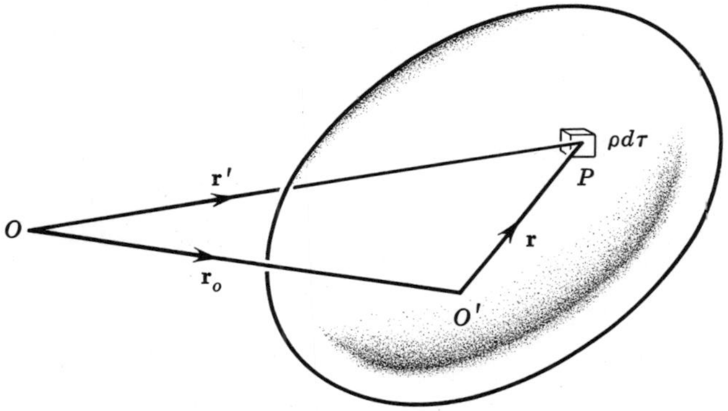

FIG. 39.1. An illustration of Eq. 39.2 concerning the energy of a charge distribution in an external field

(*cf.* Eq. 38.2). In this expression the operator ∇_0 is the gradient with differentials taken with respect to the coordinates of O', $\Phi(\mathbf{r}_0)$ is the potential at the point O', $\mathbf{r}$ is the vector from O' to P; and $\Phi(\mathbf{r})$ is the potential at the point P. Putting this expression for $\Phi(\mathbf{r})$ into Eq. 39.1, we have for the energy of the charge distribution

$$W = \sum_{n=0} (1/n!)\iiint(\mathbf{r}\cdot\nabla_0)^n \rho(\mathbf{r})\Phi(\mathbf{r}_0)d\tau. \qquad (39.3)$$

Expanding the scalar product $\mathbf{r}\cdot\nabla_0$ and writing out the power series for a few terms, we find

$$W = q\Phi(\mathbf{r}_0)+\mathbf{M}\cdot\nabla_0\Phi(\mathbf{r}_0)+\tfrac{1}{2}\mathbf{Q}:\nabla_0\nabla_0\Phi(\mathbf{r}_0)+\text{higher terms.} \qquad (39.4)$$

Here q is the total charge in the region, $\mathbf{M}$ is the first moment, and $\mathbf{Q}$ the second moment tensor, defined just as in §38. Note, however, that the integrations in this equation are over the position $\mathbf{r}$, so that the subscripts must be omitted from Eq. 38.12 and Eq. 38.14 in deriving it.

Suppose, as is generally the case, that the external field $\Phi(\mathbf{r}_0)$ can be restricted to solutions of Laplace's equation; and suppose in addition that the charge distribution has an axis of symmetry, the z-axis, so that $Q_{xx}=Q_{yy}$. Then we have the following:

$$\mathbf{Q}:\nabla_0\nabla_0\Phi = Q_{xx}\partial^2\Phi/\partial x_0\partial x_0+Q_{yy}\partial^2\Phi/\partial y_0\partial y_0+Q_{zz}\partial^2\Phi/\partial z_0\partial z_0;$$

the nondiagonal terms disappear because of the symmetry. Using $Q_{yy}=Q_{xx}$ and Laplace's equation, this reduces easily to

$$\mathbf{Q}:\nabla_0\nabla_0\Phi = 2Q^*\partial^2\Phi/\partial z_0\partial z_0 \qquad (39.5)$$

where Q^* is the strength defined in Eq. 38.17. For such charge distributions the energy in an external field depends only on the quadrupole strength and the second derivative of the external field in the direction of the symmetry axis. From Eq. 39.4, if there is no net charge and no dipole moment:

$$W = Q^*\partial^2\Phi/\partial z_0^2. \qquad (39.6)$$

This formula is basic to most practical attempts to measure the quadrupole strengths of nuclei.

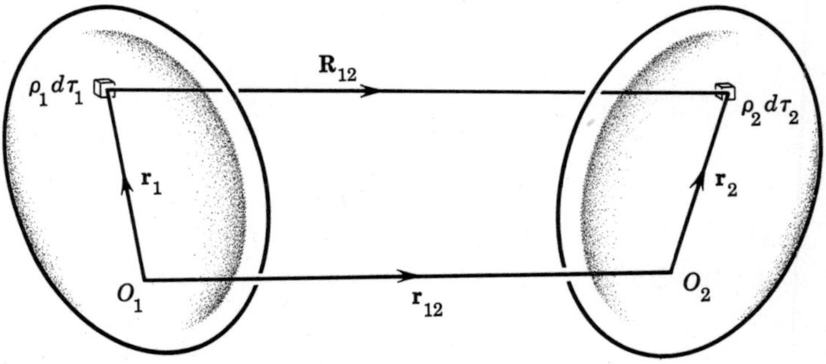

FIG. 39.2. An illustration of the discussion of the interaction energy between two charge distributions, Eqs. 39.7 and 39.10

Consider two charge distributions on nonoverlapping regions, in fact separated by distances that are fairly large compared with the dimensions of either region; we wish to find the energy of either distribution in the field produced by the other distribution. Suppose that both charge distributions can be expanded in Taylor series with respect to some arbitrary point in their respective regions (see Fig. 39.2). We can write down the energy of the second distribution in terms of its multipole structure and the field at O_2 due to the first distribution; from Eq. 39.4 it is clearly

$$W = q_2\Phi(\mathbf{r}_{12})+\mathbf{M}_2\cdot\nabla_2\Phi(\mathbf{r}_{12})+\tfrac{1}{2}\mathbf{Q}_2:\nabla_2\nabla_2\Phi(\mathbf{r}_{12})+\cdots. \qquad (39.7)$$

Here we have taken the point O_1 as the origin of coordinates. Now we can express the potential Φ at O_2 due to the first charge distribution in terms of the distance r_{12} and the multipole structure of the first distribution; from Eq. 38.15 we have

$$4\pi\epsilon\Phi(\mathbf{r}_{12}) = q_1/r_{12} - \mathbf{M}_1 \cdot \nabla_2(1/r_{12}) + \tfrac{1}{2}\mathbf{Q}_1 : \nabla_2\nabla_2(1/r_{12}) + \cdots. \quad (39.8)$$

It is convenient here to change the differentiations to the other end of the vector $\mathbf{r}_{12}$:

$$\nabla_1 = -\nabla_2. \quad (39.9)$$

Putting Eq. 39.8 into Eq. 39.7 and writing the result out in the operator notation, we have

$$4\pi\epsilon W = (q_2 + \mathbf{M}_2 \cdot \nabla_2 + \tfrac{1}{2}\mathbf{Q}_2 : \nabla_2\nabla_2)(q_1 + \mathbf{M}_1 \cdot \nabla_1 + \tfrac{1}{2}\mathbf{Q}_1 : \nabla_1\nabla_1)(1/r_{12}), \quad (39.10)$$

higher-order terms having been omitted.

In multiplying out this operator product, the first term is the familiar Coulomb term $q_2 q_1/r_{12}$; then there are two charge-dipole terms $q_2\mathbf{M}_1 \cdot \nabla_1(1/r_{12})$ and $q_1\mathbf{M}_2 \cdot \nabla_2(1/r_{12})$ which decrease with distance with an inverse square dependence. The dipole-dipole term is $\mathbf{M}_2\mathbf{M}_1 : \nabla_1\nabla_2(1/r_{12})$, and it decreases with the cube of the distance. The charge-quadrupole terms also decrease with the cube of the distance. The dipole-quadrupole terms decrease with the fourth power of the distance, and the quadrupole-quadrupole term decreases with the fifth power of the distance. The higher the negative power of the dependence on distance, the shorter the *range* of the interaction.

It is worth looking at Eq. 39.10 with the thought that the operators are effective field source strengths, or generalized charges:

$$q^* = q + \mathbf{M} \cdot \nabla + \tfrac{1}{2}\mathbf{Q} : \nabla\nabla + \cdots. \quad (39.11)$$

Then the field produced by such a source, given by Eqs. 39.8 and 39.9, would be

$$4\pi\epsilon\Phi(\mathbf{r}) = q^*(1/r) \quad (39.12)$$

and the interaction energy of two such sources would be given by

$$4\pi\epsilon W_{12} = q_2^* q_1^*(1/r_{12}). \quad (39.13)$$

The idea that a source of a field can be an operator rather than a simple quantity is not familiar in classical physics, but has become commonplace in modern nuclear theory.

To see more concretely what these very formal expressions mean, we may consider two simple distributions that are easily worked out in detail. First, consider a charge q distributed uniformly around a circle of radius a in the x-y plane, and a charge $-q$ located at its center.

This distribution obviously has zero net charge and zero dipole moment, but it does have a quadrupole strength. The tensor components are

$$Q_{xx} = Q_{yy} = \tfrac{1}{2}qa^2,\ Q_{zz} \text{ and all others zero.} \tag{39.14}$$

The numerical quadrupole strength—the distribution is symmetrical—is

$$Q^* = -\tfrac{1}{4}qa^2. \tag{39.15}$$

We can regard this example as a first approximation to an oblate spheroid of positive charge, with a very small oblateness, neutralized by a concentric sphere of negative charge.

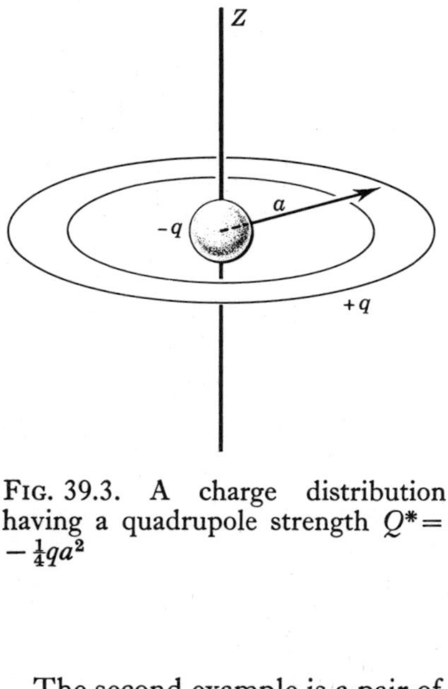

FIG. 39.3. A charge distribution having a quadrupole strength $Q^* = -\frac{1}{4}qa^2$

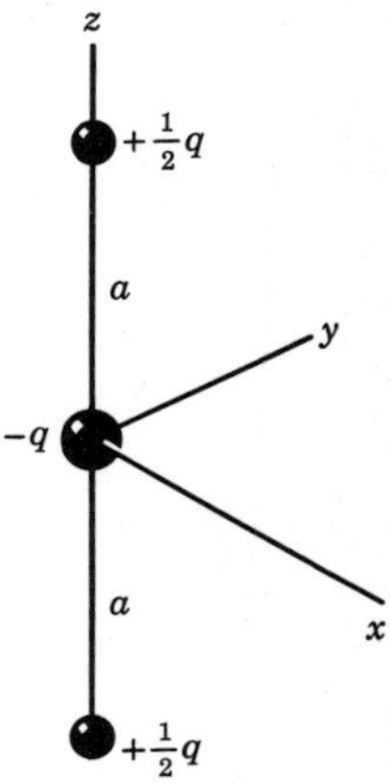

FIG. 39.4. A charge distribution having quadrupole strength $Q^* = +\frac{1}{2}qa^2$

The second example is a pair of charges, each $+\frac{1}{2}q$ at positions a and $-a$ on the z-axis, neutralized by a charge $-q$ at the origin. This again has zero net charge and zero dipole moment, while its quadrupole tensor is

$$Q_{zz} = qa^2, \text{ all others zero.} \tag{39.16}$$

The quadrupole strength is

$$Q^* = \tfrac{1}{2}qa^2. \tag{39.17}$$

This example can be regarded as a first approximation to a prolate spheroid of positive charge, with very small prolateness, neutralized by a concentric sphere of negative charge.

These two cases are typical of the quadrupole strengths observed in atomic nuclei, and the formulas for their interactions are of practical interest. The energy of either type of quadrupole in a field of potential Φ is given by Eq. 39.6:

$$W_{\text{oblate}} = -\tfrac{1}{4}qa^2\partial^2\Phi/\partial z^2; \quad W_{\text{prolate}} = \tfrac{1}{2}qa^2\partial^2\Phi/\partial z^2. \qquad (39.18)$$

In determining the energy of interaction of two such quadrupoles, we can use Eq. 39.5 in Eq. 39.10, with the particular values of Q^* in each case. For simplicity we shall discuss only quadrupoles with parallel axes of symmetry. We then have

$$4\pi\epsilon W_{12} = Q_1{}^*Q_2{}^*(\partial^2/\partial z_1{}^2)(\partial^2/\partial z_2{}^2)(1/r_{12}). \qquad (39.19)$$

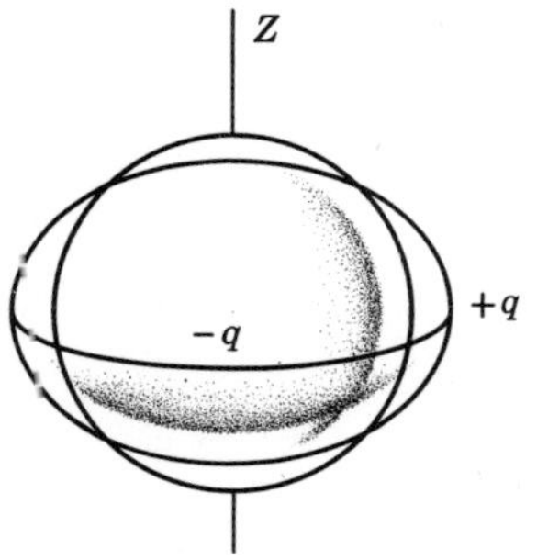

FIG. 39.5. A charge distribution with a negative quadrupole strength Q^*. It consists of a negative charged sphere balanced by an oblate spheroid of positive charge

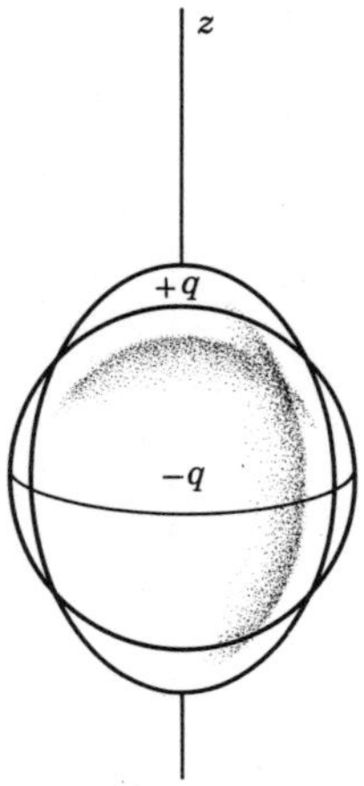

FIG. 39.6. A charge distribution with a positive quadrupole strength Q^*. It consists of a negative charged sphere balanced by a prolate spheroid of positive charge

Writing

$$r_{12}{}^2 = (x_1-x_2)^2+(y_1-y_2)^2+(z_1-z_2)^2, \qquad (39.20)$$

it is easy to prove that Eq. 39.19 is equivalent to

$$4\pi\epsilon W_{12} = 9Q_1{}^*Q_2{}^*(1/r_{12}{}^5)[1-10(z_1-z_2)^2/r_{12}{}^2+35(z_1-z_2)^4/3r_{12}{}^4]. \qquad (39.21)$$

When the two quadrupoles lie in the same equatorial plane, $z_1 = z_2$ and their interaction energy is $9Q_1{}^*Q_2{}^*/r_{12}{}^5$; when they are coaxial, and $r_{12} = z_1 - z_2$, it becomes $24Q_1{}^*Q_2{}^*/r_{12}{}^5$. The interaction is quite strongly dependent on direction and mutual orientation of the quadrupoles, but it is at all times inversely porportional to the inverse fifth power of the distance between them.

It should be emphasized that all the above calculations have assumed

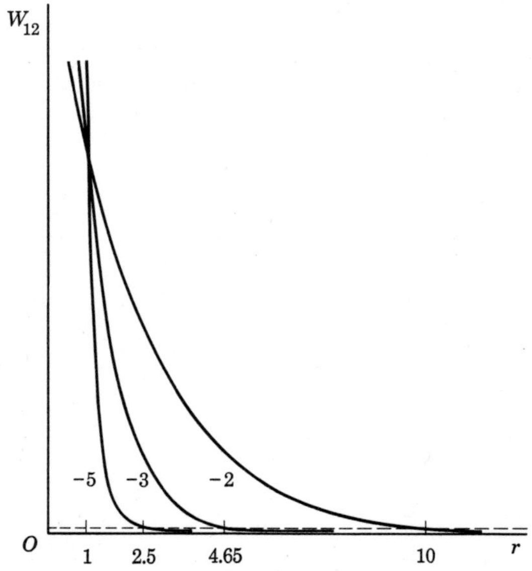

FIG. 39.7. The relative ranges of interaction forces varying with distance according to various powers of the radial separation. An r^{-5} force decreases by a factor of 100 for an increase in radius by a factor of $2\frac{1}{2}$. The inverse square force drops by a similar ratio with an increase in radius by a factor of ten

that neither of the two charge distributions is influenced by the proximity of the other. If induced charges occur, one region may be polarized by the field of the other, the quadrupole moment itself becomes a function of the field, and the calculations become much more complicated.

As we have seen, the interaction energy of two permanent dipoles varies as the inverse cube of the distance; if the two charge distributions become polarized only through their mutual interactions, possessing no moments in isolation, their interaction energy can be shown to vary as the inverse sixth power of the distance; this term in their mutual energy is known as the *van der Waals energy*.

Similarly, the interaction energy of two permanent quadrupoles varies, as we have seen, as the inverse fifth power of the distance; if the two distributions possess only mutually induced quadrupole moments, their interaction energy can be shown to depend on distance to the inverse tenth power. Although the van der Waals energy is attractive (negative), the induced quadrupole–quadrupole energy is positive (repulsion). The range of interaction is roughly defined as the distance within which the magnitude of the energy is comparable with the mean energy of thermal motion; the higher the negative power of distance involved, the shorter the range. Thus, the induced quadrupole repulsion is of a much shorter range than the van der Waals attraction. A combination of these two terms is usually sufficient to account for most of the gas imperfections observed experimentally.

Exercise IV.39

39.1. Develop Eq. 39.4 in detail from Eq. 39.3.
39.2. Check Eq. 39.5 in detail.
39.3. Prove Eq. 39.9.
39.4. Write the interaction energy between two charge distributions in the form

$$4\pi\epsilon W_{12} = \int\cdots\int (1/r_{12})\rho_1\rho_2\, d\tau_1 d\tau_2,$$

where the integrals are taken over the two separate regions. Use the Taylor series in the form:

$$1/R_{12} = \sum_n (1/n!)[(\mathbf{r}_2-\mathbf{r}_1)\cdot\nabla_2]^n(1/r_{12}),$$

where $\mathbf{R}_{12}=\mathbf{r}_{12}+\mathbf{r}_1-\mathbf{r}_2$, hence deriving Eq. 39.10.
39.5. Prove Eqs. 39.14 and 39.15.
39.6. Prove Eqs. 39.16 and 39.17.
39.7. List the dependence upon distance of the forces of attraction or repulsion between the various types of multipole-charge distributions.
39.8. Find the force exerted on each other by two quadrupoles like that illustrated in Fig. 39.3, when they are coaxial at a distance r_{12}.
39.9. Find the force exerted on each other by two quadrupoles like that shown in Fig. 39.4, when coaxial at distance r_{12}.
39.10. A nuclear magneton is $\frac{1}{2}\hbar e/mc = 0.505\times10^{-23}$ erg per gauss. The proton magnetic moment is 2.793 nuclear magnetons. The magnetic moment of an electron is 0.928×10^{-20} erg per gauss. Find the difference in energy between parallel and antiparallel orientations of the electron and proton in the H-atom, assuming the electron to be in the first Bohr orbit, the plane of the orbit being perpendicular to the magnetic moments. Find the frequency of the radiation corresponding to this energy difference. (1 gauss $=10^{-4}$ weber per meter^2, $\hbar = h/2\pi$.)
39.11. The electrical quadrupole moment of the deuterium nucleus is 0.002738 proton barn, and its magnetic dipole moment is 0.957355 nuclear magneton. The barn is 10^{-24} cm^2. Compare the interaction energies between the deuterium nucleus and an electron in the first Bohr orbit of heavy hydrogen due (a) to the electrical charge-charge interaction, (b) to the magnetic dipole-dipole interaction, and (c) to the electrical quadrupole-charge interaction.

Assume the orbit to be perpendicular to both dipole moments, and the magnetic dipole moment of the nucleus to be parallel to the axis of symmetry of its quadrupole.

39.12. The van der Waals interaction between two atoms may be written in the form $V(r) = -e_0(r_0/r)^6$. For helium $r_0 = 2.87$ angstrom, $e_0/k = 20.44°K$, and for argon $r_0 = 3.85$ angstrom, $e_0/k = 240°K$. Here k is Boltzmann's constant. Find in each gas the induced electrical dipole moments when the separation of a pair of atoms is r_0.

39.13. Consider helium or argon at N.T.P. and find the mean van der Waals energy per liter.

IV.40 The Electromagnetic Field.

The equations of the previous few sections apply equally well to electrostatic or to magnetostatic fields. However this parallelism between the two fields is highly artificial, and disappears as soon as time variations are included. Thus, *Faraday's law of induction* states that any change of the flux of *magnetic induction* **B** through any fixed loop induces an electromotive force around the loop:

$$\partial/\partial t \iint \mathbf{B} \cdot d\boldsymbol{\sigma} = -\oint \mathbf{E} \cdot d\mathbf{r}. \tag{40.1}$$

The surface integral is taken over any surface bounded by the loop, and the line integral is taken around the loop. For Eq. 40.1 to be consistent with Lenz' law, the positive directions of $d\mathbf{r}$ and $d\boldsymbol{\sigma}$ must be chosen so that the vector product $d\mathbf{r} \times d\boldsymbol{\sigma}$ (both positive) is directed outward from the loop.

Applying Stokes' theorem to the integral on the right, we can derive the differential form of *Faraday's law*:

$$\nabla \times \mathbf{E} = -\partial \mathbf{B}/\partial t. \tag{40.2}$$

Faraday's law has two important consequences. First, the electric field strength is no longer a conservative field when the magnetic flux varies in time: in fact, it turns out eventually that energy can flow between electric and magnetic forms through time varying fields. The second consequence is that no magnetic poles can exist; to see this, we take the divergence of Eq. 40.2, and note that $\nabla \cdot \nabla \times \mathbf{E} = 0$ identically, so that

$$(\partial/\partial t)\nabla \cdot \mathbf{B} = 0.$$

The divergence of **B** is therefore necessarily independent of time at every point in space, and the only value that can reasonably satisfy this condition is zero. It is assumed therefore that **B** is always solenoidal:

$$\nabla \cdot \mathbf{B} = 0. \tag{40.3}$$

All apparent magnetic poles occur in pairs, positive and negative, and are due to electric currents; no free poles exist in the electromagnetic field theory.

As another direct result of Faraday's law, a charge moving through a magnetic field experiences a force in addition to that due to the existence of any electric field. The force acting, per unit volume, on material having charge density ρ and producing a current density $\mathbf{J}$ by its motion, is found to be

$$\mathbf{F} = \rho\mathbf{E}+\mathbf{J}\times\mathbf{B}. \qquad (40.4)$$

Because of this force, any material placed in a field may become both electrically and magnetically polarized. Electric polarization is due to the relative displacement of positive and negative charges in the atoms or molecules of the material, and it is measured by the polarization vector $\mathbf{P}$ defined as the dipole moment per unit volume at the point of interest. Magnetic polarization is due to induced currents and to re-orientation of electron spins within atoms; it is also measured by the net magnetic dipole moment per unit volume, represented by the vector symbol $\mathbf{M}$.

When polarization is not uniform in space, a net charge density is produced, called the *induced charge*. At surfaces of discontinuity, it appears as a surface charge, and within a uniform material it appears as a body charge density which can be proved equal to

$$\rho_{\text{induced}} = -\nabla\cdot\mathbf{P}. \qquad (40.5)$$

This induced charge is naturally a source of electric field and has to be included in the charge density appearing on the right of Poisson's equation, and in Eq. 34.14, *viz.*: $\nabla\cdot\mathbf{E}=\rho/\epsilon$. It is therefore convenient to define a new vector, called (for historical reasons only) *electric displacement*

$$\mathbf{D} = \epsilon\mathbf{E}+\mathbf{P}, \qquad (40.6)$$

which has the property

$$\nabla\cdot\mathbf{D} = \rho_{\text{free}}. \qquad (40.7)$$

On the right side of this, we exclude the induced charge: by "free" we mean only that part of the net charge density that exists independently of the inducing field. In electrically neutral materials, whose charge is entirely induced, the electric displacement vector is solenoidal:

$$\nabla\cdot\mathbf{D} = 0. \qquad (40.8)$$

In many materials the polarization $\mathbf{P}$ is directly proportional to the field strength $\mathbf{E}$, so that we can define a new constant ϵ' depending on the kind of material, such that

$$\mathbf{D} = \epsilon'\mathbf{E}. \qquad (40.9)$$

The ratio ϵ'/ϵ is called the *dielectric constant* of the material: it is a pure dimensionless number. The vector $\mathbf{D}$ has dimensionality $[Q/M^2]$ and is the same as that of polarization.

Similarly magnetic polarization induces a magnetic polarity density also given by the divergence of the polarization vector:

$$\rho^m_{\text{induced}} = -\nabla \cdot \mathbf{M}. \tag{40.10}$$

But because we know that **B** is solenoidal (Eq. 40.3), it is convenient to define a vector for the magnetic displacement differently from the definition of **D**. Thus we let

$$\mathbf{H} = (1/\mu)\mathbf{B} - \mathbf{M}. \tag{40.11}$$

Note the opposite sign compared with that used in Eq. 40.6. This definition of the *magnetic vector* **H** is such that the divergence of **H** is equal to the induced magnetic polarization—in fact, the only magnetic polarity that can exist:

$$\nabla \cdot \mathbf{H} = \rho^m_{\text{induced}}. \tag{40.12}$$

It is not customary to call **H** the magnetic displacement in spite of its rough analogy with **D**. In many materials **M** is proportional to **B**, so that a new constant can be defined such that

$$\mathbf{B} = \mu' \mathbf{H}. \tag{40.13}$$

The ratio μ'/μ is called the *permeability* of the material and is also a pure dimensionless number. The dimensionality of **H** is the same as that of **M**, pole strength per unit area, or dipole moment per unit volume [Q/SM] (see also Eq. 34.10). The practical unit of magnetic induction **B** is the weber per meter^2, which equals one kilogram per coulomb second, and that of the magnetic vector **H** is the ampere turn per meter [Q/SM].

Finally we note *Ampere's law*, which states that either an electric current or a changing electric field **D** gives rise to a magnetic field, and thus causes **H** to become a nonconservative field:

$$\iint (\mathbf{J} + \partial \mathbf{D}/\partial t) \cdot d\boldsymbol{\sigma} = \oint \mathbf{H} \cdot d\mathbf{r}. \tag{40.14}$$

The term $\partial \mathbf{D}/\partial t$ was added by Maxwell to Ampere's law, to make the law consistent with the conservation of electric charge. Thus, in the absence of singularities, we may use Stokes' theorem on the right side of Eq. 40.14 and deduce the differential form of the law:

$$\nabla \times \mathbf{H} = \partial \mathbf{D}/\partial t + \mathbf{J}. \tag{40.15}$$

Taking the divergence of this equation, the left side vanishes identically and we are left with

$$(\partial/\partial t)\nabla \cdot \mathbf{D} + \nabla \cdot \mathbf{J} = 0. \tag{40.16}$$

On the other hand, taking the partial time derivative of Eq. 40.7, we have

$$(\partial/\partial t)\nabla\cdot\mathbf{D} = \partial\rho_{\text{free}}/\partial t,$$

which, combined with Eq. 40.16, yields

$$\partial\rho_{\text{free}}/\partial t+\nabla\cdot\mathbf{J} = 0. \tag{40.17}$$

This is equivalent to the equation of conservation of charge, Eq. 15.3, if we can identify the current as due to the motion of free charges:

$$\mathbf{J} = \rho_{\text{free}}\mathbf{v}, \qquad (40.18)^*$$

where $\mathbf{v}$ is the velocity of the charge. Incidentally, $\partial\mathbf{D}/\partial t$ is often called the *displacement current.*

This section has presented a brief review of the basic laws of the electromagnetic field; it is presumed that the student will already be more or less familiar with more technical detail than has been included here.

Exercise IV.40

40.1. Derive Eq. 40.5 for the induced charge density from the definition of polarization.

40.2. What are "lines of magnetic induction"? Prove that, because of Eq. 40.3, these lines are necessarily closed loops.

40.3. Prove that the potential at a point in a polarizable material is given by

$$4\pi\epsilon\Phi(\mathbf{r}) = \iiint(1/|\mathbf{r}-\mathbf{r}'|)[\rho_{\text{free}}(\mathbf{r}')-\nabla\cdot\mathbf{P}(\mathbf{r}')]d\tau'.$$

40.4. From Eq. 40.4 derive the transformation equation for the electric field vector relative to reference systems moving with uniform relative velocities:

$$\mathbf{E}' = \mathbf{E}+\mathbf{v}\times\mathbf{B}.$$

Show that, when v/c is sufficiently small the companion transformation,

$$\mathbf{B}' = \mathbf{B}-\epsilon\mu\mathbf{v}\times\mathbf{E}$$

is consistent with the invariance of Eq. 40.2:

$$\nabla\times\mathbf{E}' = -\partial\mathbf{B}'/\partial t.$$

IV.41 The Electromagnetic Wave Equations.

It is convenient here to review the set of field equations generally known as *Maxwell's equations*. In the rational system of units and in the order we have derived them, they are:

(40.2): $$\nabla\times\mathbf{E} = -\partial\mathbf{B}/\partial t. \tag{41.1}$$

(40.3): $$\nabla\cdot\mathbf{B} = 0. \tag{41.2}$$

* When the current is carried by several types of ion, Eq. 40.18 must be replaced by: $\mathbf{J} = \sum_n \rho_n\mathbf{v}_n$, where $\mathbf{v}_n$ is the mean velocity of the nth type of ion.

From Eqs. 40.7 and 40.9,

$$\nabla \cdot \mathbf{E} = \rho_{\text{free}}/\epsilon'. \tag{41.3}$$

From Eqs. 40.13 and 40.15,

$$\nabla \times \mathbf{B} = \mu'(\mathbf{J} + \epsilon' \partial \mathbf{E}/\partial t). \tag{41.4}$$

We may also include the equation for body force,

(40.4): $$\mathbf{F} = \rho \mathbf{E} + \mathbf{J} \times \mathbf{B}, \tag{41.5}$$

the identification of current with transport of free charge

(40.18): $$\mathbf{J} = \rho_{\text{free}}\, \mathbf{v}, \tag{41.6}$$

and the empirical Ohm's law of conduction

$$\mathbf{J} = \sigma \mathbf{E}, \tag{41.7}$$

where σ is called the *conductivity* of the material.

Maxwell derived the electromagnetic wave equation from these as follows. Take the curl of Eq. 41.1 and use Eq. 6.21 on the left side. Interchange the order of differentiation on the right, then make use of Eq. 41.3 on the left and of Eq. 41.4 on the right, to obtain

$$\nabla^2 \mathbf{E} - \epsilon' \mu' \partial^2 \mathbf{E}/\partial t^2 = \mu' \partial \mathbf{J}/\partial t + \nabla(\rho_{\text{free}}/\epsilon'). \tag{41.8}$$

In vacuo, where there is no charge density or current vector, we have

$$\nabla^2 \mathbf{E} - \epsilon\mu \partial^2 \mathbf{E}/\partial t^2 = 0. \tag{41.9}$$

This is the vector wave equation corresponding to the velocity of propagation:

$$c = 1/(\epsilon\mu)^{1/2}. \tag{41.10}$$

When charge and current are present, their variations, appearing on the right of Eq. 41.8, act as sources of the wave field.

Again, take the curl of Eq. 41.4 and the time derivative of Eq. 41.1; it is then easy to show that

$$\nabla^2 \mathbf{B} - \epsilon' \mu' \partial^2 \mathbf{B}/\partial t^2 = -\mu' \nabla \times \mathbf{J}. \tag{41.11}$$

In vacuo, where $\mathbf{J}$ is zero, we again have the simple wave equation, this time for the magnetic induction $\mathbf{B}$:

$$\nabla^2 \mathbf{B} - \epsilon\mu \partial^2 \mathbf{B}/\partial t^2 = 0, \tag{41.12}$$

with the same speed of propagation as with the electric vector. In the presence of a current, the curl of the current density appearing on the right of Eq. 41.11 is a source for this magnetic wave field. The numerical value of the speed of propagation in vacuo is given by Eqs. 34.4 and 34.7:

$$c = 3 \times 10^8 \text{ M/S}. \tag{41.13}$$

The *refractive index*, *n*, of a transparent material is the ratio of the speed in vacuo to the speed in the material, so

$$n = (\epsilon'\mu'/\epsilon\mu)^{1/2}. \tag{41.14}$$

In static fields we found it convenient to introduce a scalar potential because the fields were individually conservative. We have seen now that in general this conservation is violated in time-dependent fields. It is nevertheless possible and convenient to define potential functions of a more general type in terms of Maxwell's equations. Thus, from Eq. 41.2, we see that a vector function **A** exists such that

$$\mathbf{B} = \nabla \times \mathbf{A}. \tag{41.15}$$

Putting this into Eq. 41.1 yields the equation

$$\nabla \times (\mathbf{E} + \partial \mathbf{A}/\partial t) = 0, \tag{41.16}$$

which shows that the vector $\mathbf{E} + \partial \mathbf{A}/\partial t$ represents a conservative field even when **E** does not. A scalar function Φ exists such that

$$\mathbf{E} + \partial \mathbf{A}/\partial t = -\nabla\Phi. \tag{41.17}$$

In fact, Φ reduces to the ordinary electrostatic potential when the field is independent of time, $\partial \mathbf{A}/\partial t = 0$. The two potentials **A** and Φ are sufficient to characterize the electromagnetic field in the general case.

It is not difficult to find a wave propagation equation for **A** and for Φ. Write Eq. 41.15 into Eq. 41.11 and interchange the order of the differentiation on the left, we find:

$$\nabla \times \nabla^2\mathbf{A} - \epsilon'\mu'\nabla \times \partial^2\mathbf{A}/\partial t^2 = -\mu'\nabla \times \mathbf{J}. \tag{41.18}$$

This is satisfied by the following wave equation:

$$\nabla^2\mathbf{A} - \epsilon'\mu'\partial^2\mathbf{A}/\partial t^2 = -\mu'\mathbf{J}. \tag{41.19}$$

There is, however, an arbitrary term in the source function on the right, because any function with a zero curl can be added to this equation and it would still satisfy Eq. 41.18. In empty space **A** satisfies the same wave equation as the vectors **E** and **B** found previously:

$$\nabla^2\mathbf{A} - \epsilon\mu\partial^2\mathbf{A}/\partial t^2 = 0. \tag{41.20}$$

Again, use Eq. 41.17 to eliminate **E** from Eq. 41.8. Collecting the two terms in **A**, we compare them with the time derivative of Eq. 41.19 and find the result:

$$\nabla\nabla^2\Phi - \epsilon'\mu'\nabla(\partial^2\Phi/\partial t^2) = -\nabla(\rho_{\text{free}}/\epsilon'). \tag{41.21}$$

This is satisfied by the following wave equation:

$$\nabla^2\Phi - \epsilon'\mu'\partial^2\Phi/\partial t^2 = -\rho_{\text{free}}/\epsilon'. \tag{41.22}$$

Again there remains an arbitrary constant in the source function, which could be added to this result and still satisfy Eq. 41.21: we assume this constant to be zero so that Eq. 41.22 will reduce to Poisson's equation in the static case.

The arbitrary function involved in the source function for the vector potential in Eq. 41.19 exists essentially because of its definition in Eq. 41.15. We can add to **A** the gradient of any scalar function:

$$\mathbf{A}' = \mathbf{A} + \nabla f \tag{41.23}$$

and still find the same magnetic induction:

$$\mathbf{B}' = \nabla\times\mathbf{A}' = \nabla\times\mathbf{A} = \mathbf{B}. \tag{41.24}$$

It is convenient to restrict this arbitrariness by accepting vector potentials only if they satisfy in addition the equation:

$$\nabla\cdot\mathbf{A} = 0, \tag{41.25}$$

so that the arbitrary function f is restricted to satisfy Laplace's equation

$$\nabla^2 f = 0. \tag{41.26}$$

The empirically significant Maxwell equations are independent of the function f, a circumstance that is described as *gauge invariance* of Maxwell's field equations. Detailed applications of these concepts can be found in more advanced texts on electromagnetic theory.

Exercise IV.41

41.1. Derive Eq. 41.8 from Eqs. 41.1 and 41.4.

41.2. Derive Eq. 41.11 from Eqs. 41.1 and 41.4. Write out explicitly the three equations contained in Eq. 41.11.

41.3. Derive Eq. 41.19 from Eqs. 41.15 and 41.11. Write out explicitly the three component equations contained in Eq. 41.19.

41.4. Derive Eq. 41.21 from Eqs. 41.17 and 41.19.

41.5. Show that the following functions, representing plane propagated waves, satisfy, respectively, Eqs. 41.9 and 41.12:

$$\mathbf{E} = \mathbf{E}_0 \sin[(2\pi/\lambda)(x-ct)],$$
$$\mathbf{B} = \mathbf{B}_0 \sin[(2\pi/\lambda)(x-ct)].$$

where λ is an arbitrary wavelength, $\mathbf{E}_0$ and $\mathbf{B}_0$ are arbitrary constant vectors. Show that if

$$\mathbf{E}_0 = \mathbf{j}E_0 \quad \text{and} \quad \mathbf{B}_0 = \mathbf{k}B_0,$$

with E_0 and B_0 constants, Eq. 41.1 is satisfied when $E_0 = cB_0$.

41.6. Show that the plane wave vector potential

$$\mathbf{A} = \mathbf{A}_0 \cos[(2\pi/\lambda)(x-ct)]$$

satisfies Eq. 41.20. Prove that, if $\mathbf{A}_0 = A_1\mathbf{j} + A_2\mathbf{k}$ it also satisfies Eq. 41.25. Find the constants A_1 and A_2 for this to satisfy Eq. 41.15 when the vector $\mathbf{B}$ is given by the last exercise. Put the function $\mathbf{A}$ so determined into Eq. 41.17 and find the scalar potential Φ.

IV.42 Energy and Stress in the Electromagnetic Field.

In elementary physics we are familiar with the calculation of the work done in charging a condenser and with the idea that this work produces energy of the field in the space between the conductors. We are also familiar with the similar calculation of the work done in building up a current in an induction coil against its self inductance and with the idea that this work produces energy of the magnetic field in the space inside the solenoid. In fact, it can be proved quite generally that the work required to build up any electromagnetic field can be regarded as producing energy in space, the density of this energy being exactly that calculated in the elementary examples mentioned, *viz.*:

$$W = \tfrac{1}{2}(\mathbf{E}\cdot\mathbf{D} + \mathbf{B}\cdot\mathbf{H}), \tag{42.1}$$

being measured in rational units, joules per cubic meter.

This expression for the energy density is valid even when the fields are nonstatic, and therefore conservation of energy demands that energy can flow through space to compensate for changes that occur at any given point. We shall develop the equation for the conservation of electromagnetic energy in space from Maxwell's equations. Take the scalar product of Eq. 41.4 with $\mathbf{E}$, and of Eq. 41.1 with $\mathbf{H}$, and combine the two results:

$$\mathbf{E}\cdot\partial\mathbf{D}/\partial t + \mathbf{H}\cdot\partial\mathbf{B}/\partial t = -\mathbf{E}\cdot\mathbf{J} + \mathbf{E}\cdot\nabla\times\mathbf{H} - \mathbf{H}\cdot\nabla\times\mathbf{E}. \tag{42.2}$$

It is not difficult to prove that this is equivalent to

$$\partial W/\partial t = -\mathbf{E}\cdot\mathbf{J} - \nabla\cdot(\mathbf{E}\times\mathbf{H}), \tag{42.3}$$

with of course W being given by Eq. 42.1. This equation may be integrated over some finite region in space, and the result can be read as stating: the rate of increase of electromagnetic energy in the region equals the rate of flow of energy in through the surface less the rate of production of Joule heat in the region. The Joule heat is produced at the rate $\iiint \mathbf{E}\cdot\mathbf{J}\,d\tau$. The other term must then give the flow in through the surface. We have from Gauss' integral theorem

$$\iiint \nabla\cdot(\mathbf{E}\times\mathbf{H})\,d\tau = \iint (\mathbf{E}\times\mathbf{H})\cdot d\boldsymbol{\sigma}, \tag{42.4}$$

where $d\boldsymbol{\sigma}$ is the outward drawn normal vector surface element. We therefore are obliged to interpret the vector

$$\mathbf{S} = \mathbf{E} \times \mathbf{H} \tag{42.5}$$

as the flow of energy per unit area per unit time at the surface: it is known as *Poynting's vector.*

It is clear that it is not necessary to have time-varying fields in order to have transport of electromagnetic energy through space. Consider a voltaic cell driving a steady current through a resistance. It is easy to calculate the Poynting vector (a) in the space between the plates of the battery, and (b) near the surface of the resistance at some distance from the battery. It is found that the Poynting vector flows away from the battery into the surrounding space, and returns to the resistance through the surface. The rate at which electromagnetic energy enters any element of the resistance is just equal to the Joule heat rate in that element. It is worth emphasizing that the energy for Joule heat is not carried through the conductor from the battery, but through the sur-

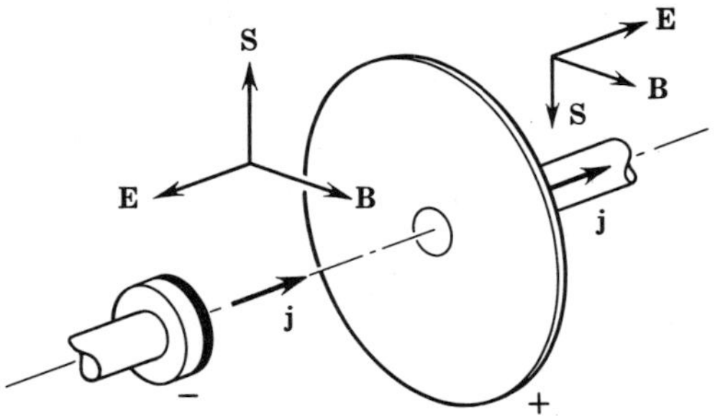

FIG. 42.1. The flow of energy from a battery through the field into an external conducting resistance

rounding field: the electrons whose motion carries the current absorb energy from the field, and give it up by collisions with the atoms of the conductor.

This interpretation of the process has not been proved unique, but it appears to be the simplest possible one that can be carried through under all conceivable circumstances. It is instructive to try to imagine an experimental arrangement in which a conductor is shielded in such a way as to prevent any Poynting flux from reaching it, and so to prevent the arrival of any Joule heat through the surrounding medium. It will invariably be found that a complete shield must short circuit the conductor so that no current can flow through it.

We have seen in an exercise, Ex. 41.5, that in plane electromagnetic waves the electric and magnetic vectors are perpendicular to each other and to the direction of propagation, and the amplitudes are related by

$$E_0 = cB_0. \tag{42.6}$$

Putting this relation into Eq. 42.1 for the density of energy in the field, we find that the magnitude of the density is proportional to

$$W_0 = \epsilon E_0^2. \tag{42.7}$$

Also, in this case, Eq. 42.5 becomes

$$S_0 = E_0 H_0 = E_0^2/\mu c. \tag{42.8}$$

The ratio of energy flux S to energy density W is then

$$S_0/W_0 = 1/\epsilon\mu c = c. \tag{42.9}$$

In the plane electromagnetic wave in vacuo, therefore, the energy flows along with the same speed of propagation as the wave phase. The energy density at any fixed point varies with the frequency of the wave, but the peaks of energy are propagated through space and no violation of conservation can occur.

Consider an isotropic and homogeneous dielectric material containing free charge density ρ_f and subject to a static electric field $\mathbf{E}$. The force per unit volume exerted on the dielectric by the field is

$$\mathbf{f} = \rho_f \mathbf{E}. \tag{42.10}$$

That it is the free charge alone, and not the total charge (including the induced charge), that occurs in this equation is not at once obvious. To derive the expression one has to use the energy principle in the form: rate of loss of energy by the field equals the rate of working by the field on the dielectric material, the energy being given by Eq. 42.1. If the dielectric is not homogeneous, so that its dielectric capacity is a function of position, it can be shown that the body force includes terms depending on the variations of ϵ' with position, and possibly also with the density of the material; the last terms give rise to electrostriction effects. We shall omit these additional terms in our discussion and work simply with Eq. 42.10.

The body force in a continuous material is a consequence of stress: it is in fact equal to the divergence of the stress matrix (see Eq. 12.13). We can therefore think of the electric field causing stresses in the dielectric material, and we can find an expression for the stress matrix in terms of the electric vector. From Eq. 42.10 and Eq. 40.7, we have

$$\mathbf{f} = \mathbf{E}\nabla\cdot\mathbf{D}, \tag{42.11}$$

provided the dielectric capacity is a constant. This can be expressed in the equivalent form, as a divergence:

$$\mathbf{f} = \nabla \cdot \epsilon'(\mathbf{EE} - \tfrac{1}{2}\mathrm{E}\mathbf{E}^2), \tag{42.12}$$

where E is the unit (identity) matrix. This is equivalent to Eq. 12.13 if we take the stress matrix to be

$$\mathbf{T}_{\mathrm{el}} = \epsilon'(\mathbf{EE} - \tfrac{1}{2}\mathrm{E}\mathbf{E}^2). \tag{42.13}$$

It turns out incidentally that, although we have derived this assuming ϵ' to be a constant, the result is true even for variable ϵ', except that the electrostriction term is not included.

When the field is not static, terms due to the magnetic fields have to be added, and these are found to be given by a matrix closely analogous with Eq. 42.13, *viz.*:

$$\mathbf{T}_{\mathrm{mag}} = (1/\mu')(\mathbf{BB} - \tfrac{1}{2}\mathrm{E}\mathbf{B}^2). \tag{42.14}$$

The total stress matrix

$$\mathbf{T} = \mathbf{T}_{\mathrm{el}} + \mathbf{T}_{\mathrm{mag}} \tag{42.15}$$

has been thought of here as a condition of the material in the field. But we note at once that the matrix does not vanish in vacuo; it merely takes the form

$$\mathbf{T}_{\mathrm{vac}} = \epsilon(\mathbf{EE} - \tfrac{1}{2}\mathrm{E}\mathbf{E}^2) + (1/\mu)(\mathbf{BB} - \tfrac{1}{2}\mathrm{E}\mathbf{B}^2), \tag{42.16}$$

where ϵ and μ are the vacuum values introduced in §34. It appears, therefore, that the stress is an essential property of the *field* rather than of the material in the field. The material merely modifies the stress matrix by its effect on the parameters ϵ' and μ'.

From Eq. 12.2 we know the force per unit area across any surface in a continuum under stress. Where a surface separates two dielectrics that have different dielectric capacities, it is clear that the force across the surface from one side will differ from that produced on the other side, the lack of balance yielding a net force on the surface between the dielectrics. It can be shown from Eq. 42.13 that the force is a tension when the electric field is perpendicular to the surface and a pressure when the field is parallel to the surface. These are just the effects stated by Faraday in his original concept of tubes of force exerting tension longitudinally and pressure laterally.

We have seen that the body force is given by the divergence of the stress matrix; the question arises, on what does this body force act in the vacuum? It is clear that $\nabla \cdot \mathbf{T}_{\mathrm{vac}}$ need not be zero, so that a force is then acting on every element of the "empty" field. In prerelativity physics, this question was answered by ascribing inertia to the ether;

it was believed that stress cannot exist without a substantial medium in which it can operate. Today we have to abandon this explanation in favor of a more abstract interpretation; we know that nothing more substantial than energy itself is needed as a vehicle for inertia. The stresses and body force in the vacuum are required to maintain the transport of energy in the field. The divergence of the total stress matrix must in fact equal the body force on any material in the field plus the force exerted on the energy in the field—*i.e.* the rate of change of the momentum in the field. We can write the divergence of the stress matrix in the form:

$$\nabla\cdot\mathbf{T} = \mathbf{E}\nabla\cdot\mathbf{D}+\epsilon'(\mathbf{E}\cdot\nabla\mathbf{E}-\tfrac{1}{2}\nabla\mathbf{E}^2)+(1/\mu')(\mathbf{B}\cdot\nabla\mathbf{B}-\tfrac{1}{2}\nabla\mathbf{B}^2). \quad (42.17)$$

As an exercise it is easy to show that

$$\mathbf{A}\cdot\nabla\mathbf{A}-\tfrac{1}{2}\nabla\mathbf{A}^2 = -\mathbf{A}\times\nabla\times\mathbf{A}, \quad (42.18)$$

where $\mathbf{A}$ is any vector function. Applying this to both $\mathbf{E}$ and $\mathbf{B}$ in the above expression, and making use of Maxwell's equations, Eqs. 41.1 and 41.4, we arrive at the following result:

$$\nabla\cdot\mathbf{T} = \rho_f\mathbf{E}+\mathbf{J}\times\mathbf{B}+\partial(\mathbf{D}\times\mathbf{B})/\partial t. \quad (42.19)$$

The first two terms we recognize immediately as being the body force on the material in the field—the electrostatic and the magnetic effects. The last term is then to be identified with the rate of change of the momentum in the field itself. Therefore we shall ascribe a *momentum density* $\mathbf{M}$ to the field:

$$\mathbf{M} = \mathbf{D}\times\mathbf{B} = \epsilon'\mu'\mathbf{S} = \mathbf{S}/c^2, \quad (42.20)$$

which is closely related with the Poynting vector $\mathbf{S}$. In fact, the momentum density is equal to the energy flow density divided by c^2. This relation, which is equivalent to making mass density equal to energy density divided by c^2, is now regarded, through the theory of relativity, as a universal property of energy.

Exercise IV.42

42.1. Derive in detail Eq. 42.3 from Eq. 42.2.

42.2. Consider a long rod of resistivity ρ, radius a, carrying a current of density J per unit area of cross section. Use Ohm's law for the electric field just outside the surface of the rod, and Ampere's law for the magnetic field just outside the surface of the rod. Calculate the total rate of flow into the rod, per unit length, from the electromagnetic energy of the field. Compare this with the Joule heating in the rod.

42.3. Derive Eq. 42.12 from Eq. 42.11. Hint: use $\mathbf{E} = -\nabla\Phi$.

42.4. Show that the components of the stress tensor, Eq. 42.13 are:

$$\mathbf{T}_{\mathrm{el}} = \epsilon' \begin{Bmatrix} \frac{1}{2}(E_x^2 - E_y^2 - E_z^2) & E_xE_y & E_xE_z \\ E_xE_y & \frac{1}{2}(E_y^2 - E_z^2 - E_x^2) & E_yE_z \\ E_xE_z & E_yE_z & \frac{1}{2}(E_z^2 - E_x^2 - E_y^2) \end{Bmatrix} \tag{42.21}$$

42.5. Diagonalize the matrix $\mathbf{T}_{\mathrm{el}}$, and show that it takes the form

$$T_{11} = -T_{22} = -T_{33} = \tfrac{1}{2}\epsilon' E^2.$$

42.6. Write down all the components of the matrix $\mathbf{T}_{\mathrm{mg}}$, and diagonalize it. Do $\mathbf{T}_{\mathrm{el}}$ and $\mathbf{T}_{\mathrm{mg}}$ become diagonal in the same coordinate system?

42.7. Derive Eq. 42.17 from Eq. 42.13 and Eq. 42.14.

42.8. Prove Eq. 42.18.

IV.43 The Lagrangian Form of Electromagnetic Field Laws. We saw in §26 that Hamilton's variation principle was applicable to continuous media and mentioned there that this ensures its applicability to the behavior of fields. We have just seen in the last section that an energy density, a momentum density vector, and a stress matrix can all be defined as functions of position in space in terms of the electromagnetic field variables. In every essential the electromagnetic field can be treated as a mechanical system, and we can expect that Hamilton's variation principle is applicable here. In other words, we should be able to set up a suitable Lagrangian function of the field variables, such that Hamilton's variation principle, Eq. 26.4, shall result in the equations of motion of the field, *viz.*: Maxwell's equations. We know of no general rule for the construction of the Lagrangian function. We might be tempted to guess that it is the difference between the kinetic and potential energy of the field, but we are in difficulties here because we have not clearly defined the meaning of "kinetic" and "potential" energies in the field. Actually the correct Lagrangian has been discovered to be:

$$\mathscr{L} = \tfrac{1}{2}(\mathbf{H}\cdot\mathbf{B} - \mathbf{E}\cdot\mathbf{D}), \tag{43.1}$$

which yields the proper equations for the vacuum field. We could here interpret $\frac{1}{2}\mathbf{E}\cdot\mathbf{D}$ as the potential energy density, and $\frac{1}{2}\mathbf{H}\cdot\mathbf{B}$ as the kinetic energy density, and indeed this interpretation has some advantages.

Hamilton's variation principle demands that we make variations in the field variables and find those field variables that maximize the Lagrangian integral, thus:

$$\delta \iiiint \mathscr{L}\, dx\,dy\,dz\,dt = 0. \tag{43.2}$$

We have, however, seen that the six variables E_m and B_m ($m=1, 2, 3$) are in a sense redundant, because the four variables Φ, A_1, A_2, A_3, the four potentials of the electromagnetic field, are sufficient in fact to define the field completely. Thus the vectors **E** and **B** are defined in terms of Φ and **A** by the equations

$$\left.\begin{aligned}\mathbf{E} &= -\nabla\Phi - \partial\mathbf{A}/\partial t \\ \mathbf{B} &= \nabla\times\mathbf{A}\end{aligned}\right\}. \tag{43.3}$$

(See Eqs. 41.17 and 41.15.) Two sets of Maxwell's equations are then simple identities stemming from these definitions, *viz.*: Eqs. 41.1 and 41.2. The variation principle provides a single equation from which the other Maxwell equations can be deduced, *viz.*: Eqs. 41.4 and 41.3. To make this derivation, we have to express the Lagrangian in terms of the potentials as field variables, thus:

$$\mathscr{L} = \tfrac{1}{2}(1/\mu)(\nabla\times\mathbf{A})^2 - \tfrac{1}{2}\epsilon(\nabla\Phi + \partial\mathbf{A}/\partial t)^2. \tag{43.4}$$

In §26 we set up the Lagrangian equations of motion, Eq. 26.12, for a field with only one variable, u, and its four partial derivatives. Here there are four independent variables, and it is not difficult to prove that this has simply the effect of providing a formally similar equation for each of the independent variables. Thus, the generalization of Eq. 26.12 that we need here is:

$$\partial\mathscr{L}/\partial A_m = (\partial/\partial t)(\partial\mathscr{L}/\partial\dot{A}_m) + \sum_n (\partial/\partial x_n)(\partial\mathscr{L}/\partial A_{m,n}), \tag{43.5}$$

$$\partial\mathscr{L}/\partial\Phi = (\partial/\partial t)(\partial\mathscr{L}/\partial\dot{\Phi}) + \sum_n (\partial/\partial x_n)(\partial\mathscr{L}/\partial\Phi_{,n}). \tag{43.6}$$

In Eq. 43.5 we have written $\dot{A}_m$ for $\partial A_m/\partial t$ and $A_{m,n}$ for $\partial A_m/\partial x_n$, and it is understood that m can have any of the three values 1, 2, 3: there are three equations corresponding to the three field variables A_m. In Eq. 43.6 we have also used the dot notation for the time derivative, and $\Phi_{,n}$ is an abbreviation for $\partial\Phi/\partial x_n$.

For the particular Lagrangian density shown in Eq. 43.4, there is no explicit dependence on the potentials themselves—it depends only on their derivatives. Therefore, the left-hand sides of Eqs. 43.5 and 43.6 are all zero. In fact, it is not a difficult exercise to prove that the three equations in Eq. 43.5 are equivalent to Maxwell's equation Eq. 41.4 in the form

$$\nabla\times\mathbf{B} = \epsilon\mu\partial\mathbf{E}/\partial t, \tag{43.7}$$

which is correct for the vacuum field. Also Eq. 43.6 reduces simply to

$$\nabla\cdot\mathbf{E} = 0, \tag{43.8}$$

also true for the vacuum field. The Lagrangian formulation presented here thus interprets these Maxwell equations as the equations of motion of the electromagnetic field. This derivation further implies that the Maxwell equations are invariant for coordinate transformations (*cf.* the discussion of the invariance of the Lagrangian equations of motion in §25 and §27).

So far we have discussed only the Lagrangian for the vacuum field. It is also possible to set up a Lagrangian for the field in material media carrying free charge density and current. The resulting equations of motion then turn out to be Maxwell's equations with the proper modifications due to the material, the charge, and the current terms. The Lagrangian density is found to be:

$$\mathscr{L} = \tfrac{1}{2}(\mathbf{B}\cdot\mathbf{H}-\mathbf{E}\cdot\mathbf{D})+(\mathbf{J}\cdot\mathbf{A}-\rho\Phi). \tag{43.9}$$

The part of the Lagrangian depending on the charge and current sources may be thought of as the difference between a kinetic energy density $\mathbf{J}\cdot\mathbf{A}$ and a potential energy density $\rho\Phi$; the full Lagrangian density is then the sum of a field Lagrangian having the same form as Eq. 43.1 and this new part depending on the charge and current. We leave it as an exercise to derive Maxwell's equations, Eq. 41.4 and 41.3, by using Eqs. 43.5 and 43.6 with the Lagrangian in Eq. 43.9. Note that the charge and current densities are not varied.

The Lagrangian formulation of the electromagnetic field laws is at the basis of modern quantum field theories.

Exercise IV.43

43.1. Show that Eq. 43.5 reduces to Eq. 43.7 and that Eq. 43.6 reduces to Eq. 43.8 when the Lagrangian density is given by Eq. 43.4.

43.2. Show that with Eq. 43.9 for the Lagrangian density, Eq. 43.5 becomes the Maxwell equation Eq. 41.4 and that Eq. 43.6 becomes Eq. 41.3. Note: use the proper modification of Eq. 43.4 for the field part of the Lagrangian in Eq. 43.9.

43.3. Derive Eqs. 43.5 and 43.6 from the variation principle by following a generalization of the method of §26 to several independent variables.

IV.44 The Homogeneous Wave Equation. In the next few sections we shall discuss methods of solving the scalar wave equation, Eq. 41.22. Any one of the wave equations obeyed by vector variables, Eqs. 41.8, 41.11, or 41.19, can be regarded as a set of three component equations, each of which possesses the same type of solutions as the scalar wave equation. The additional problem involved in solving the vector equation is to find three solutions that together form a vector function satisfying the other conditions on the vector field. For instance, the solutions of Eq. 41.19 must satisfy also the solenoidal condition, Eq. 41.25.

We start with the scalar wave equation in empty space:

$$\nabla^2\Phi = (1/c^2)\partial^2\Phi/\partial t^2 \tag{44.1}$$

and seek only oscillatory solutions of the form:

$$\Phi = U(x, y, z)e^{i\omega t}. \tag{44.2}$$

Substituting this into Eq. 44.1 and canceling the time factor, we find Helmholtz' equation in U:

$$\nabla^2 U + k^2 U = 0, \tag{44.3}$$

where

$$k = \omega/c.$$

This equation can be solved by separation of the variables, and in Cartesian coordinates we let

$$U(x, y, z) = X(x)Y(y)Z(z). \tag{44.4}$$

Substituting this into Eq. 44.3 yields separate equations in the three coordinates:

$$\left.\begin{aligned} d^2X/dx^2 + k_1{}^2 X &= 0 \\ d^2Y/dy^2 + k_2{}^2 Y &= 0 \\ d^2Z/dz^2 + k_3{}^2 Z &= 0 \end{aligned}\right\}. \tag{44.5}$$

where the three separation constants are related:

$$k_1{}^2 + k_2{}^2 + k_3{}^2 = k^2. \tag{44.6}$$

The solutions of Eq. 44.3 can therefore be expressed in the form

$$U = U_0 \sin(k_1 x + a_1) \sin(k_2 y + a_2) \sin(k_3 z + a_3), \tag{44.7}$$

where the k's are restricted only by Eq. 44.6 and the phase angles a_1, a_2, and a_3 are completely arbitrary. Multiplying this expression by the time factor to obtain Φ in Eq. 44.2, we see that the solution is a standing wave pattern. By algebraic manipulation of the solution, it can be expressed as the superposition of progressive waves with their directions of propagation parallel to vectors whose components are $(\pm k_1, \pm k_2, \pm k_3)$; these directions can be pictured as being derived from the original vector $\mathbf{k} = (k_1, k_2, k_3)$ by reflection from the rectangular walls of an enclosure.

If we ask that the potential Φ be zero permanently over the walls of the rectangular enclosure, all the phase angles must be set zero and the wave numbers k_1, k_2, and k_3 must be restricted to the following values:

$$k_j L_j = n_j \pi \quad (j = 1, 2, 3), \tag{44.8}$$

where n_1, n_2, and n_3 are integers; and L_1, L_2, and L_3 are the linear dimensions of the enclosure. The standing waves have frequencies restricted by Eq. 44.6, thus:

$$\omega^2/c^2 = k^2 = \pi^2(n_1{}^2/L_1{}^2 + n_2{}^2/L_2{}^2 + n_3{}^2/L_3{}^2). \tag{44.9}$$

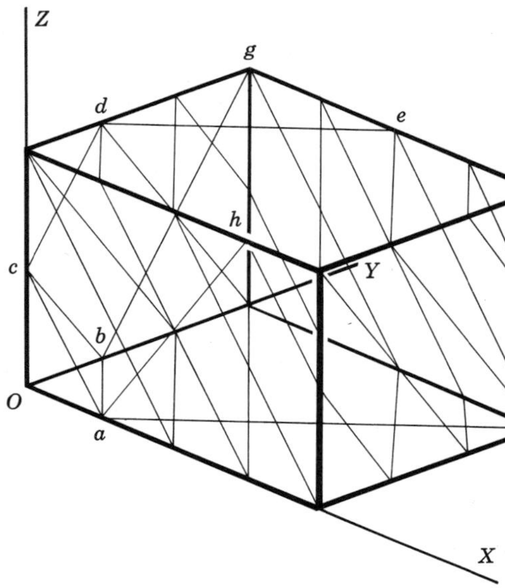

FIG. 44.1. One family of nodal planes (4, 3, 2) for the function in Eq. 44.7 when the **n**-vector of Eq. 44.9 has components (4, 3, 2). The plane marked *acdefa* is one member of another family produced by reflecting *abc* in the $X-Z$ plane, *viz.*: $(-4, 3, -2)$. The plane marked *abgh* is a member of a third family, *viz.*: $(-4, -3, 2)$. There exists a fourth family not drawn in the diagram: $(4, -3, -2)$

It is worth while to compare the present results with the discussion of the standing waves in a rectangular membrane given in §10. The theory of electromagnetic resonators and the propagation of electromagnetic waves along wave guides take off from this point, but we shall not pursue the subject here.

By expressing the Laplacian operator in Eq. 44.3 in terms of spherical polar coordinates (see Ex. 8.11), we can solve the equation in the form

$$U = R(r)T(\theta)F(\phi). \tag{44.10}$$

Substituting this into Eq. 44.3 and manipulating the terms, we can separate out the functions of ϕ from the others and find

$$F(\phi) = e^{im\phi}, \tag{44.11}$$

where m must be an integer for the function to be a single-valued one. The function of θ also separates out in the same way as with Laplace's equation (see §35) with Legendre polynomials:

$$T(\theta) = T_{n,m}(\cos\theta) \tag{44.12}$$

as before. The radial equation, however, is not the same as in Laplace's equation; it turns out to be

$$d^2R/dr^2+(2/r)(dR/dr)+[k^2-n(n+1)/r^2]R = 0. \tag{44.13}$$

This can be reduced to Bessel's equation by making the transformation of variables:

$$\begin{aligned} x &= kr, \\ X(x) &= x^{1/2}\, R(r), \end{aligned} \tag{44.14}$$

after which Eq. 44.13 becomes

$$(1/x)(d/dx)(x dX/dx)+[1-(n+\tfrac{1}{2})^2/x^2]X = 0. \tag{44.15}$$

This is *Bessel's equation*; it has generally well-behaved solutions if n is an integer, in which case the solutions are known as *Bessel's functions* of order $n+\frac{1}{2}$:

$$X(x) = J_{n+1/2}(x) \quad \text{or} \quad J_{-n-1/2}(x). \tag{44.16}$$

The subscript can be either positive or negative, but if it is negative the function has an infinity at the origin; if our physical problem calls for well-behaved functions at the origin, we just leave out the functions with negative indices.

Collecting all the information together, we have for the most general solution obtainable in this way:

$$U(r, \theta, \phi) = \sum_{n,m}\sum A_{n,m} r^{-1/2} J_{n+1/2}(kr) T_{n,m}(\cos\theta) e^{im\phi}. \tag{44.17}$$

Together with the time factor, as in Eq. 44.2, the final solution of the wave equation is

$$\Phi = U(r, \theta, \phi) e^{i\omega t}. \tag{44.18}$$

We may still superpose many solutions like this with different frequencies, *i.e.* different values of k or ω, and by doing so we can in principle match an arbitrary time-dependence of the field at a source or boundary.

As with the rectangular enclosure, we can here ask for the standing wave solutions in a spherical enclosure in which the function vanishes at all times on the sphere of radius a. These are obtained simply by adjusting the parameter k in Eq. 44.17 so that ka is a node of the Bessel function. This in fact leaves us with a discrete spectrum of permitted k-values, one set for each value of n in the equation:

$$J_{n+1/2}(ka) = 0. \tag{44.19}$$

It is worth while at this point to review the work in §10 on the normal modes of the circular membrane. Here we have had to use a somewhat different set of Bessel functions, but the general method is quite similar in the two problems.

As another example in which the spherical polar coordinates are useful, we shall consider oscillatory solutions of the wave equation extending to infinity in all directions, originating from a specified type of singularity as source at the origin, namely, *dipole radiation.* We note that the general solution, Eq. 44.17, contains angle factors that we can adjust to match the dipole solution, Eq. 36.7, of Laplace's equation. Thus we take $m=0$ in Eq. 44.17, and all A_n zero when n is even, and obtain

$$U = \sum_{\text{odd } n} A_n r^{-1/2} J_{n+1/2}(kr) P_n(\cos\theta), \tag{44.20}$$

which has the right angular dependence for the dipole field. To make the radial dependence correct we have to restrict the choice of n more severely. In fact, it is known that

$$\text{Lim}_{x\to 0}: \qquad x^{-1/2} J_{-1\frac{1}{2}}(x) \to 1/x^2. \tag{44.21}$$

Therefore, the solution

$$U = Ar^{-1/2} J_{-1\frac{1}{2}}(kr) P_1(\cos\theta) \tag{44.22}$$

behaves asymptotically as $r\to 0$ like the dipole solution of Laplace's equation, provided the dipole is vanishingly short. We can add any term with a zero at the origin provided it has the right angular dependence, and this is provided by the Bessel function with positive index in Eq. 44.20, and the desired solution is

$$U = [AJ_{-1\frac{1}{2}}(kr) + BJ_{1\frac{1}{2}}(kr)] r^{-1/2} \cos\theta. \tag{44.23}$$

Multiplying this by the time factor $e^{i\omega t}$ to give the complete solution of the wave equation, we get an outward propagated wave that looks like an oscillating dipole of frequency ω at the origin.

At very great distances from the source, we have

$$\text{Lim}_{x\to\infty}: \qquad \left.\begin{aligned} (\tfrac{1}{2}x)^{1/2} J_{1\frac{1}{2}}(x) &\to -\cos x \\ (\tfrac{1}{2}x)^{1/2} J_{-1\frac{1}{2}}(x) &\to -\sin x \end{aligned}\right\}, \tag{44.24}$$

and so the solution Eq. 44.23 approaches the form

$$\begin{aligned} U &= (A\cos kr + B\sin kr)(1/r)\cos\theta \\ &= U_0 \cos(kr+\alpha)(1/r)\cos\theta. \end{aligned} \tag{44.25}$$

With the time factor included, this gives a propagated wave with a phase velocity $c=\omega/k$, whose squared amplitude decreases with the square of the distance from the source. The function

$$\Phi = U_0(1/r)\cos\theta \times \cos(kr+\alpha) e^{i\omega t} \tag{44.26}$$

is the standard form of dipole radiation.

Exercise IV.44

44.1. Derive Eq. 44.5 in detail from Eqs. 44.3 and 44.4.

44.2. Develop Eq. 44.7 in terms of exponential functions with imaginary exponents. Hence show that the solution can be regarded as a superposition of progressive waves propagated in directions parallel to the vectors whose components are $(\pm k_1, \pm k_2, \pm k_3)$. Discuss these directions in terms of reflections at rectangular boundaries parallel to the coordinate planes.

44.3. Obtain Eq. 44.13 from Eqs. 44.3 and 44.10 as outlined in the text.

44.4. Use Eq. 44.9 to find the lowest frequency of electromagnetic resonance waves in a hollow cubic metal cavity with sides of 1 cm.

44.5. Develop the solutions of Eq. 44.3 in cylindrical polar coordinates when the potential function is permanently zero over the surface of a hollow circular cylinder.

IV.45 The Inhomogeneous Wave Equation. We shall consider the wave equation with a source function that is a simple oscillatory function of time, but an arbitrary function of position in space:

$$\nabla^2\Phi-(1/c^2)\partial^2\Phi/\partial t^2 = -(\rho/\epsilon)e^{i\omega t}, \tag{45.1}$$

where ρ is some function of position only. The solution of equations with more general time-dependent sources may be found by combining solutions of equations with simple harmonic sources. We assume that the potential Φ has the form $Ue^{i\omega t}$, as in Eq. 44.2, and substitute this into Eq. 45.1 to find the Helmholtz equation for U:

$$\nabla^2 U+k^2 U = -\rho/\epsilon, \tag{45.2}$$

with $k=\omega/c$.

Suppose the source function can be expressed as a linear combination of some more convenient functions, in the form

$$\rho(\mathbf{r}) = \sum_s a_s g_s(\mathbf{r}), \tag{45.3}$$

where a_s are constants, and $g_s(\mathbf{r})$ are the new functions. Let $G_s(\mathbf{r})$ be the solutions of the equations

$$\nabla^2 G_s+k^2 G_s = -g_s/\epsilon. \tag{45.4}$$

If we can choose the functions $g_s(\mathbf{r})$ strategically, it may be possible to write down the functions $G_s(\mathbf{r})$ easily enough. If now we multiply Eq. 45.4 by a_s and then sum all the s-values, we find, noting the linearity of the Laplacian operator:

$$\nabla^2\sum_s a_s G_s+k^2\sum_s a_s G_s = -(1/\epsilon)\sum_s a_s g_s = -\rho/\epsilon. \tag{45.5}$$

Comparing this with Eq. 45.2, we note that the function

$$U(\mathbf{r}) = \sum_s a_s G_s(\mathbf{r}) \tag{45.6}$$

is a solution of the inhomogeneous equation.

Green's method of solving the inhomogeneous equation, Eq. 45.2, consists in a particular choice of the functions $g_s(\mathbf{r})$, namely, the Dirac delta functions with their peaks located at various points $\mathbf{r}_s$:

Green's source functions:

$$g_s(\mathbf{r}) = \delta(\mathbf{r}-\mathbf{r}_s). \tag{45.7}$$

To satisfy Eq. 45.3, the coefficients a_s take on the values

$$a_s = \rho(\mathbf{r}_s)d\tau_s, \tag{45.8}$$

$\rho(\mathbf{r}_s)$ being the value of the source function $\rho(\mathbf{r})$ at the position $\mathbf{r}_s$, and $d\tau_s$ a small volume element around the point $\mathbf{r}_s$. We let the points $\mathbf{r}_s$ become indefinitely more crowded together, and the volume elements $d\tau_s$ get smaller, so that Eq. 45.3 becomes an integral:

$$\rho(\mathbf{r}) = \iiint \rho(\mathbf{r}_s)\delta(\mathbf{r}-\mathbf{r}_s)d\tau_s, \tag{45.9}$$

which is an identity from the definition of the Dirac delta function. With Green's source functions, the function G_s is a solution of the equation (see Eq. 45.4):

$$\nabla^2 G_s(\mathbf{r})+k^2 G_s(\mathbf{r}) = -\delta(\mathbf{r}-\mathbf{r}_s)/\epsilon \tag{45.10}$$

and the solution, Eq. 45.6, of the inhomogeneous equation is, from Eq. 45.8:

$$U(\mathbf{r}) = \iiint \rho(\mathbf{r}_s)G_s(\mathbf{r})d\tau_s. \tag{45.11}$$

To make the notation more explicit, we generally write $G(\mathbf{r}, \mathbf{r}_s)$ for $G_s(\mathbf{r})$ and so write

$$U(\mathbf{r}) = \iiint \rho(\mathbf{r}_s)G(\mathbf{r}, \mathbf{r}_s)d\tau_s \tag{45.12}$$

for the solution in terms of Green's function $G(\mathbf{r}, \mathbf{r}_s)$.

The central problem is now to find the solution of Green's equation, Eq. 45.10, for Green's function $G(\mathbf{r}, \mathbf{r}_s)$. If we compare Eq. 45.2 with Poisson's equation, Eq. 34.17, and the solution, Eq. 45.12, with the solution of Poisson's equation, Eq. 37.1, we see that we have already found Green's function for the special case when $k=0$:

$$k = 0: \qquad G(\mathbf{r}, \mathbf{r}_s) = (1/4\pi\epsilon)(1/|\mathbf{r}-\mathbf{r}_s|). \tag{45.13}$$

Rewriting Poisson's equation in the present notation, we say that Eq. 45.13 is the solution of Green's equation:

$$\nabla^2 G(\mathbf{r}, \mathbf{r}_s) = -\delta(\mathbf{r}-\mathbf{r}_s)/\epsilon. \tag{45.14}$$

Remember that $(1/r)$, and therefore Eq. 45.13, is a solution of Laplace's equation except at the origin where the singularity exists. The

Dirac delta function takes care of the singularity at the origin in Eq. 45.14.

The dimensionality of the problem happens to make a significant difference to Green's functions, and we shall consider one, two, and three-dimensional problems separately. It is clear that Eq. 45.13 is Green's function for Poisson's equation in three dimensions; and before undertaking the study of Helmholtz' equation in one or more dimensions, we shall first consider a problem that results in a one-dimensional Poisson equation.

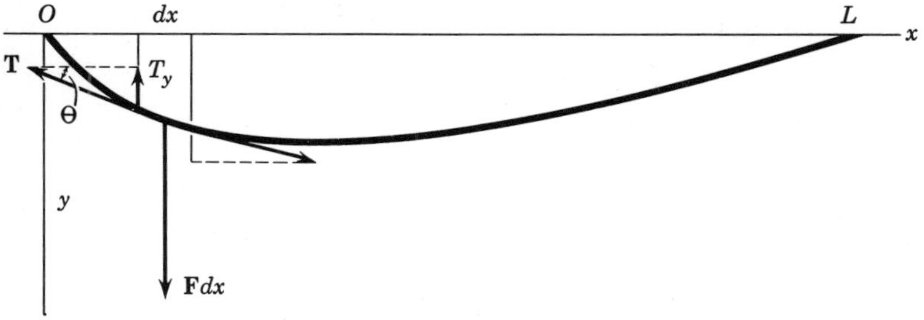

FIG. 45.1. String under continuous load **F** per unit length, Eq. 45.15, where $T_y = T dy/dx$ and $Fdx = -dT_y$

Let a flexible string of length L be stretched by a tension T, and a continuous load $F(x)$ per unit length act transversely to produce a curve $y(x)$ for the shape of the string. Provided the load is nowhere too great, it is easy to prove that the shape of the string must satisfy the differential equation:

$$d^2y/dx^2 = -(1/T)F(x). \tag{45.15}$$

This is a one-dimensional Poisson equation with a source function proportional to $F(x)$, and we proceed to use Green's method to solve it. Green's function is a solution of the equation

$$d^2G/dx^2 = -\delta(x-x_s)/T, \tag{45.16}$$

and the analog of Eq. 45.12 for the solution of Eq. 45.15 is

$$y(x) = \int F(x_s)G(x, x_s)dx_s. \tag{45.17}$$

Green's function evidently represents the shape of the string when one pulls it to one side at the point x_s with a concentrated force of unit magnitude. Eq. 45.16 tells us that Green's function has a zero second derivative except at the point x_s, where the second derivative is infinite. Therefore $G(x, x_s)$ is a linear function of x, with a break in slope at the point x_s. To find this break in slope precisely, we can integrate Eq. 45.16 over a short interval surrounding the break point: $x_s - \epsilon \leqslant x \leqslant x_s + \epsilon$, to obtain

$$dG/dx \Big|_{x_s-\epsilon}^{x_s+\epsilon} = -(1/T)\int_{x_s-\epsilon}^{x_s+\epsilon} \delta(x-x_s)dx = -1/T. \qquad (45.18)$$

This tells us that the gradient dG/dx makes a discontinuous jump at x_s through an amount $-1/T$. In the given problem, we also want the solution that vanishes at both ends of the string, where $x=0$ and $x=L$.

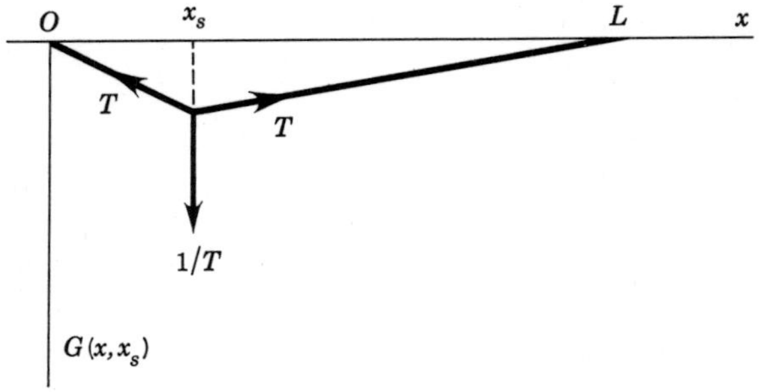

FIG. 45.2. Green's function for the loaded string, Eq. 45.19

By inspection we can see that the following solution satisfies both the boundary conditions and has the right discontinuity at the load point:

$$\left.\begin{array}{ll} 0 \leqslant x \leqslant x_s: & G(x, x_s) = x(L-x_s)/LT \\ x_s \leqslant x \leqslant L: & G(x, x_s) = x_s(L-x)/LT \end{array}\right\}. \qquad (45.19)$$

Using this Green's function in Eq. 45.17 yields the general solution for an arbitrary load function $F(x)$:

$$y(x) = (1/T)\int_0^x F(x_s)[x_s(L-x)/L]dx_s + (1/T)\int_x^L F(x_s)[x(L-x_s)/L]dx_s. \qquad (45.20)$$

Note especially that, in the first integral here, $x_s < x$, so we have to use the second part of Green's function from Eq. 45.19, while in the second integral, it is the first part of Green's function because $x_s > x$.

Exercise IV.45

45.1. Write out the three-dimensional delta function $\delta(\mathbf{r}-\mathbf{r}_s)$ as the product of three one-dimensional delta functions $\delta(x-x_s)$, $\delta(y-y_s)$, and $\delta(z-z_s)$, and prove Eq. 45.9 by using the property given in Eq. 3.27 in each of the three dimensions.

45.2. Prove that the shape of a uniformly loaded string is

$$y = (W/T)x(L-x)/2,$$

where L is the length, W is the load per unit length, and T is the tension in the string.

45.3. If a number of discrete loads are added to the string, note the linearity of the equation and show that the correct solution is obtained by adding the individual solutions for the separate loads. Work out the shape of the string under a uniform load W per unit length plus a concentrated load LW at the center.

45.4. Consider the wave equation with a time dependent source

$$\nabla^2\Phi-(1/c^2)\partial^2\Phi/\partial t^2 = -\rho(\mathbf{r}, t)/\epsilon. \tag{A}$$

Let $\rho(\mathbf{r}, t)$ be periodic in time with period τ and be represented by a Fourier series (*cf.* Eq. 3.23):

$$\rho(\mathbf{r}, t) = \sum_n A(\mathbf{r})(1/\tau)^{1/2}\, e^{2\pi nit/\tau}. \tag{B}$$

Prove that the solution is $\Phi=\sum_n U_n(r)\, e^{2\pi nit/\tau}$, where $U_n(\mathbf{r})$ is a solution of the Helmholtz equation:

$$\nabla^2U_n+(2\pi n/c\tau)^2U_n = -A_n(\mathbf{r})(1/\tau)^{1/2}/\epsilon. \tag{C}$$

45.5. A cable having a mass of 1.0 gram per cm length is stretched between two points at the same level 30 meters apart; the tension is adjusted to raise the mid-point of the cable to a point 20 cm below the level of the two ends. Find the tension.

A steady wind exerts a uniform horizontal load perpendicular to its length on the cable equal to one half its weight. What is the position of the center of the cable, and what is its shape?

If the tension in the cable, when undisturbed, is one half its breaking strength, what is the maximum uniform horizontal load that may be applied without breaking the cable?

IV.46 The Inhomogeneous Wave Equation, *continued*. Consider again the loaded string discussed in the last section, and let the load depend harmonically on time at all points:

$$\partial^2y/\partial x^2-(1/c^2)\partial^2y/\partial t^2 = -(1/T)F(x)e^{i\omega t}. \tag{46.1}$$

If we seek oscillatory solutions having the same frequency as the load:

$$y(x, t) = y(x)e^{i\omega t}, \tag{46.2}$$

we find the Helmholtz equation for the amplitude as a function of position:

$$d^2y(x)/dx^2+k^2y(x) = -F(x)/T, \tag{46.3}$$

with $k=\omega/c$. Green's function for this problem is the solution of

$$d^2G(x, x_s)/dx^2+k^2G(x, x_s) = -\delta(x-x_s)/T \tag{46.4}$$

and the solution of Eq. 46.3 is then

$$y(x) = \int G(x, x_s)F(x_s)dx_s. \tag{46.5}$$

From the discussion of §45, we expect Green's function to be discontinuous at the point x_s, which is the point of application of the concentrated load in Eq. 46.4. To find this discontinuity, we integrate Eq. 46.4 over a small interval including the point x_s:

$$dG/dx\Big|_{x_s-\epsilon}^{x_s+\epsilon} + k^2 \int_{x_s-\epsilon}^{x_s+\epsilon} G(x, x_s)dx = -1/T. \tag{46.6}$$

From physical considerations, the discontinuity in G is entirely in its derivative, and the magnitude is continuous across the point of application of the load because the string does not fracture there. It follows that the second term on the left of the last equation goes to zero when ϵ is taken indefinitely small. Therefore this equation states that the gradient of G is discontinuous by the amount $-1/T$ in passing over the point x_s. We now have the function G satisfying the equation

$$d^2G/dx^2 + k^2G = 0, \text{ except at } x = x_s, \tag{46.7}$$

and at x_s it must have the discontinuity described by the equation

$$(dG/dx)_{x_s+\epsilon} - (dG/dx)_{x_s-\epsilon} = -1/T. \tag{46.8}$$

It is not difficult to verify that the following function satisfies all of these conditions and may be accepted as Green's function for the problem:

$$\left.\begin{array}{lll} x < x_s: & G(x, x_s) = -(i/2kT)e^{+ik(x-x_s)} \\ x > x_s: & G(x, x_s) = -(i/2kT)e^{-ik(x-x_s)} \end{array}\right\}. \tag{46.9}$$

We have not set up any boundary values at the ends of the string, and if we accept Eq. 46.9 as Green's function, we shall have to be content to treat the string as having an infinite length. If the string had a finite length and we required definite conditions at the ends, Green's function for the problem would be different from Eq. 46.9, although it would still have to satisfy Eqs. 46.7 and 46.8.

Combining Green's function with the time factor $e^{i\omega t}$, we see that the concentrated load at the point x_s produces a wave traveling to the left in the region $x < x_s$, and to the right in the region $x > x_s$; this is just what we would have expected intuitively.

Putting Eq. 46.9 into Eq. 46.5 gives the complete solution to the wave motion on an infinite string under the action of an oscillatory distributed load:

$$y(x, t) = -(i/2kT)e^{i\omega t}\left[\int_{-\infty}^{x} F(x_s)e^{-ik(x-x_s)}dx_s + \int_{x}^{\infty} F(x_s)e^{ik(x-x_s)}dx_s\right]. \tag{46.10}$$

In the first integral here, $x > x_s$, so we have to use the second part of

Green's function in Eq. 46.9, while in the second integral we use the first part of Green's function because $x<x_s$. When we know the amplitude of the oscillatory load as a function of position, $F(x_s)$, we can solve the problem by evaluating the integrals in Eq. 46.10.

It is easy to set up an interesting problem that leads to a two-dimensional Helmholtz equation. Thus let the flexible membrane discussed in §10 be subjected to a transverse force $F(x, y, t)$ per unit area, so that it suffers transverse displacements $u(x, y, t)$ that obey the differential equation:

$$\partial^2u/\partial x^2+\partial^2u/\partial y^2-(1/c^2)\partial^2u/\partial t^2 = -F(x, y, t)/S, \qquad (46.11)$$

where S is the tension in the membrane. Let the load be simple harmonic in the time, writing

$$F(x, y, t)/S = f(x, y)e^{i\omega t}, \qquad (46.12)$$

and seek solutions of Eq. 46.11 having the form

$$u(x, y, t) = \Phi(x, y)e^{i\omega t}. \qquad (46.13)$$

Then Φ satisfies the inhomogeneous Helmholtz equation:

$$\partial^2\Phi/\partial x^2+\partial^2\Phi/\partial y^2+k^2\Phi = -f(x, y), \qquad (46.14)$$

with $k=\omega/c$. The equation for Green's function for this problem is then

$$\partial^2G/\partial x^2+\partial^2G/\partial y^2+k^2G = -\delta(x-x_s)\delta(y-y_s). \qquad (46.15)$$

Here G is a function of the form $G(\mathbf{r}, \mathbf{r}_s)$, where $r^2=x^2+y^2$, *etc.*, the radius being in the plane of x, y. We shall consider the problem in polar coordinates and rewrite Eq. 46.15 in the form

$$(1/r)(d/dr)(rdG/dr)+k^2G = -\delta(\mathbf{r}-\mathbf{r}_s). \qquad (46.16)$$

This equation is homogeneous except at the singularity $\mathbf{r}=\mathbf{r}_s$. To find the singularity in G at this point, we may integrate Eq. 46.16 over a circle of radius r about $\mathbf{r}_s$ as center, obtaining

$$2\pi r(dG/dr)+k^2\int 2\pi rGdr = -1. \qquad (46.17)$$

We may guess that the second term becomes negligible when $r\to 0$, and so we would have

$$\text{Lim: } r\to 0: \qquad 2\pi(dG/dr) = -1/r$$

or

$$G\to -(1/2\pi)\ln r. \qquad (46.18)$$

The function G then has a logarithmic infinity at the source point. This verifies that the second term in Eq. 46.17 is in fact negligible.

Finally we need a solution of the Helmholtz equation, Eq. 46.16, with zero on the right-hand side except at the singularity. The homogeneous Helmholtz equation here is in fact a Bessel equation of zero order, and the solution that behaves like Eq. 46.18 turns out to be:

$$G(\mathbf{r}, \mathbf{r}_s) = \tfrac{1}{4}i[J_0(kr)+iN_0(kr)], \tag{46.19}$$

where J_0 and N_0 are the Bessel functions of the first and second kind, respectively, and r is the distance from source point to field point, $|\mathbf{r}-\mathbf{r}_s|$. When r is large this solution approaches the following:

$$r \to \infty: \qquad G \to \tfrac{1}{2}(2\pi kr)^{-1/2} \exp[-i(kr+\tfrac{1}{4}\pi)]. \tag{46.20}$$

When multiplied by the time factor $e^{i\omega t}$, this becomes a propagated wave. At intermediate distances the wave form is complicated: there is always a wake trailing behind the wave front whose shape changes during propagation.

The solution of Eq. 46.11 with oscillatory source function is now

$$u(\mathbf{r}, t) = u_0 e^{i\omega t} 2\pi \int f(\mathbf{r}_s) G(\mathbf{r}, \mathbf{r}_s) r_s dr_s. \tag{46.21}$$

This solution goes out to infinity. If we had imposed boundary conditions at a finite distance, the equation for Green's function would have had these boundary conditions and Green's function would have been different.

Let us finally return to the three-dimensional inhomogeneous wave equation satisfied by the electrical potential, Eq. 45.1, and the corresponding Helmholtz equation, Eq. 45.2. The equation for Green's function in this case is Eq. 45.10, and we wish to find the nature of the singularity in G at the source point $\mathbf{r}_s$. To do so we may integrate the equation over the volume of a small sphere of radius R with its center at the singularity $\mathbf{r}_s$, and allow R to approach zero after the integration. We have to start from Eq. 45.10:

$$\nabla^2 G(\mathbf{r}, \mathbf{r}_s) + k^2 G(\mathbf{r}, \mathbf{r}_s) = -\delta(\mathbf{r}-\mathbf{r}_s)/\epsilon$$

and obtain

$$\iiint \nabla^2 G d\tau + k^2 \iiint G d\tau = -1/\epsilon. \tag{46.22}$$

We shall guess that the second term on the left is negligible when $R \to 0$. The first term transforms by Gauss' theorem into a surface integral over the surface of the sphere, and we assume G to be symmetrical so that the integral becomes asymptotically $4\pi R^2 dG/dr$. Thus Eq. 46.22 yields

$$dG/dR = -1/4\pi R^2 \epsilon. \tag{46.23}$$

We expect, then, that as $R \to 0$, $G \to 1/4\pi R\epsilon$. In this case the second term in Eq. 46.22 does indeed become negligible as was assumed. The solution of Helmholtz' equation that approaches Eq. 46.23 in the limit of small R is:

$$G(\mathbf{r}, \mathbf{r}_s) = (1/4\pi\epsilon)(1/R)e^{-ikR}, \tag{46.24}$$

with $R = |\mathbf{r} - \mathbf{r}_s|$.

The corresponding solution of the inhomogeneous wave equation is now

$$\Phi(\mathbf{r}, t) = (1/4\pi\epsilon)e^{i\omega t}\iiint \rho(\mathbf{r}_s)(1/R)e^{-ikR}d\tau_s. \tag{46.25}$$

Exercise IV.46

46.1. Verify that Eq. 46.9 satisfies Eqs. 46.7 and 46.8.

46.2. In Eq. 46.10, let $F(x) = W$, a constant, over a finite length of the string between $x=0$ and $x=a$, and zero elsewhere. Find the amplitude of the wave propagated in the region $x > a$. What happens when the wavelength $\lambda = 2\pi/k$ equals (i) twice the distance a, (ii) just the distance a?

46.3. Verify that Eq. 46.24 satisfies Eq. 45.10.

46.4. Give the detailed arguments culminating in Eq. 46.21, and the similar argument culminating in Eq. 46.25, omitted from the discussion in the text.

46.5. As an application of Eq. 46.25, let the field point $\mathbf{r}$ be taken as the origin at the center of a hollow sphere of radius R_0, and let the source density be zero inside this sphere, and depend only on distance from the center outside the sphere, according to the formula

$$\rho(\mathbf{r}_s) = (1/R)e^{-aR} \ (R > R_0)$$

where a is a constant. Find the potential at the center. Discuss the physical meaning of the results in the two limiting cases given by letting a approach either zero or infinity.

IV.47 The General Boundary Value Problem. Early in this chapter we proved a uniqueness theorem for solutions of the Laplace equation; and we have been implicitly using uniqueness theorems for other more general equations, without proof, throughout the later sections of the chapter, relying mostly on physical intuition for their validity. The general theory of uniqueness under boundary conditions is a well-developed subject in pure mathematics, but it would take us too far afield from our present course to go thoroughly into the subject here. It is, however, quite essential to have some idea about the problem, and we present here, without any pretense at proof, a summary of the uniqueness theorems that apply to the differential equations of interest in theoretical physics.

It is necessary to consider two types of boundary, called the open and the closed boundary respectively. In the open type of boundary the region of interest extends indefinitely in one or more dimensions, and there is no specification of the solution in these directions. The closed type of boundary completely surrounds the region in which the solution is sought, and the boundary conditions are specified in all directions.

It is also necessary to consider three types of boundary conditions:

Dirichlet conditions: the value of the solution is given over the boundary.

Neumann conditions: the normal gradient of the solution is given over the boundary.

Cauchy conditions: both value and normal derivative of the solution are given over the boundary.

For certain differential equations the uniqueness of the solution can be assured equally well either by Dirichlet or Neumann boundary conditions; for others neither Dirichlet nor Neumann conditions are sufficient to determine the solution, and the stronger Cauchy condition is required. The differential equations may be classed under three heads:

Elliptic equations: here the differential operators are homogeneous in the second partial derivatives—they all occur with the same sign; examples are Laplace's and Poisson's equations and Helmholtz' equation.

Hyperbolic equations: these have one of the second-order derivatives opposite in sign to the others, an example being the wave equation.

Parabolic equations: the derivative with respect to one of the variables occurs only in first order, *e.g.* the diffusion equation. For more general equations, a stricter definition of the above categories is necessary, but the above suffices for the equations we have used. The following table summarizes the uniqueness theorems that apply to the several types of boundary, boundary conditions, and differential equations.

TABLE OF UNIQUENESS THEOREMS

<table>
<tr><th>Boundary Condition</th><th>Type of Boundary</th><th>Hyperbolic Equations</th><th>Elliptic Equations</th><th>Parabolic Equations</th></tr>
<tr><td>Dirichlet or Neumann</td><td>open</td><td colspan="2">insufficient specification</td><td>unique solution</td></tr>
<tr><td>Dirichlet or Neumann</td><td>closed</td><td>therefore many solutions</td><td>unique solution</td><td rowspan="2">over-</td></tr>
<tr><td>Cauchy</td><td>open</td><td>unique solution</td><td>specification</td></tr>
<tr><td>Cauchy</td><td>closed</td><td colspan="3">generally no solutions are possible</td></tr>
</table>

From this table we can see for instance that the uniqueness theorem for the diffusion equation, which is parabolic, requires the use of an open boundary over which either the value or the normal gradient is specified. Consider the equation with one space dimension and time as the independent variables:

$$a^2\partial^2\rho/dx^2 = \partial\rho/\partial t. \tag{47.1}$$

The domain of the solution is a two-dimensional region with x and t as the dimensions. The boundary is left open in one time direction, *i.e.* the future, but includes all x-values between two stated limits, finite or infinite. If the x-boundary extends over the whole x-axis at an initial time, $t=0$, the initial value of the function $\rho(x, t)$ at $t=0$ must be prescribed for the Dirichlet condition, while the rate of change, $\partial\rho/\partial t$, must be prescribed at all values of x, at $t=0$, for the Neumann condition. The future values of $\rho(x, t)$ for all values of x and t are then determined by the unique solution.

If the boundary extends only over a finite distance on the x-axis, in the range $a \leqslant x \leqslant b$, then either (Dirichlet conditions) the function must be specified over all x between a and b and also at $x=a$ and $x=b$ for all times, or (Neumann conditions) the time derivative of the function must be specified at $t=0$ over x between a and b, and the space derivative $\partial\rho/\partial x$ must be specified at all times along $x=a$ and $x=b$. The solution is then determined at all times for intermediate positions.

The uniqueness theorem for the wave equation, which is hyperbolic, requires Cauchy conditions on an open boundary. Consider the simple case

$$c^2\partial^2\Phi/\partial x^2 = \partial^2\Phi/\partial t^2.$$

The boundary is like the one just discussed for the diffusion equation, but the conditions must be the stronger ones: we must know both the function and its time derivative on the x-axis at the initial time in order to determine the solution uniquely for all time in the future.

If in either of the above examples we attempted to use a closed boundary and to specify the function both at the initial time and at some future time, we would in general find no solution that would fit such conditions.

The theorem that a differential equation possesses a unique solution does not guarantee that the solution is easy to determine in practice. It happens that in this introductory text all the equations we have met are relatively elementary, and their solutions are well known. However in more advanced work the physicist is frequently faced with differential equations whose solutions are not well known; some indication of this was given in the section on hydrodynamics where the exact differential equations of motion were seen to be nonlinear, the linear approximation being used instead. In nonlinear equations the solutions may suffer from instability. By this is meant that a vanishingly small change in boundary values may induce a catastrophic change in the solution. When this is the case it is impossible to find a unique solution because, of course, the boundary values are always prescribed by experimental data, and these are subject to errors and uncertainties. The best that one can expect in such a situation is to find a whole family of possible solutions consistent with the boundary information available and discuss the next step in terms of probability theory. The study of these instabilities is a fascinating and relatively new field of applied mathematics.

Exercise IV.47

47.1. In the light of the fact that many basic laws of physics are nonlinear differential equations, discuss the implications of the instability of their solutions for the classical theory of deterministic causality.

References

The following texts contain useful material for collateral reading with the contents of this chapter.

Classical Electricity and Magnetism, W. K. H. Panofsky and M. Phillips, Addison-Wesley Publishing Co., Inc., Reading, Mass., 1955.

Electrodynamics, A. Sommerfeld, Academic Press, Inc., New York, 1952.

Foundations of Potential Theory, O. D. Kellog, Dover Publications, Inc., New York, 1953.

Methods of Mathematical Physics, H. and B. S. Jeffreys, Cambridge University Press, 1956.

Methods of Theoretical Physics, P. M. Morse and H. Feshbach, McGraw-Hill Book Co., Inc., New York, 1953.

Theoretical Physics; Electromagnetism, F. W. Constant, Addison-Wesley Publishing Co., Inc., Reading, Mass., 1958.

Chapter V

THE THEORY OF RELATIVITY

Chapter V

THE THEORY OF RELATIVITY

V.48 The Principle of Invariance and the Concept of Absolute Velocity. One of the basic principles of theoretical physics is the assertion that general laws shall be expressed in a form that is true for all observers. This may be called the principle of invariance and is regarded as the formal expression of the scientist's conviction that the physical world has an existence outside of and independent of the observer's cognition of it. We must here state the principle of invariance more objectively.

The first example of the principle of invariance discussed in this text was in the theory of elasticity: we required both sides of the equation expressing Hooke's law in an isotropic solid to transform in the same way—*viz.* the tensor transformation—so that the equation would remain true in all coordinate systems derivable by pure rotations from one another. It is clear, however, that we were not worried here about different *observers* agreeing with each other; we were merely making sure that the law came out the same for all possible points of view, whether an observer were present or not. Throughout Chapter II on continuum theory of matter we kept this form of the principle of invariance in mind: the laws of physics are to remain invariant under all pure rotations in coordinate space. In Chapter III we extended the principle of invariance to include more general transformations—namely, those between generalized coordinates both in configuration space and in momentum space, and finally for the contact transformations in phase space. There was no suggestion here either that any of these generalized coordinate systems would have to be associated with some human observer. The invariance of the laws of motion in the Hamiltonian variational form, under contact transformations, was arranged purely as a mathematical convenience in handling complicated problems; it was an operational rule of the technique and had nothing essential to do with any connection with human observers. In Chapter IV on field theory, we have not discussed invariance explicitly, but we should have noticed that Maxwell's laws were expressed in vector form so that they are in fact invariant for all coordinate rotations. We have seen that it is possible to express the laws of electromagnetism in a Hamiltonian form, and so to make them invariant under the more general contact transformations.

In the light of these considerations, we shall revise the statement made at the opening of this section, leaving out the anthropomorphic term "observer": the principle of invariance states only that the general laws of physics shall be expressed in a form that is invariant under as wide a class of coordinate transformations as possible.

In elementary mechanics another aspect of the principle of invariance is significant, *viz.* the invariance of the laws of motion with respect to transformations among coordinate systems having relative uniform motion with one another, the so-called *inertial systems*. In classical mechanics one assumes the transformation relations:

$$t' = t; \quad x' = x - vt; \quad y' = y; \quad z' = z. \tag{48.1}$$

It can be seen that this transformation leaves Newton's laws of motion unchanged in form. Thus from

$$F_1 = md^2x/dt^2 \tag{48.2}$$

one derives

$$F_1' = m'd^2x'/dt'^2, \tag{48.3}$$

provided that

$$F' = F \tag{48.4}$$

and

$$m' = m. \tag{48.5}$$

Force and mass are unchanged by relative uniform motion on Newton's theory. No uniform motion of the system makes any difference to Newton's laws, so no purely mechanical experiments could ever determine anything more than relative motion between physical objects. So far as Newton's laws of motion are concerned, the question of the existence of an absolute standard of rest or of uniform motion was quite meaningless. The same however cannot be asserted of Maxwell's laws of the electromagnetic field. One can see this in several ways. For instance, consider plane electromagnetic waves transmitted in the x-direction according to the equation:

$$\partial^2\Phi/\partial x^2 = (1/c^2)\partial^2\Phi/\partial t^2. \tag{48.6}$$

The transformation of Eq. 48.1 implies that

$$\partial^2/\partial x'^2 = \partial^2/\partial x^2; \quad \partial^2/\partial t'^2 = \partial^2/\partial t^2 + 2v\partial^2/\partial x\partial t + v^2\partial^2/\partial x^2. \tag{48.7}$$

Taking Φ as scalar, independent of relative velocity, it is easy to prove that Eq. 48.6 transforms into

$$(c-v)^2\partial^2\Phi/\partial x'^2 = \partial^2\Phi/\partial t'^2. \tag{48.8}$$

Therefore the relative motion of the two systems affects the velocity of propagation of the waves just as if they were being transmitted by some material medium through which the observers were moving. The true velocity of propagation should be observed only when the measurement is made in a system at rest in the transmitting medium. Relative velocity v through the medium should show up in the velocity of propagation as it does in Eq. 48.8.

Because of these results, it was hoped that by measuring the speed of light from stellar sources relative to the earth at different seasons of the year, we could determine the motion of the earth through the ether, and hence the absolute velocity of the earth in space. As is well known, between 1881, when Michelson first devised a sufficiently precise method, and 1933, when D. C. Miller reported a very slight effect, repeated experiments were conducted by interferometric methods capable of detecting effects of an order one hundred times smaller than the expected one. The concensus is that no effect has been detected that can be attributed to motion through the ether. This direct attack was not the only method used. In the discussion given in Chapter IV, concerning the theory of the electromagnetic field, it was seen that the velocity of propagation of light is $c=(\epsilon\mu)^{-1/2}$. The two parameters ϵ and μ are scale factors in the expressions for the inverse square law for the electrostatic and magnetostatic fields, respectively. One would therefore expect that motion relative to the preferred frame of reference, by changing the apparent velocity of light, would disturb the inverse square law in some way. Many experimental attempts were made to detect this disturbance, based in detail on Maxwell's equations, but all failed completely.

The ether hypothesis was eventually abandoned, because it had to be endowed with many mutually contradictory properties. The paradox still remained, however, that relative velocity between two reference systems ought to show some effect on the wave propagation. Such an effect has never been found. Early efforts to explain this failure lead to the *Fitzgerald-Lorentz contraction hypothesis.* According to this, all samples of matter, including rigid bodies and our standard measuring instruments, suffer a contraction in their linear dimensions in the direction of their motion through the preferred frame of reference. When, for example, in the Michelson-Morley experiment we attempt to find the difference in optical path length of two rays of light, one transmitted perpendicular and the other parallel to the velocity v, the contraction in the latter direction is such as to annul *identically* the otherwise expected effect. This hypothesis was so subtle in its mathematical detail that *all* the expected effects of the motion through the preferred frame of reference could be annulled identically. Yet the

Fitzgerald-Lorentz contraction, by its very universal nature, could never be directly observed or verified. This status of the preferred frame of reference was particularly absurd: its existence could be granted so long as we also admitted other effects which, themselves unobservable, ensured that the preferred frame shall also remain undetectable. The only logical course was to deny the existence of the preferred frame in the first place, and to admit that the electromagnetic field laws are not affected by relative motion, but are in principle true—as they are in practice!—for all inertial systems, with the same universal value for the velocity of light. Taking this inevitable step, we must reject the transformations of Eq. 48.1. Lorentz then discovered the required modifications of Eq. 48.1 that do leave the electromagnetic field laws invariant. We shall discuss these transformations in detail shortly. Einstein gave the Lorentz transformations a physical meaning. He also pointed out that Newton's laws of motion are not invariant for the Lorentz transformations, so that mechanical experiments could lead to a discovery of a preferred frame of reference if Newton's laws are true. Einstein was able to modify Newton's laws of motion to make them invariant under the Lorentz transformations, and so ensured the nonexistence of a preferred frame of reference both for electromagnetic and mechanical experiments.

In our discussion of these developments we shall not follow their historical sequence. Instead we shall give a modernized version of Minkowski's four-dimensional geometry, as the approach most in harmony with the general point of view of the present text.

Exercise V.48

48.1. The Fitzgerald-Lorentz hypothesis stated that the lengths of all material bodies in motion through the ether contracted in their direction of motion in the ratio $(1-v^2/c^2)^{1/2}:1$, where v is the velocity through the ether. A body is spherical in shape when at rest. Find its shape when in motion.

48.2. A spherical conductor of radius a contains one electronic charge. Find the energy of its electromagnetic field outside the sphere by integrating the energy density (a) when the sphere is at rest, and (b) when the sphere is distorted by motion with speed v. Find the difference in energy as a function of v when v^2/c^2 is negligible compared with unity, and when v^2/c^2 is not neglected.

V.49 The Four-Dimensional Algebra of Space-Time. It will be recalled that in Chapter I we gave the aim of mathematical physics as the construction of abstract mathematical objects for the description of the results of physical measurements. The success of this scheme depends on the correct interpretation of the physical results and their proper identification with the appropriate mathematical objects. Thus, time, mass, and energy were identified with scalar invariants; displacement, velocity, and force with 3-vectors; stress, strain, and linear transformations with 3-by-3 tensors. If these identifications are correct, observable relationships among the physical objects must be predicted by the algebraic properties of the corresponding mathematical objects.

We now interpret the dilemma posed by the negative results of the Michelson-Morley experiments as a failure of the physical quantity called velocity to behave exactly like the mathematical object called a vector. The sum of two parallel mathematical vectors of magnitude c and v is by definition a mathematical vector of magnitude $v+c$. The physical vector that results from measuring the velocity of light, added to the physical vector that results from measuring the velocity of a moving reference system, yields still the same velocity of light; these vectors do not behave like mathematical vectors, and therefore we must proceed to construct, or invent, a new mathematical object that describes the physical situation more precisely.

In our remarks near the end of §2 in Chapter I, we hinted that space and time must be joined in a four-dimensional vector. We can see qualitatively why this should be so. The basic elements of the physical world are *events* that occur in space and time, and the data required to locate an event, or to differentiate between two events,

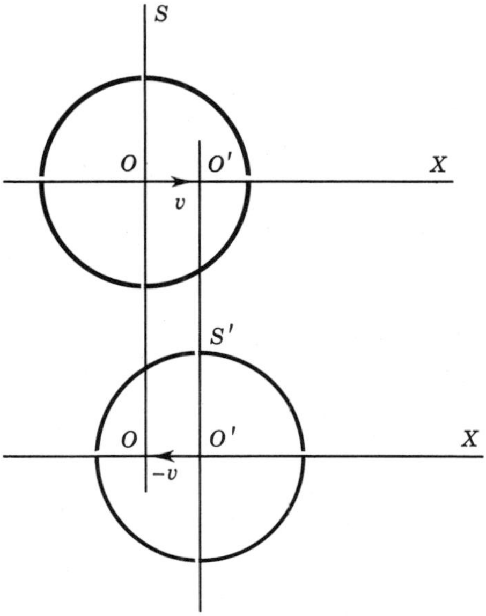

FIG. 49.1. A single pulse of radiation drawn from two relatively moving points of view

consists of four numbers, t, x, y, and z in the familiar notation. The essence of Minkowski's theory is that the four numbers must be treated as an ordered set of four components of a vector, and that a suitable algebra of such four-dimensional vectors must be constructed to describe their observed relationships. The separation of the four items into one set of three space coordinates, treated as a mathematical 3-vector, and one scalar time, is definitely in conflict with the observed behavior of the velocity of light. In fact, a more careful examination of the behavior of light waves leads to a definite clue to the appropriate scheme for combining time and space in one four-dimensional algebra.

Consider two reference systems S and S' having a relative velocity v along their common x-axis, their origins coinciding with each other at the time $t=0$ when they both start counting time. Let a source of light at the common origin emit a sharp pulse at the time $t=0$. Then the accepted fact is that this one pulse of light, as it propagates, will be measured as a spherical wave front on both systems, traveling out with the speed c in all directions. Each agrees that any motion of the source makes no difference to the speed c. But on system S the center of the sphere is located at O, the origin of the system S, while on system S' the sphere has its center at O' which is moving relative to the point O. How can these two conflicting descriptions of the same phenomena be reconciled? The analytic expressions for the wave front given on the two systems S and S' are, respectively:

$$\left.\begin{aligned} c^2t^2 &= x^2+y^2+z^2 \\ \text{and} \qquad c^2t'^2 &= x'^2+y'^2+z'^2 \end{aligned}\right\}, \qquad \begin{matrix}(49.1)\\(49.2)\end{matrix}$$

i.e. the equation of a sphere moving out with speed c in all directions; (t, x, y, z) are the time and coordinates measured by S, whereas (t', x', y', z') are the time and coordinates as measured by S'. If we try the Newtonian transformation of Eq. 48.1 on these equations, it is at once obvious that they are mutually inconsistent. If we look for a more general linear transformation that reduces to Eq. 48.1 when v approaches zero, and also makes Eq. 49.1 consistent with Eq. 49.2, there is a unique result, namely:

$$\left.\begin{aligned} x' &= (x-vt)/(1-v^2/c^2)^{1/2}, & y' &= y \\ t' &= (t-vx/c^2)/(1-v^2/c^2)^{1/2}, & z' &= z \end{aligned}\right\}. \qquad (49.3)$$

When $v \ll c$ these are practically identical with Eq. 48.1, but when v becomes comparable with c they become very different; in particular it is clear that something spectacular would happen if v were to become equal to c: both x' and t' could become infinite. These transformations were discovered by H. A. Lorentz and are known by his name. Before discussing the physical interpretation of the Lorentz transformations, we shall consider their mathematical implications. It is clear at once that time cannot be considered a scalar under a Lorentz transformation, and it is also clear that the three space coordinates do not behave like a classical 3-vector, except in the limit when v/c is so small that the term $1-v^2/c^2$ can be equated practically to unity.

One of the first steps to make in finding the proper way of combining time and space into a four-dimensional vector algebra would be to find the scalar length for the 4-vector. In direct analogy with the definition of length elaborated in Chapter I, Eq. 2.4, one might expect that the square length of the space-time vector would be something like

$$c^2t^2+x^2+y^2+z^2,$$

the sum of the squares of the four components. The velocity of light is used to multiply the time to make this component have the same dimensionality of length as the other components; the velocity of light is used rather than any other velocity because, experimentally it is a universal constant. The Lorentz transformation however does not leave this quantity invariant—it is not a scalar and therefore not satisfactory for the length. Instead we can see that the following expression is invariant for the Lorentz transformation:

$$s^2 = c^2t^2-(x^2+y^2+z^2). \tag{49.4}$$

This could be defined as the square length of the interval whose four components are ct, x, y, z.

As in §2 a subscript notation is more convenient, and we shall represent the interval ct, x, y, z by the ordered set of four real numbers:

$$\mathbf{x} = (x_0, x_1, x_2, x_3). \tag{49.5}$$

The subscript zero is used for the time-component:

$$x_0 = ct, \quad x_1 = x, \quad x_2 = y, \quad x_3 = z. \tag{49.6}$$

The sum of two 4-vectors is defined precisely as in Eq. 2.1:

$$\mathbf{A}+\mathbf{B} \equiv (A_0+B_0, A_1+B_1, A_2+B_2, A_3+B_3) \tag{49.7}$$

Scalar products are defined differently, thus:

$$\mathbf{A}\cdot\mathbf{B} \equiv A_0B_0-(A_1B_1+A_2B_2+A_3B_3) \tag{49.8}$$

The vector product will not be used until we discuss matrix theory in four dimensions. The difference between Eq. 49.8 and the more obvious generalization of Eq. 2.2, where all the signs would be positive, sets the time-like component apart from the three space components and has the mathematical consequence that the length of a vector can be zero without any of the four components being zero. Such a

4-vector is called a null vector. Thus from Eq. 49.8 with $\mathbf{A}=\mathbf{B}$, we have a null vector $A \equiv \sqrt{\mathbf{A}\cdot\mathbf{A}}=0$ if

$$A_0^2 = A_1^2+A_2^2+A_3^2. \tag{49.9}$$

Comparing this with Eqs. 49.1 and 49.2, we see that the length of the interval between emission and absorption of a light pulse is zero. If A_0^2 is less than $A_1^2+A_2^2+A_3^2$, then A is imaginary, and the vector is called *spacelike*; while if A_0^2 is greater than $A_1^2+A_2^2+A_3^2$, it has a real length and is called *time-like*. If two events are separated by a space-like interval it would be necessary to travel at a speed faster than light to be present at both.

The angle between two 4-vectors could be defined just as in Eq. 2.5, with $\cos\theta=(\mathbf{A}\cdot\mathbf{B})/AB$. But because of the negative signs in the scalar product it is possible for this quantity to be greater than unity, so that the angle would often be imaginary. It is in fact more convenient to define the angle by means of the hyperbolic cosine, thus:

$$\cosh\phi = (\mathbf{A}\cdot\mathbf{B})/AB, \tag{49.10}$$

where ϕ is defined as the hyperbolic angle between the vectors $\mathbf{A}$ and $\mathbf{B}$. This always gives a real value to the angle between any two space-like vectors, or between any two time-like vectors; between a time-like and a space-like vector the angle is always imaginary.

It is convenient to introduce four unit vectors in analogy with Eq. 2.6:

$$\mathbf{i}_0 = (1, 0, 0, 0);\quad \mathbf{i}_1 = (0, 1, 0, 0);\quad \mathbf{i}_2 = (0, 0, 1, 0);\quad \mathbf{i}_3 = (0, 0, 0, 1). \tag{49.11}$$

From the scalar product defined in Eq. 49.8 and the length defined in Eq. 49.4, or equivalently:

$$A^2 = A_0^2-(A_1^2+A_2^2+A_3^2), \tag{49.12}$$

these are all unit vectors, and they satisfy the identities:

$$\mathbf{i}_\alpha\cdot\mathbf{i}_\beta = 0,\ \alpha \neq \beta;\quad \mathbf{i}_0\cdot\mathbf{i}_0 = 1;\quad \mathbf{i}_m\cdot\mathbf{i}_m = -1\quad (m = 1, 2, 3). \tag{49.13}$$

The length of $\mathbf{i}_0$ is unity, the length of $\mathbf{i}_m$ $(m=1, 2, 3)$ is $\sqrt{-1}$. Any 4-vector $\mathbf{A}$ can be expressed in terms of these unit vectors:

$$\mathbf{A} = \sum_\alpha A_\alpha \mathbf{i}_\alpha \tag{49.14}$$

where the Greek subscripts take all four values 0, 1, 2, 3.

To make use of this 4-vector algebra in the problems of relativity it is necessary to proceed with transformation theory, generalized from the treatment given in §4 for three dimensions. Let us introduce a new set of unit vectors $\mathbf{i}_\alpha'$ that satisfy the same conditions of Eq. 49.13. Any one vector $\mathbf{F}$ can be expressed in terms of either system of units:

$$\mathbf{F} = \sum_\alpha F_\alpha \mathbf{i}_\alpha = \sum_\alpha F_\alpha' \mathbf{i}_\alpha', \tag{49.15}$$

the analogy to Eq. 4.1. We take the scalar product of this equation with $\mathbf{i}_\beta'$, recall the negative signs in Eq. 49.13, and have

$$\sum_\alpha F_\alpha \mathbf{i}_\alpha \cdot \mathbf{i}_0' = F_0'; \quad \sum_\alpha F_\alpha \mathbf{i}_\alpha \cdot \mathbf{i}_m' = -F_m' \quad (m = 1, 2, 3). \tag{49.16}$$

At first we shall discuss only the simplest kind of transformation. Thus consider a plane rotation in one "plane" containing only time and one space dimension. The transformation Eq. 49.15 then reads

$$F_0 \mathbf{i}_0 + F_1 \mathbf{i}_1 = F_0' \mathbf{i}_0' + F_1' \mathbf{i}_1'; \tag{49.17}$$

$F_2 = F_2'$, $F_3 = F_3'$ and $\mathbf{i}_2' = \mathbf{i}_2$, $\mathbf{i}_3' = \mathbf{i}_3$. We take the scalar product of Eq. 49.17 with $\mathbf{i}_0'$ and with $\mathbf{i}_1'$, respectively, obtaining:

$$F_0 \mathbf{i}_0 \cdot \mathbf{i}_0' + F_1 \mathbf{i}_1 \cdot \mathbf{i}_0' = F_0', \quad F_0 \mathbf{i}_0 \cdot \mathbf{i}_1' + F_1 \mathbf{i}_1 \cdot \mathbf{i}_1' = -F_1'. \tag{49.18}$$

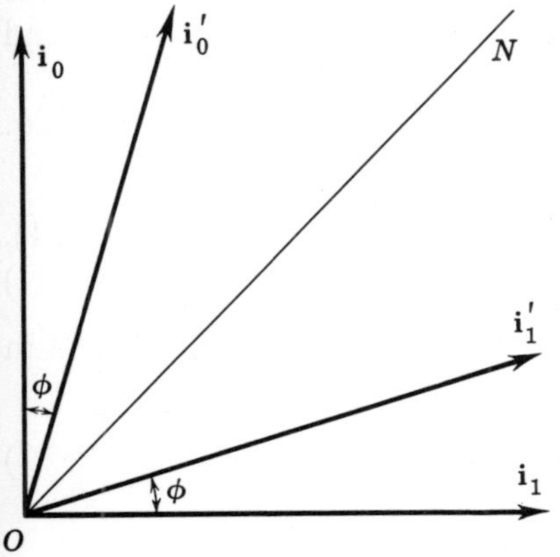

FIG. 49.2. The hyperbolic rotation of space-time, Eqs. 49.20 and 49.21

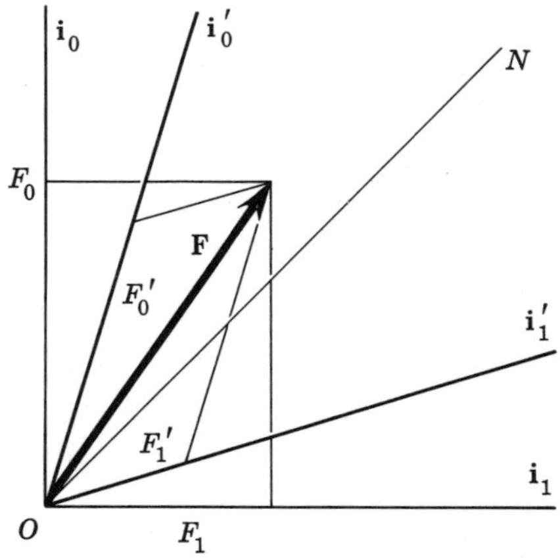

FIG. 49.3. An illustration of Eq. 49.27, the Lorentz transformation of the vector $\mathbf{F}$

Putting Eq. 49.18 into the right-hand side of Eq. 49.17, we have

$$F_0 \mathbf{i}_0 + F_1 \mathbf{i}_1 = \mathbf{i}_0'(F_0 \mathbf{i}_0 \cdot \mathbf{i}_0' + F_1 \mathbf{i}_1 \cdot \mathbf{i}_0') - \mathbf{i}_1'(F_0 \mathbf{i}_0 \cdot \mathbf{i}_1' + F_1 \mathbf{i}_1 \cdot \mathbf{i}_1'). \tag{49.19}$$

This must be an identity for any vector $\mathbf{F}$. Let $F_0 = 1$, $F_1 = 0$, then the identity becomes

$$\mathbf{i}_0 = \mathbf{i}_0'(\mathbf{i}_0 \cdot \mathbf{i}_0') - \mathbf{i}_1'(\mathbf{i}_0 \cdot \mathbf{i}_1'). \tag{49.20}$$

Again, let $F_0=0$, $F_1=1$, and we have instead

$$\mathbf{i}_1 = \mathbf{i}_0'(\mathbf{i}_1\cdot\mathbf{i}_0')-\mathbf{i}_1'(\mathbf{i}_1\cdot\mathbf{i}_1'). \tag{49.21}$$

Eqs. 49.20 and 49.21 are the transformation relations for the unit vectors. Take the scalar product of the two equations, and the scalar product of each equation with itself, using Eq. 49.13, both for the original and the new unit vectors, and we find:

$$\left.\begin{aligned} (\mathbf{i}_0\cdot\mathbf{i}_0')^2-(\mathbf{i}_0\cdot\mathbf{i}_1')^2 &= 1 \\ (\mathbf{i}_0\cdot\mathbf{i}_0')(\mathbf{i}_1\cdot\mathbf{i}_0')-(\mathbf{i}_0\cdot\mathbf{i}_1')(\mathbf{i}_1\cdot\mathbf{i}_1') &= 0 \\ (\mathbf{i}_1\cdot\mathbf{i}_0')^2-(\mathbf{i}_1\cdot\mathbf{i}_1')^2 &= -1 \end{aligned}\right\}. \tag{49.22}$$

Introducing the angles between the unit vectors, from the definition in Eq. 49.10, thus:

$$\left.\begin{aligned} \cosh\phi_{00} &= \mathbf{i}_0\cdot\mathbf{i}_0', & \cosh\phi_{11} &= -\mathbf{i}_1\cdot\mathbf{i}_1' \\ \cosh\phi_{01} &= (\mathbf{i}_0\cdot\mathbf{i}_1')/\sqrt{-1}, & \cosh\phi_{10} &= (\mathbf{i}_1\cdot\mathbf{i}_0')/\sqrt{-1} \end{aligned}\right\}, \tag{49.23}$$

we find

$$\left.\begin{aligned} \cosh^2\phi_{00} &= 1-\cosh^2\phi_{01} \\ \cosh\phi_{00}\cosh\phi_{10} &= -\cosh\phi_{01}\cosh\phi_{11} \\ \cosh^2\phi_{11} &= 1-\cosh^2\phi_{10} \end{aligned}\right\} \tag{49.24}$$

from Eq. 49.22. We conclude that

$$\left.\begin{aligned} \cosh\phi_{01} &= \pm\sqrt{-1}\sinh\phi_{00} \\ \cosh\phi_{10} &= \pm\sqrt{-1}\sinh\phi_{11} \end{aligned}\right\} \tag{49.25}$$

and then that

$$\begin{aligned} \tanh\phi_{00} &= \pm\tanh\phi_{11}, \quad \text{or} \\ \phi_{00} &= \pm\phi_{11}. \end{aligned} \tag{49.26}$$

The transformation can therefore be expressed in terms of a single angle ϕ written for ϕ_{00}. Using Eq. 49.23 in Eq. 49.18, the latter transformation becomes:

$$\left.\begin{aligned} F_0' &= \cosh\phi\cdot F_0\pm\sinh\phi\cdot F_1 \\ -F_1' &= \mp\sinh\phi\cdot F_0-\cosh\phi\cdot F_1 \end{aligned}\right\}. \tag{49.27}$$

To make use of this purely mathematical result of the algebra, we shall write

$$\tanh\phi = v/c, \tag{49.28}$$

where v is a relative velocity of two reference systems, and c the velocity of light, as before, and note that

$$\left.\begin{aligned} \cosh\phi &= 1/(1-v^2/c^2)^{1/2} \\ \sinh\phi &= (v/c)/(1-v^2/c^2)^{1/2} \end{aligned}\right\}. \tag{49.29}$$

Then let **F** be the interval (ct, x, y, z) in the familiar notation, and we find the transformation relation of Eq. 49.27 becomes

$$\left.\begin{aligned} t' &= (t \pm vx/c^2)/(1-v^2/c^2)^{1/2}, \\ x' &= (x \pm vt)/(1-v^2/c^2)^{1/2}, \\ y' &= y, \qquad z' = z. \end{aligned}\right\} \qquad (49.30)$$

These are exactly the Lorentz transformation equations, which we have already noted are needed to explain the constancy of the velocity of light. We may therefore conclude tentatively that our four-dimensional algebra is correct for space time. Relative velocity of two reference systems is connected with a hyperbolic rotation in the algebra, through Eq. 49.28. A more searching examination is required before we finally conclude that this is true. Before proceeding with this examination in the next section, we shall point out here that because the interval length is invariant under the Lorentz transformations, such transformations are incapable of changing a time-like vector into a space-like vector or vice versa—because this would involve altering the length from real to imaginary. Thus the time-like vectors are permanently separated from space-like vectors by the null vectors.

Exercise V.49

49.1. Use Eq. 49.3 to deduce Eq. 49.2 from Eq. 49.1.

49.2. Prove that the expression in Eq. 49.4 is invariant under the transformation of Eq. 49.3.

49.3. Find the angle between each pair of the unit vectors in Eq. 49.11 from the definition Eq. 49.10.

49.4. Find the angle between the following pairs of 4-vectors:

(a) (2, 1, 0, 0) and (2, 0, 0, 0);
(b) (2, 1, 0, 0) and (2, −1, 0, 0);
(c) (2, 0, 1, 0) and (2, 0, 0, 0);
(d) (2, 0, 1, 0) and (2, 0, 0, 1);
(e) (1, −2, 0, 0) and (1, 2, 0, 0);
(f) (1, 0, 0, 0) and (1, 1, 0, 0);
(g) (1, −1, 0, 0) and (1, 1, 0, 0);
(h) (0, 1, 0, 0) and (0, 1, 1, 0).

Which of these vectors are space-like, which time-like, and which null?

49.5. Consider a transformation of the form

$$F_1\mathbf{i}_1 + F_2\mathbf{i}_2 = F_1'\mathbf{i}_1' + F_2'\mathbf{i}_2'; \quad F_0' = F_0; \quad F_3' = F_3.$$

Follow the same kind of reasoning as that from Eqs. 49.17 to 49.22 and prove that it represents an ordinary space rotation in the x-y plane.

V.50 Kinematic Implications of the Lorentz Transformations. We have seen that the 4-vector algebra developed in the last section results in the Lorentz transformation, when it is assumed that space and time measurements are to be described in terms of this algebra. The hypothesis that this description is legitimate will be checked by comparing other implications of the Lorentz transformations with experiment. In the present section we develop some of the more direct implications for purely kinematic phenomena.

Consider a small displacement in space and time, the separation between two events that occur at nearby places within a short interval of time. The four components of the vector joining these two events are to be measured in a reference system S and written in either of the alternative notations:

$$(cdt, dx, dy, dz) \quad \text{or} \quad (dx_0, dx_1, dx_2, dx_3), \quad dx_0 = cdt. \tag{50.1}$$

The separation between the same two events is to be measured also in another reference system S' moving relative to S, the results being written either

$$(cdt', dx', dy', dz') \quad \text{or} \quad (dx_0', dx_1', dx_2', dx_3'), \quad dx_0' = cdt'. \tag{50.2}$$

Regarding these four numbers as a 4-vector, as defined in §49, we deduce that the Lorentz transformation holds in the form:

$$dx_0' = (dx_0 - vdx_1/c)/(1-v^2/c^2)^{1/2} \quad \text{or} \quad dt' = (dt - vdx/c^2)/(1-v^2/c^2)^{1/2}; \tag{50.3}$$

$$dx_1' = (dx_1 - vdx_0/c)/(1-v^2/c^2)^{1/2} \quad \text{or} \quad dx' = (dx - vdt)/(1-v^2/c^2)^{1/2}. \tag{50.4}$$

The x-axis of each system is taken parallel to their relative velocity $\mathbf{v}$. We now consider several special cases. First let the two events occur at the same time but at different places, as observed in the system S; then from Eq. 50.3:

$$dt = 0, \quad dt' = -vdx'/c^2. \tag{50.5}$$

Two events that are simultaneous in system S are not simultaneous in system S', because dt' is not zero if the events do not coincide completely in S. We say that simultaneity is relative.

Secondly, let the two events occur at the same place but at different times, as observed in the system S; then

$$dx = 0, \quad dt' = dt/(1-v^2/c^2)^{1/2}, \quad dx' = -vdt' = -vdt/(1-v^2/c^2)^{1/2}, \tag{50.6}$$

from Eqs. 50.3 and 50.4. Events that occur at a fixed place in system S occur at different places in system S', and appear to have a greater time separation to S' than they do to S. Let the two events occur at the same place in S', then from Eq. 50.4, $dx'=0$ means $dx=vdt$, and Eq. 50.3 leads to

$$dx' = 0, \quad dt = dt'/(1-v^2/c^2)^{\frac{1}{2}}. \tag{50.7}$$

Comparing this with Eq. 50.6, we see that the situation is completely symmetrical between the two reference systems. We may summarize it as follows: Any periodic process, in particular a clock, appears to go slow when observed from a reference system through which the clock is moving.

Next consider a rigid rod of length $dx = L_0$ at rest in the system S. Its length is independent of the times at which the positions of its ends are marked. But the length as measured on the system S', moving relative to the rod, must be obtained by marking the two ends simultaneously for the system S'. This means that the marking of the two ends of the rod by S' are events for which $dt' = 0$. From Eq. 50.3 this means that $dt = v\,dx/c^2$, and so Eq. 50.4 gives

$$dx' = dx(1 - v^2/c^2)^{1/2}. \tag{50.8}$$

The length of the rod measured in system S' is less than that measured by the system in which it is at rest. This is the Lorentz contraction

$$L(v)' = L(0)(1 - v^2/c^2)^{1/2}. \tag{50.9}$$

On the other hand, if the rod were at rest in system S', having L_0' length, it must be measured by S by simultaneous markings, these events having the time interval $dt = 0$. Eq. 50.4 then gives directly

$$dx' = dx/(1 - v^2/c^2)^{1/2}, \tag{50.10}$$

or the Lorentz contraction in the form

$$L(v) = L(0)'(1 - v^2/c^2)^{1/2}. \tag{50.11}$$

Again the situation is completely symmetrical between the two systems S and S', and we can summarize the result by stating that the length of a rod as measured in a moving system is less than its length at rest. It does not make any difference which system the rod happens to rest in, this system will record the maximum length; this maximum length is called the *proper length* of the rod.

It is also convenient to use the term *proper time* for the time interval between two events that is measured by a system in which the events occur at the same place. From Eq. 50.6 the proper time is the smallest value that the time interval will be assigned among various reference systems.

The proper length and the proper time are scalars for the Lorentz transformation; they are defined to have a single value for a given pair of events, independent of the reference system we happen to be using.

The proper time between a pair of events exists only when the

4-vector joining the events is time-like. The invariant interval corresponding to the vector in Eq. 50.1, or Eq. 50.2, is evidently

$$ds^2 = c^2dt^2 - dx^2 - dy^2 - dz^2. \tag{50.12}$$

Putting $dx = dy = dz = 0$ in this, so that the time would be the proper time $d\tau$, we see that

$$ds^2 = c^2d\tau^2. \tag{50.13}$$

It is therefore possible to rewrite Eq. 50.12 in terms of the scalar $d\tau$ instead of ds, thus:

$$c^2d\tau^2 = c^2dt^2 - dx^2 - dy^2 - dz^2,$$

and therefore

$$(d\tau/dt)^2 = 1 - v^2/c^2, \tag{50.14}$$

where we have written v for the velocity whose components are dx/dt, dy/dt, and dz/dt; this is the velocity with which an observer moves who is present at both events.

We shall now consider the addition of two velocities. Let a point move with speed u in the x-direction relative to system S. What will be the speed of this point as measured in system S'? We write it by definition:

$$u' = dx'/dt' \tag{50.15}$$

and use the Lorentz transformation for the dx' and dt' on the right. Thus from Eqs. 50.3 and 50.4, we have at once

$$u' = (dx - vdt)/(dt - vdx/c^2) = (u - v)/(1 - uv/c^2). \tag{50.16}$$

If the relative motion of the two observers remains constant and we let u approach the speed of light c, we find that u' also approaches the speed of light, independent of the relative velocity v:

$$u \to c: \qquad u' \to (c - v)/(1 - v/c) = c. \tag{50.17}$$

It is therefore not possible to decrease the speed of a particle approaching one with speed c by moving away from it at any possible speed.

Consider a transversely moving object with speed $dx/dt = 0$ and

$$u = dy/dt. \tag{50.18}$$

We ask for the transverse speed $u' = dy'/dt'$ in the system S'. We simply use Eq. 50.3 for dt' and note that $dy' = dy$, so that

$$u' = u(1 - v^2/c^2)^{1/2}. \tag{50.19}$$

Transverse motion is reduced by the relative motion of the reference system.

It is clear that velocity as ordinarily defined is not a true vector.

We can, however, define a 4-vector that plays the role of velocity in our theory. Thus, any 4-vector, multiplied or divided by a scalar, is still a 4-vector. In particular, we may divide the 4-vector displacement (cdt, dx, dy, dz) by the scalar proper time $d\tau$ and obtain a 4-vector that corresponds to velocity:

4-vector velocity:

$$(cdt/d\tau,\ dx/d\tau,\ dy/d\tau,\ dz/d\tau),$$

or

$$\mathbf{V} = (dt/d\tau)(c,\ v_x,\ v_y,\ v_z), \tag{50.20}$$

where v_x, v_y, v_z are the three components of velocity as ordinarily measured. From Eq. 50.14 this is a 4-vector of magnitude given by Eq. 49.12 as

$$V = (1-v^2/c^2)^{-1/2}(c^2-v^2)^{1/2} = c. \tag{50.21}$$

Thus all velocities have the same magnitude in the 4-vector algebra; they differ only in their directions in the four-dimensional sense. One might say that an object having a high speed in the ordinary sense is

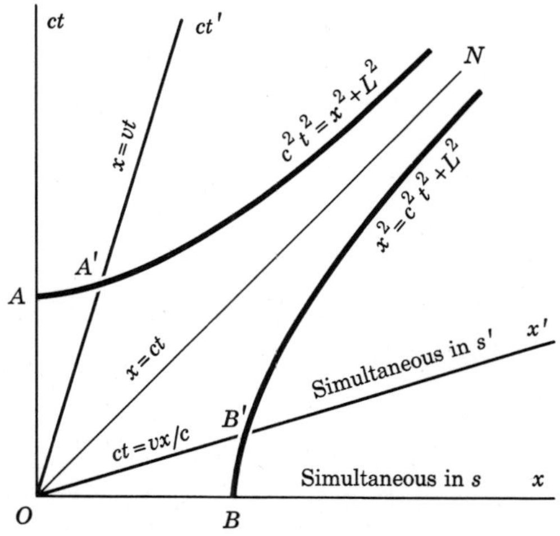

FIG. 50.1. An illustration of the discussion leading to Eqs. 50.22 and 50.23

pointing away from the time-direction toward the null direction, where $v=c$. An object with no velocity in the ordinary sense has a 4-vector velocity pointing directly along the time axis—it is moving into the future with speed c! By increasing the ordinary speed in a space-like direction, one can slow down the motion into the future; this is another, somewhat picturesque way of speaking about the time contraction due to relative motion.

We must notice that the time contraction is insensitive to the sign of the relative motion in space. Suppose a passenger on an interplanetary projectile could be shot away at high speed approaching that of light, be brought to rest after some time, and returned at high speed again. Since the clocks, and all periodic processes, on the moving system, when measured against the standards of the system at rest, appear to go slow, the traveling observer should return aged less than a twin observer who remained at rest. On the other hand, the time contraction effect is reciprocal: the traveler could as well take himself as at rest and assert that the journey was made by the other observer, who should then be the younger when they meet again. This paradox is resolved when we note that an absolute change in velocity is experienced by the real traveler, and he is not justified in asserting that the journey was made by the other. Their relative ages depend on the detailed history of the accelerations experienced by the traveler, but it can be shown that both observers will finally agree about their relative ages on being brought together again. For this and other features of the Lorentz transformation, a graphical construction due to Minkowski is highly illuminating. In the first diagram Fig. 50.1, we have two perpendicular axes representing ct and x, ignoring the dimensions y, z. The line O-ct' is in the direction $x = vt$ to represent the history of the moving observer S'. The line O-x' is in the direction $c^2t = vx$ because this makes $t' = 0$ in the Lorentz transformation Eq. 49.3: all events on the line O-ct' are simultaneous for the S' system. The angle between O-ct and O-ct' is equal to the angle between O-x and O-x'. The Lorentz transformation looks more like a strain than a rotation in this diagram.

The line O-N is in the direction $x = ct$ and represents the path of a light pulse emitted at the origin. Any straight line drawn from O into the angle between O-ct and O-N represents a possible history of a material particle moving to the right. Any straight line drawn from O into the angle between O-x and O-N could not represent the history of a particle, but a system could be found such that events on this line would appear simultaneous, with O. The next few diagrams, together with their captions, illustrate the reciprocal nature of the Lorentz contraction of lengths and times. They are based on the following theorem from analytical geometry. Draw the two hyperbolas:

$$\text{(a) } c^2t^2 - x^2 = L^2 \quad \text{and} \quad \text{(b) } x^2 - c^2t^2 = L^2, \tag{50.22}$$

where L is the length of some standard measuring rod. Refer these same curves to oblique coordinates, with the axes O-ct' and O-x', but with the length of the rod given by the intersections of the hyperbolas

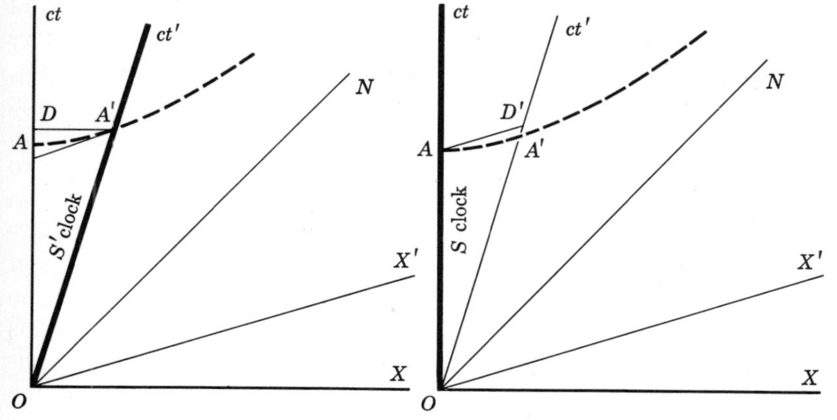

FIG. 50.2. A comparison of the time intervals recorded by two clocks. On the left, the clock is at rest in S', giving unit interval OA'. The event simultaneous with A' according to S is the event D, and the interval OD is greater than the unit time OA for S by the Lorentz factor. On the right, the clock is at rest in S and gives unit interval OA. The event simultaneous with A according to S' is D', and OD' is greater than OA', the unit for S'. Thus the S clock runs slow to S', and the S' clock runs slow to S

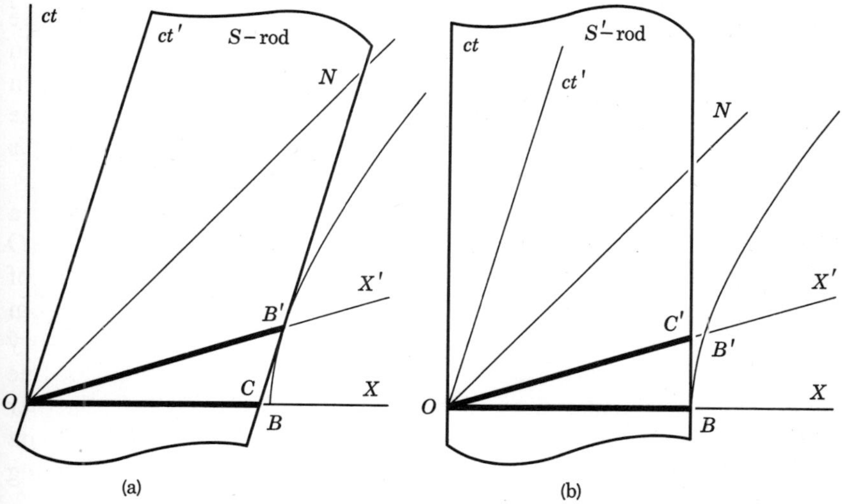

FIG. 50.3. A comparison of the histories of two measuring rods of unit length. On the left, the rod is at rest in S'; its proper length is OB', one unit on the S' system; its apparent length relative to S is OC, which is less than the S-unit OB by the Lorentz contraction. On the right, the rod is at rest in S; its proper length is OB, one unit on the S-system; its apparent length in the S' system is OC', which is less than the S' unit by the Lorentz contraction. Note the complete reciprocity in the situation.

with these axes; then their equations on the new system of coordinates are identical:

$$c^2t'^2 - x'^2 = L^2 \quad \text{and} \quad x'^2 - c^2t'^2 = L^2. \tag{50.23}$$

The invariance of the four-dimensional "length" is thus duplicated in this graphical construction.

Figure 50.5 illustrates the interplanetary space traveler. The line O-P is his outward bound path along his own time axis O-ct'. On the inertial reference system O-ct, O-x the journey out takes three years, but on the moving inertial reference system, O-ct', O-x', it takes only

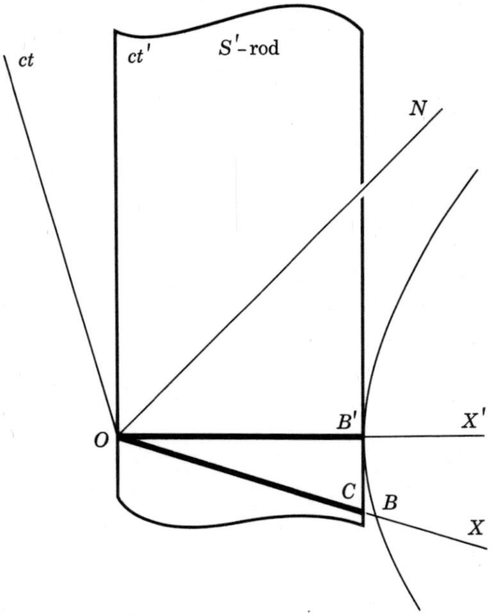

FIG. 50.4. The same situation as in (a) of Fig. 50.3, but drawn from the S' point of view, who regards himself as at rest, with S moving to the left

two years. At P the traveler changes to the return inertial system and reaches home at Q in another two years of his own time, finding a total elapsed time of six years in the stationary system. This summary has neglected the effects of the three acceleration events experienced by the traveler: first at O where he is projected into the orbit, second at P where his motion is reversed, and finally at Q to bring him to rest at home. The exact effect of these events on his time measuring system cannot be estimated on the basis of the theory so far developed and must wait until the theory of acceleration and gravitation have been elaborated. However, we can say that whatever effects they are, whether adding to or decreasing the age of the traveler, they are independent of the total length of the journey traveled at uniform speed. On the other hand, the difference in age due to the latter is proportional to the length of the journey and can be made as large as we please by extending the journey out into space. Thus we can ignore the effects of the accelerations in first approximation for a long enough trip.

It is instructive to examine the events on Earth from the traveler's system. The diagram shows events A and B on Earth that are reckoned

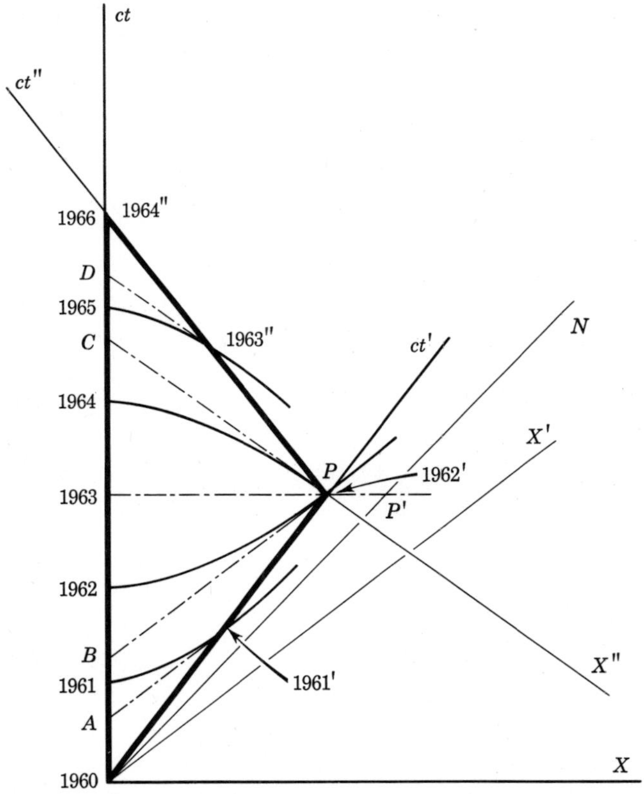

FIG. 50.5. The Minkowski diagram for an interplanetary space traveler

to be simultaneous, respectively, with the end of the first and second years on the traveler's system. At the end of his second year, the traveler switches systems, and in this new system the event C on Earth is reckoned to be simultaneous with P. The time difference between B and C, as reckoned by the traveler, is just the time taken for the speed reversal of his orbit. If he makes the journey long enough, as we have just seen, this time can be neglected. Thus all the terrestrial events occurring between *B* and *C* are practically simultaneous with the turning event in the traveler's history. On the outward trip, the Earth seems to be staying young compared with the traveler; at *B* the Earth has only experienced about a year and a half of time, whereas at the turning point *P*, the traveler is already two years older. Also on the return trip the time for the traveler is two years, but the time between *C* and *Q* on Earth is only another year and a half. It is the telescoping effect between *B* and *C* that ages the Earth compared with the traveler when the whole process is added up. Before his first two years are finished, the events later than *B* are still in the future for the traveler, but as soon as he starts his third year, using the return inertial

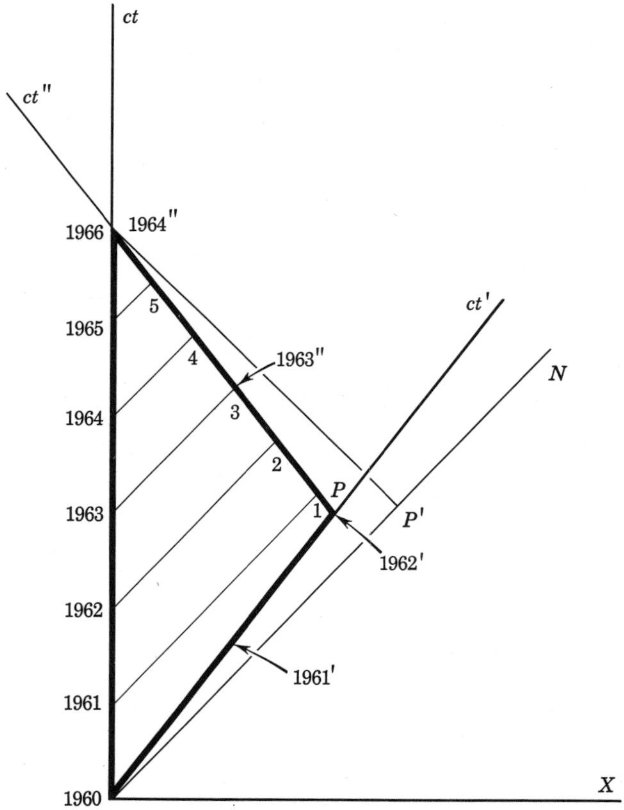

FIG. 50.6. Radio signals reach the space traveler with the speed of light. New Year messages from Earth reach him at the points numbered 1,2,3,4,5 for the terrestrial years 1961–1965

system, the events before C are already in the past. Taking into account the finite (if small) time occupied by the accelerating process of speed reversal at P, we can think of the traveler switching his inertial system continually, with his space axis O-x' moving over to the new direction (from parallel to O-x', to P-x'' in the diagram) in a continuous fashion. The events picked out on Earth by the traveler as simultaneous with his turning events will then pass continuously from B to C. He will regard all these events as practically simultaneous with P.

This last statement is not to be understood as meaning that the traveler will actually "see" the Earth aging in the course of a few moments. What he actually sees is conveyed to him across space by light signals. As the next diagram, Fig. 50.6, shows, the traveler will not see the end of the first terrestrial year until after he has started on the return trip, early in his third year in space. Thereafter the terrestrial calender will be seen to turn over at the rate of some $2\frac{1}{2}$ times a year on the traveler's calendar. The sudden telescoping of events between B and C will not be directly visible; but, when the traveler calculates back and attempts to locate these events on his two time systems, he will find no place for them except at the moment of transfer; they could be placed before P on his returning system, or after P on his outward system, but he was not using these systems at those times.*

It may be noted that by making the space traveler move at speeds very nearly equal to the speed of light, this telescoping time effect can be made to embrace almost the entire history of the stationary system. In this sense all the events between O and Q on Earth are simultaneous with a single event marked as P' on Fig. 50.6. In general, all events on the Minkowski diagram joined by space-like vectors are simultaneous in some reference system.

* From the point of view of general relativity (see §54) no "twin paradox" ever arises, but a full discussion from this point of view is not possible within the limits of the present text. Such a discussion would proceed somewhat as follows. Identical (*e.g.* atomic) clocks carried by two observers record the generalized integrated interval ($\int ds$) of their respective histories. If their paths intersect at events A and B, their relative ages at A and B are determined by the difference in integrated interval of the two paths. These intervals depend only on the gravitational fields through which they pass and do not depend on what reference systems may be used. Whether we describe the experience of a traveler by means of the curved metric of gravitational theory, or by locally tangential Galilean frames, the results are the same—an example of the general principle of equivalence. The inertial frames set up for the traveler in the above discussion are in fact these tangential Galilean frames. They naturally form a nonaffine system if extended out into spacetime from their points of tangency and are useful only in the immediate neighborhood of the traveler himself. They give his correct age, but result in ambiguity if he attempts to use them to discuss the detailed history of events on the Earth while he is far out in space.

Exercise V.50

50.1. Solve Eqs. 50.3 and 50.4 for dx and dt as functions of dx' and dt'.

50.2. A particle moves with speed u in a direction bisecting the angle between O-x and O-y. Find the direction of its motion in a reference system moving with speed v along O-x, using Eqs. 50.17 and 50.19.

50.3. Derive Eq. 50.23 from Eq. 50.22 using theorems of analytical geometry, as mentioned in the text.

50.4. What is the speed of the space traveler in Fig. 50.5?

50.5. Draw Fig. 50.3b from the point of view of system S'.

50.6. The shape of a circular object at rest in system S is

$$x^2+y^2 = a^2.$$

What is its shape as measured in the system S' moving with speed v in the x-direction?

50.7. A disc of radius a rotates about its axis with angular velocity ω radians per second. Considering that its perimeter suffers the Lorentz contraction, what is the ratio of its perimeter to its diameter as a function of ω? What is the maximum possible angular velocity set by the contraction? For a more detailed discussion of this problem, which takes into account the possible effects of the accelerations involved in the rotation, see §54.

V.51 Dynamical Implications of the Lorentz Transformation. Let us assume that a mass m_0 can be assigned to a particle. It is to be a measurement by an observer at rest relative to the particle and is called the *proper mass*; by definition it is a scalar quantity. The *4-vector momentum* is then defined as the product of the proper mass and the 4-vector velocity of Eq. 50.20:

$$\mathbf{p}^* = m_0 d\mathbf{s}/d\tau = m_0(dt/d\tau)(c, v_x, v_y, v_z). \qquad (51.1)$$

We shall interpret the three-space-like components of this vector as the relativistic expression of the ordinary three-dimensional momentum; we note that these components are not simply m_0 times the 3-vector velocity components but that instead of m_0 there is an effective mass m:

$$m = m_0(dt/d\tau) = m_0(1-v^2/c^2)^{-1/2}, \qquad (51.2)$$

where of course v is the speed of the particle relative to the observer who ascribes the mass m to it. The effective mass may increase without limit as v approaches c.

We may also define a 4-vector force by equating it to the rate of change of the 4-vector momentum, but the rate must be taken with respect to the proper time to ensure that this force shall be a 4-vector. Thus, from Eq. 51.1 we may write for the force:

$$\mathbf{F}^* = (d/d\tau)m_0(dt/d\tau)(c, v_x, v_y, v_z). \qquad (51.3)$$

If it happens that m_0 is independent of time, we may write this from Eq. 50.20 in the form

$$\mathbf{F}^* = m_0 d\mathbf{V}/d\tau. \tag{51.4}$$

Now we have seen that $\mathbf{V}$ is a vector of constant magnitude, Eqs. 50.20 and 50.21, so that any variation in $\mathbf{V}$ is necessarily orthogonal to $\mathbf{V}$:

$$d\mathbf{V}\cdot\mathbf{V} = 0, \tag{51.5}$$

and it follows from Eq. 51.4 that, when m_0 is constant,

$$\mathbf{F}^*\cdot\mathbf{V} = 0. \tag{51.6}$$

The three space-like components of Eq. 51.3 read:

$$(1-v^2/c^2)^{1/2}F_n^* = (d/dt)(mv_n) \quad (n = 1, 2, 3). \tag{51.7}$$

Eq. 51.3 is verbally identical with Newton's law of motion—force equals rate of change of momentum—but neither of the words "force" and "momentum" here has its classical meaning. The classical force is determined by the time rate of change with respect to the system through which the particle is moving, not the proper time used in Eq. 51.3. The expression on the right of Eq. 51.7 would be more closely analogous with the classical expression. It is customary, therefore, to call this expression the classical relative force:

$$n = 1, 2, 3: \quad F_n \equiv (1-v^2/c^2)^{1/2}F_n^* = (d/dt)(mv_n). \tag{51.8}$$

Also the quantities mv_n are called the three components of relative momentum. The law of motion contains, as the three component equations in 51.8, the statement: relative force equals (d/dt)(relative momentum). Eq. 51.6 states in general terms that the rate of doing work on a particle of fixed proper mass by a 4-vector force $\mathbf{F}^*$ is always zero. The proper energy of such a particle is therefore a constant—it can depend only on the proper mass.

The work done by the relative force, which is what we would measure physically, is not zero, and we can easily compute it as follows. Let the particle start from rest and be accelerated along the x-axis to a final speed v_0. Then the work done by the relative force $\mathbf{F}$ is

$$\begin{aligned} W(v_0) &= \int_0^{v_0} F dx = \int_0^{v_0} Fv dt = \int_0^{v_0} v d(mv) \\ &= \Big| vm_0 v/(1-v^2/c^2)^{1/2} \Big|_0^{v_0} - \int_0^{v_0} m_0 v/(1-v^2/c^2)^{1/2} dv \\ &= m_0 v_0^2/(1-v_0^2/c^2)^{1/2} + \Big| m_0 c^2 (1-v^2/c^2)^{1/2} \Big|_0^{v_0} \\ &= m_0 c^2/(1-v_0^2/c^2)^{1/2} - m_0 c^2 = (m-m_0)c^2. \end{aligned} \tag{51.9}$$

Thus the increase in energy—the work done—is proportional to the increase in relative mass due to the motion. The general relation

$$E = mc^2 \tag{51.10}$$

for the relative energy is consistent with this result and can in fact be derived by an independent argument dealing with the collision processes between different masses. It is the famous Einstein energy-mass relation.

To a first approximation Eq. 51.9 is equivalent to the classical kinetic energy. Expanding the factor $1/(1-v^2/c^2)^{1/2}$ yields:

$$E(v)-E(0) = \tfrac{1}{2}m_0v^2+(3/8)m_0v^4/c^2+\cdots. \tag{51.11}$$

So when $v \ll c$, it reduces to the classical expression for the kinetic energy. Equation 51.10 implies further that the relative energy can change not only due to motion, but also due to changes in the proper mass. In fact, this is fundamental in the explanation of the mass defects of atomic nuclei and accounts for the energy source of the stars, including the sun, and for the energy produced in fission and fusion reactions in modern nuclear engineering.

Returning for a moment to Eq. 51.4, we see now that the time-like component of this equation can be written in the form

$$c(1-v^2/c^2)^{1/2}F_0^* = (d/dt)(mc^2). \tag{51.12}$$

The time-like component of $\mathbf{F}^*$ is thus the proper-time derivative of the relative energy divided by c:

$$F_0^* = (d/d\tau)(E/c). \tag{51.13}$$

The time-like component of the 4-vector momentum is also E/c, and for this reason the vector $\mathbf{p}^*$ is sometimes called the *energy-momentum vector.*

The procedure we have used above for setting up the relativistic form of Newton's equations of motion was based on the invariance postulate: the equations must be invariant for Lorentz transformations between relatively moving reference systems. We ensured this by stating Eq. 51.3 in the form of a 4-vector equation. It also turns out that the same equations can be put into a Lagrangian form so that they are invariant also under a wide class of contact transformations. Thus, we seek a Lagrangian function L such that the relativistic equations of motion become the consequences of the variation principle of Eq. 24.2:

$$\delta\int Ldt = 0. \tag{51.14}$$

From the point of view of relativity, this variation equation must be Lorentz invariant; in other words, the quantity Ldt must be either a scalar, a vector, etc., in the four-dimensional algebra. Since it is but one quantity, it must be a scalar. In four dimensions, however, dt is not a scalar, but rather the proper time $d\tau$ of Eq. 50.14 is the scalar: $d\tau = dt(1-v^2/c^2)^{1/2}$. Thus Ldt will be scalar if $L/(1-v^2/c^2)^{1/2}$ is a scalar. It is clear that

$$L = -m_0c^2(1-v^2/c^2)^{1/2} \tag{51.15}$$

satisfies this condition. It also turns out to yield the right equations of motion, derived from Eq. 24.15:

$$F_n = \partial L/\partial x_n = (d/dt)[m_0v_n/(1-v^2/c^2)^{1/2}] \quad (n = 1, 2, 3), \tag{51.16}$$

which are identical with Eq. 51.8.

The Hamiltonian function associated with this Lagrangian by Eq. 27.4 is:

$$H = \sum_n p_nv_n - L = m_0c^2/(1-v^2/c^2)^{1/2} = mc^2, \tag{51.17}$$

and is just the relative energy, E, of Eq. 51.10.

The relative energy E can be expressed explicitly in terms of the momentum. Classically this is of course $E_{\text{class}} = p^2/2m$. Here instead we start with the statement that the momentum p is

$$p = mv = m_0v/(1-v^2/c^2)^{1/2} \tag{51.18}$$

and that

$$E = mc^2 = m_0c^2/(1-v^2/c^2)^{1/2}. \tag{51.19}$$

From these two equations we can prove that

$$E^2 = c^2p^2 + m_0{}^2c^4. \tag{51.20}$$

When the momentum is small enough one can show of course that this reduces to the classical expression. When the momentum is large enough so that

$$p \gg m_0c, \tag{51.21}$$

then the energy approaches the value

$$E = cp. \tag{51.22}$$

The condition Eq. 51.21 can be met formally by letting $v = c$ and $m_0 = 0$; these are believed to be applicable to the photon, or quantum of radiating electromagnetic energy, so that in this case in which we know

$$E = h\nu, \tag{51.23}$$

where h is Planck's constant and ν the frequency of the radiation, there is a momentum to be associated with the photon given by Eq. 51.22 as

$$p_{\text{photon}} = h\nu/c = h/\lambda. \tag{51.24}$$

When a material particle such as an electron is given a velocity approaching c, its energy is also given by Eq. 51.22, and, as we shall see in Chapter VI, a wave is associated with the electron also, through the equations:

$$E_{\text{particle}} = h\nu, \quad p_{\text{particle}} = h/\lambda. \tag{51.25}$$

There is a close analogy between high-speed material particles and the photon.

The variational principle of Eq. 51.14, with the Lagrangian of Eq. 51.15, is susceptible of a very interesting interpretation. Thus we can write these equations in the form

$$-\delta\int m_0 c^2 d\tau = -m_0 c\,\delta\int ds = 0. \tag{51.26}$$

In words: the space-time history of a particle, passing between two given events, is the path of shortest four-dimensional interval. It is the generalized straight line in our four-dimensional geometry. The analog here with Fermat's principle in optics is suggestive again of the close association of waves and particles.

Exercise V.51

51.1. Use Eq. 51.3 and Eq. 50.20 to prove Eq. 51.6 in terms of the explicit components of the two vectors **P** and **V**.

51.2. Prove Eq. 51.16 from Eqs. 51.15, 51.14, and 24.15.

51.3. Derive Eq. 51.20 from Eqs. 51.18 and 51.19.

51.4. Prove that Eq. 51.20 reduces to the classical expression when the momentum is small enough.

51.5. Find the velocity that an electron must be given so that its momentum is ten times its rest mass times the speed of light. What is its energy at this speed?

51.6. Find the momentum and the kinetic energy of the space traveler in Ex. 50.4.

51.7. A rocket, having a total mass M_1 initially at rest, ejects fuel at constant jet velocity v relative to the rocket. After some time the final mass of the rocket has been reduced to M_2 by loss of fuel. Prove from classical mechanics that the final velocity of the rocket is

$$u = v \ln(M_1/M_2).$$

51.8. Show that the relativistic formula for the speed of the rocket in the last exercise is the same as the classical formula, the masses being reinterpreted as relativistic masses.

51.9. The space traveler of Ex. 50.4 carries 90% of his total mass in the form of fuel to be used for reversing his velocity at the time he decides to start the return to Earth. Find the jet velocity he must use.

51.10. Precise measurements of time can be made by comparison with periodic processes occurring in molecules. The ammonia clock makes use of the oscillations of the N-atom up and down through the plane of the three H-atoms. The frequency of this process is known to within one part in 10^8

at 23,870 megacycles. Let an ammonia clock be mounted on a rocket and fired out into space at the escape velocity 7 miles per sec, traveling a distance R before reversing speed and returning to Earth. Compute the minimum distance R required to yield a detectable difference in the time of flight registered on the traveling clock compared with a similar clock at rest on Earth. Consider this question purely kinematically by the method of §50, and also discuss the effects of the relative motion on the mass and force constants in the atomic oscillator.

51.11. In the linear particle accelerator Mark III, at Stanford University, electrons are given an energy 630 MeV and travel a distance 220 feet in the laboratory. Find the distance traveled by the electrons in coordinates at rest with the electrons. How precisely must the electrons be aimed to hit a target 1.0 cm in diameter at the far end of the instrument?

51.12. In the synchro-cyclotron, at Berkeley, protons were made to execute a circular path of radius 92 inches in a maximum magnetic field 15,500 gauss. Find the maximum energy of the protons in MeV, and estimate the difference due to the relativistic effect. (1 gauss $= 10^{-4}$ weber per m^2.)

V.52 Four-Dimensional Matrix Algebra for Space-Time.

To apply the theory of relativity to more complicated problems than those of simple particle mechanics, discussed in the last section, it is advisable to make use of matrix algebra in four dimensions, and we develop a notation and some theorems for this algebra here. Essentially this section is an introduction to what is often called *tensor analysis*.

The general theory of matrices outlined in §5 is valid whatever the rank of the matrices, with appropriate range to the allowed values of the subscripts. However, because of the negative signs in the scalar products used in space-time applications, Eq. 49.13, the reasoning at certain points does not go through without modification, in particular from Eq. 5.12 to Eq. 5.13. In order to make full use of the dyadic notation in space-time algebra it is necessary to introduce a *reciprocal set* of unit vectors $\mathbf{i}^\alpha$, $\alpha = 0, 1, 2, 3$; defined* by:

$$\mathbf{i}^\alpha \cdot \mathbf{i}_\beta = \delta_{\alpha\beta}, \tag{52.1}$$

or, more explicitly,

$$\mathbf{i}^0 = \mathbf{i}_0, \quad \mathbf{i}^m = -\mathbf{i}_m \quad (m = 1, 2, 3). \tag{52.2}$$

Any 4-vector in space-time can be expressed in terms of either set:

$$\mathbf{F} = \sum_\mu F_\mu \mathbf{i}_\mu = \sum_\mu F^\mu \mathbf{i}^\mu, \tag{52.3}$$

* In most mathematical texts on tensor analysis the opposite convention is used regarding the position of the subscripts and superscripts. To follow that convention one would have to go back over the whole of Chapter I and also §49 and rewrite all the subscripts as superscripts.

the quantities F^μ being called the *covariant components* of **F** and the quantities F_μ being called the *contravariant components* of **F**. From Eq. 52.2 we have

$$F^0 = F_0, \quad F^m = -F_m \quad (m = 1, 2, 3). \tag{52.4}$$

The scalar product of two vectors **F** and **G** can be expressed conveniently in terms of the covariant components of one and the contravariant components of the other vector:

$$\mathbf{F}\cdot\mathbf{G} = \sum_\mu\sum_\nu F^\mu \mathbf{i}^\mu \cdot G_\nu \mathbf{i}_\nu = \sum_\mu\sum_\nu F^\mu G_\nu \delta_{\mu\nu} = \sum_\mu F^\mu G_\mu, \tag{52.5}$$

having used Eq. 52.1. The awkward negative signs appearing in Eq. 49.8 are absorbed in the covariant components.

In order to handle matrix products similarly without trouble from the negative signs, it is convenient to use the reciprocal set of unit vectors in the symbol for the unit dyads. Thus we shall write $\mathbf{i}^\alpha\mathbf{i}_\beta$ for the dyad with a unity in the αth row and β^{th} column, other elements being zero. Instead of Eq. 5.12 we now write for the general 4×4 matrix:

$$A = \sum_\mu\sum_\nu A^\mu{}_\nu \mathbf{i}^\mu\mathbf{i}_\nu. \tag{52.6}$$

The scalar product of two matrices of this type then gives the matrix product of their components exactly as in Eq. 5.13.

At this juncture it is helpful to introduce the *dummy suffix* convention. According to this convention, if a subscript occurs twice in any expression, the summation over its allowed values is understood, and not explicitly written out. Thus, on this convention, Eq. 52.6 would be written simply as

$$A = A^\mu{}_\nu \mathbf{i}^\mu\mathbf{i}_\nu. \tag{52.7}$$

In reading the equation, and subsequent equations of this chapter, attention must be paid to detailed subscripts and the summation convention kept in mind. Confusion and error can easily result from forgetting the summation implied by the notation, but some of the equations contain so many summation signs that they are much better implied than written out explicitly.

The generalization of Eq. 5.13 now goes through as follows:

$$\begin{aligned} A\cdot B &= A^\mu{}_\nu \mathbf{i}^\mu\mathbf{i}_\nu \cdot B^\alpha{}_\beta \mathbf{i}^\alpha\mathbf{i}_\beta = A^\mu{}_\nu B^\alpha{}_\beta \mathbf{i}^\mu\mathbf{i}_\nu \cdot \mathbf{i}^\alpha\mathbf{i}_\beta \\ &= A^\mu{}_\nu B^\alpha{}_\beta \delta_{\nu\alpha} \mathbf{i}^\mu\mathbf{i}_\beta = A^\mu{}_\nu B^\nu{}_\beta \mathbf{i}^\mu\mathbf{i}_\beta, \end{aligned} \tag{52.8}$$

or

$$A\cdot B = C \quad \text{where} \quad C^\mu{}_\beta = A^\mu{}_\nu B^\nu{}_\beta. \tag{52.9}$$

The dyadic product corresponds exactly to the matrix product, as before. The trace of a matrix is the sum of its diagonal elements, and on the dummy suffix convention it is written:

$$\text{Trace } A = A^{\mu}{}_{\mu}. \tag{52.10}$$

In the dyadic notation the trace of the unit dyad is

$$\text{Trace } \mathbf{i}^{\mu}\mathbf{i}_{\nu} = \mathbf{i}^{\mu}\cdot\mathbf{i}_{\nu} = \delta_{\mu\nu}, \tag{52.11}$$

and the trace of any dyadic is

$$\text{Trace } A = A^{\mu}{}_{\nu}\mathbf{i}^{\mu}\cdot\mathbf{i}_{\nu} = \delta_{\mu\nu}A^{\mu}{}_{\nu} = A^{\mu}{}_{\mu}. \tag{52.12}$$

The double scalar product is a useful operation. This is defined by

$$\mathbf{i}^{\alpha}\mathbf{i}_{\beta}:\mathbf{i}^{\mu}\mathbf{i}_{\nu} = (\mathbf{i}^{\alpha}\cdot\mathbf{i}_{\nu})(\mathbf{i}_{\beta}\cdot\mathbf{i}^{\mu}) = \delta_{\alpha\nu}\delta_{\beta\mu}. \tag{52.13}$$

The double scalar product of two dyadics is the trace of their matrix product:

$$A:B = A^{\mu}{}_{\nu}B^{\alpha}{}_{\beta}\mathbf{i}^{\mu}\mathbf{i}_{\nu}:\mathbf{i}^{\alpha}\mathbf{i}_{\beta} = A^{\mu}{}_{\nu}B^{\nu}{}_{\mu} = \text{trace } (AB). \tag{52.14}$$

This is often called the *contracted product* of the two matrices.

The identity matrix can be expressed in terms of the unit dyads, thus:

$$E = \mathbf{i}^{\alpha}\mathbf{i}_{\alpha}. \tag{52.15}$$

For any matrix A we have the dyadic equation:

$$E\cdot A = \mathbf{i}^{\alpha}\mathbf{i}_{\alpha}\cdot A^{\mu}{}_{\nu}\mathbf{i}^{\mu}\mathbf{i}_{\nu} = A^{\mu}{}_{\nu}\mathbf{i}^{\alpha}(\mathbf{i}_{\alpha}\cdot\mathbf{i}^{\mu})\mathbf{i}_{\nu} = A^{\mu}{}_{\nu}\mathbf{i}^{\mu}\mathbf{i}_{\nu} = A,$$

thus demonstrating that E is indeed the identity.

As we have seen, a vector can be expressed in terms of either contravariant or covariant components. The vector interval or displacement in space-time is usually expressed in terms of its contravariant components alone:

$$d\mathbf{s} = dx_{\alpha}\mathbf{i}_{\alpha}. \tag{52.16}$$

All vectors expressed in terms of their contravariant components are like displacements; when expressed in terms of their covariant components they may be thought of as like forces, because scalar products between force and displacement result in a scalar.

It is necessary to generalize the differential operators that were introduced in the three-dimensional vector calculus. Thus we shall write the *four-dimensional del operator* in terms of the covariant unit vectors:

$$\Box = \mathbf{i}^{\alpha}\partial/\partial x_{\alpha}. \tag{52.17}$$

By so doing it is at once a simple formal matter to write down the *four-dimensional divergence* of a displacement-like 4-vector:

$$\Box\cdot\mathbf{F} = \mathbf{i}^{\alpha}(\partial/\partial x_{\alpha})\cdot F_{\mu}\mathbf{i}_{\mu} = \partial F_{\alpha}/\partial x_{\alpha}, \tag{52.18}$$

always remembering the dummy suffix summation convention.

We can also take the scalar product between a four-dimensional displacement $d\mathbf{s}$ and the gradient of a scalar function:

$$d\mathbf{s}\cdot\Box\Phi = \mathbf{i}_\alpha dx_\alpha\cdot\mathbf{i}^\mu\partial\Phi/\partial x_\mu = dx_\alpha\partial\Phi/\partial x_\alpha = d\Phi, \tag{52.19}$$

which is the exact analog in four dimensions of Eq. 6.11 in three dimensions. Note that in Eq. 52.19 the function Φ is a function of four independent variables instead of three.

The generalized Laplacian operator, or scalar product of the gradient operator with itself, being the product of two covariant factors, gives the negative signs again, thus:

$$\begin{aligned}\Box\cdot\Box &= \mathbf{i}^\alpha(\partial/\partial x_\alpha)\cdot\mathbf{i}^\mu(\partial/\partial x_\mu) = \mathbf{i}^\alpha\cdot\mathbf{i}^\mu(\partial^2/\partial x_\alpha\partial x_\mu)\\ &= \partial^2/\partial x_0{}^2 - \sum_{m=1}^{3}\partial^2/\partial x_m{}^2.\end{aligned} \tag{52.20}$$

This is the operator that appears in the scalar wave equation, or the *D'Alembertian operator*: in four dimensions the analog of Laplace's equation turns out to be the wave equation.

We must now discuss transformation theory again. Suppose that the original unit vectors $\mathbf{i}_\alpha$ are submitted to a Lorentz transformation (either corresponding to relative motion in some direction, or a space rotation about some axis) and that the new unit vectors are related with the original ones by

$$\mathbf{i}_\alpha' = \sum_\beta L_{\alpha\beta}\mathbf{i}_\beta = L_{\alpha\beta}\mathbf{i}_\beta, \tag{52.21}$$

or the inverse relation:

$$\mathbf{i}_\beta = L_{\beta\alpha}{}^{-1}\mathbf{i}_\alpha'. \tag{52.22}$$

We wish to find the new vectors $\mathbf{i}^{\alpha'}$ that stand in the same relation to $\mathbf{i}_\alpha'$ as the original vectors $\mathbf{i}^\alpha$ do to $\mathbf{i}_\alpha$. It is easily verified that

$$\mathbf{i}^{\beta'} = L_{\alpha\beta}{}^{-1}\mathbf{i}^\alpha, \tag{52.23}$$

or the inverse relation:

$$\mathbf{i}^\alpha = L_{\beta\alpha}\mathbf{i}^{\beta'}. \tag{52.24}$$

Here $L_{\alpha\beta}{}^{-1}$ is the $\alpha\beta^{\text{th}}$ component of the matrix reciprocal to L:

$$L_{\mu\alpha}L_{\alpha\beta}{}^{-1} = \delta_{\mu\beta}. \tag{52.25}$$

It is easy to verify that these relations result in

$$\mathbf{i}^{\alpha'}\cdot\mathbf{i}_\beta' = \delta_{\alpha\beta}, \tag{52.26}$$

as required. It is also easy to verify that the identity matrix remains of the same form in terms of the new unit vectors:

$$E = \mathbf{i}^\alpha\mathbf{i}_\alpha = \mathbf{i}^{\alpha'}\mathbf{i}_\alpha'. \tag{52.27}$$

A useful form of the transformation matrix from $\mathbf{i}^\alpha$ to $\mathbf{i}^{\alpha'}$ can be found by taking the dyad product of Eq. 52.23 by $\mathbf{i}_\beta$ as a postfactor:

$$\mathbf{i}^{\beta'}\mathbf{i}_\beta = L_{\alpha\beta}{}^{-1}\mathbf{i}^\alpha\mathbf{i}_\beta. \tag{52.28}$$

The right side of this is just the dyadic expression for the reciprocal transformation matrix, so that the equation reads simply:

$$L^{-1} = \mathbf{i}^{\beta'}\mathbf{i}_\beta. \tag{52.29}$$

(Remember always the summation convention on the dummy subscript.) Again making a dyad product of Eq. 52.21 with $\mathbf{i}^\alpha$ as a prefactor gives

$$L = \mathbf{i}^\alpha\mathbf{i}_\alpha'. \tag{52.30}$$

The proper transformation relations for both covariant and contravariant vector components are easily derived. Thus in analogy with Eq. 49.15 we have, for the contravariant components,

$$\mathbf{F} = F_\mu\mathbf{i}_\mu = F_\mu'\mathbf{i}_\mu'$$

and therefore

$$F_\nu' = F_\mu\mathbf{i}_\mu\cdot\mathbf{i}^{\nu'}. \tag{52.31}$$

It is proved in an exercise that

$$L_{\mu\nu}{}^{-1} = \mathbf{i}^{\mu'}\cdot\mathbf{i}_\nu \tag{52.32}$$

and also that

$$L_{\mu\nu} = \mathbf{i}_\mu'\cdot\mathbf{i}^\nu. \tag{52.33}$$

Eq. 52.31 therefore yields the desired transformation:

$$F_\nu' = L_{\nu\mu}{}^{-1}F_\mu. \tag{52.34}$$

For the covariant components, a similar reasoning yields the transformation:

$$F^{\nu'} = L_{\nu\mu}F^\mu. \tag{52.35}$$

The law of transformation of the components of the matrix A in Eq. 52.7 are deduced as follows:

$$A = A^\mu{}_\nu\mathbf{i}^\mu\mathbf{i}_\nu = A^\mu{}_\nu{}'\mathbf{i}^{\mu'}\mathbf{i}_\nu'. \tag{52.36}$$

Using Eq. 52.24 and Eq. 52.22 on the first form of A in this equation, we have

$$A = A^\mu{}_\nu L_{\beta\mu}\mathbf{i}^{\beta'}L_{\nu\alpha}{}^{-1}\mathbf{i}_\alpha' = L_{\beta\mu}A^\mu{}_\nu L_{\nu\alpha}{}^{-1}\mathbf{i}^{\beta'}\mathbf{i}_\alpha',$$

and comparing this with the second form in Eq. 52.36 we have the transformation relation for the coefficients:

$$A^\mu{}_\nu{}' = L_{\mu\alpha}A^\alpha{}_\beta L_{\beta\nu}{}^{-1}. \tag{52.37}$$

Thus we see that the prefactors of A transform as covariant vectors, and the postfactors as contravariant vectors.

It is apparent that, just as there are covariant and contravariant components for any given 4-vector, we may expect to find the components of a matrix capable of more than one form, with different transformation properties. The notation suggests the following four forms of the same matrix:

$$G = G^{\alpha}{}_{\beta}\mathbf{i}^{\alpha}\mathbf{i}_{\beta} = G_{\alpha\beta}\mathbf{i}_{\alpha}\mathbf{i}_{\beta} = G_{\alpha}{}^{\beta}\mathbf{i}_{\alpha}\mathbf{i}^{\beta} = G^{\alpha\beta}\mathbf{i}^{\alpha}\mathbf{i}^{\beta}. \tag{52.38}$$

Their transformation properties are easily found in exact analogy with the reasoning leading to Eq. 52.37, and they are:

$$\left.\begin{aligned} &\text{(i)} & G^{\alpha}{}_{\beta}{}' &= L_{\alpha\mu}G^{\mu}{}_{\nu}L_{\nu\beta}{}^{-1} \\ &\text{(ii)} & G_{\alpha\beta}{}' &= L_{\alpha\mu}{}^{-1}G_{\mu\nu}L_{\nu\beta}{}^{-1} \\ &\text{(iii)} & G_{\alpha}{}'^{\beta} &= L_{\alpha\mu}{}^{-1}G_{\mu}{}^{\nu}L_{\nu\beta} \\ &\text{(iv)} & G^{\alpha\beta\prime} &= L_{\alpha\mu}G^{\mu\nu}L_{\nu\beta} \end{aligned}\right\}. \tag{52.39}$$

A set of matrix components that obeys a transformation rule, such as any one of these four relations, is called a *tensor*. The rules (i) and (iii) are called *mixed tensor*, and the rules (ii) and (iv) are, respectively, pure *contravariant* and pure *covariant* tensor rules.

In Eq. 52.39 we have to think of the unit dyads $\mathbf{i}_{\alpha}\mathbf{i}_{\beta}$, $\mathbf{i}_{\alpha}\mathbf{i}^{\beta}$, and $\mathbf{i}^{\alpha}\mathbf{i}^{\beta}$ as derived from the unit dyad $\mathbf{i}^{\alpha}\mathbf{i}_{\beta}$ by using the relation Eq. 52.2 on either the prefactor or the postfactor. From Eq. 52.38 we can then derive the other forms of the components from the first form. For example, we can see that $G^{0}{}_{1} = -G_{0}{}^{1} = -G^{01} = G_{01}$. These relations are invariant under a Lorentz transformation. Under the transformation the dyads each transform according to their different rules. The new dyads are taken as the unit dyads in the new coordinate system, and the same matrix G is expressed in terms of these four different new unit dyads in Eqs. 52.39.

Provided we express a dyadic with its prefactors in contravariant form, we can apply the divergence operator to it. Thus we have

$$\square\cdot T = \mathbf{i}^{\alpha}(\partial/\partial x_{\alpha})\cdot T_{\mu}{}^{\nu}\mathbf{i}_{\mu}\mathbf{i}^{\nu} = (\partial/\partial x_{\mu})T_{\mu}{}^{\nu}\mathbf{i}^{\nu}, \tag{52.40}$$

which is a covariant vector.

A three-dimensional region in four-dimensional geometry can be represented by a vector. For example, the volume in space $dx_1dx_2dx_3$ can be represented by a time-like vector. If the volume is at rest in the reference system, its vector is parallel to the time axis, $\mathbf{i}_0$. If it is moving with speed v in the x_1-direction, it is represented by a vector parallel to $\mathbf{i}_0'$ given by the Lorentz transformation $\mathbf{i}_0' = \cosh\phi\ \mathbf{i}_0 + \sinh\phi\ \mathbf{i}_1$, where $\tanh\phi = v/c$. A three-dimensional region defined by the history of a stationary plane element dx_1dx_2 through a time dt is to be represented by a vector parallel to $\mathbf{i}_3$ of magnitude $c\,dt\,dx_1dx_2$.

Any two-dimensional region must be represented by both vectors

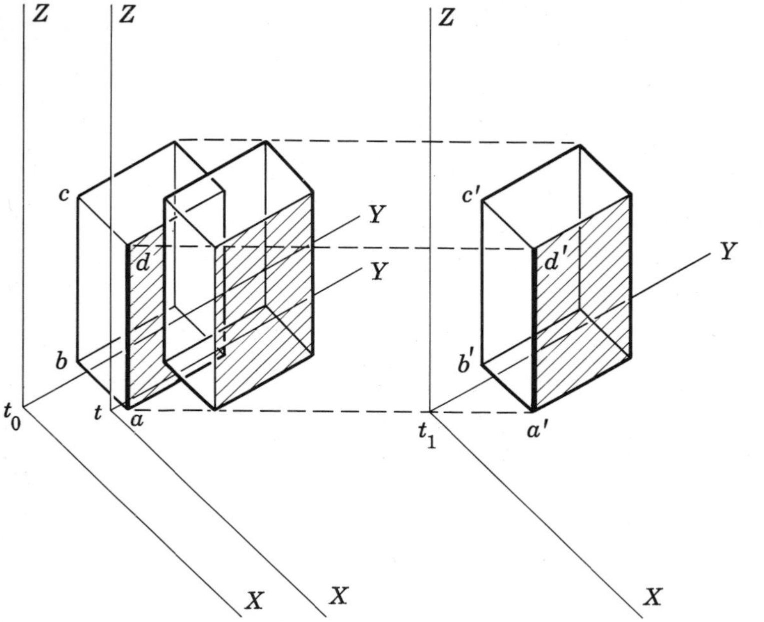

FIG. 52.1. The history of a stationary parallelopiped in space. A set of space axes is to be set up at each instant of time. Time progresses from t_0 at the left to t_1 at the right. The four-dimensional extension of this history is the three-dimensional volume of the parallelopiped multiplied by the time $t_1 - t_0$. The history of the face *abcd* to *a′b′c′d′* is one part of the hypersurface of this history

that are normal to the region. This is most obviously done by means of a dyadic compounded of both normal directions. Thus a stationary plane element dx_2dx_3 is normal both to $\mathbf{i}_0$ and $\mathbf{i}_1$. We may use the antisymmetrical dyadic for this:

$$dx_2dx_3(\mathbf{i}_1\mathbf{i}_0 - \mathbf{i}_0\mathbf{i}_1).$$

There are, however, other possible ways in which such a representation can be set up, and it is better to list the general expressions in terms of the permutation symbols. Thus we define the fourth-rank permutation symbols as follows:

$$\left.\begin{aligned} e_{\alpha\beta\gamma\delta} &= +1 \text{ when } \alpha\beta\gamma\delta \text{ is an even permutation of } 0, 1, 2, 3 \\ &= -1 \text{ when } \alpha\beta\gamma\delta \text{ is an odd permutation} \\ &= 0 \text{ when any two subscripts are equal} \end{aligned}\right\}. \qquad (52.41)$$

There are then four different "area" representations, which we can write either in terms explicitly of the permutation symbols and the unit dyads, or in terms of a "vector product" symbol; their relations are

$$\left.\begin{aligned} \mathbf{i}^\alpha \times \mathbf{i}_\beta &= e_{\alpha\beta\mu\nu}\mathbf{i}_\mu\mathbf{i}^\nu \\ \mathbf{i}_\alpha \times \mathbf{i}_\beta &= e_{\alpha\beta\mu\nu}\mathbf{i}^\mu\mathbf{i}^\nu \\ \mathbf{i}_\alpha \times \mathbf{i}^\beta &= e_{\alpha\beta\mu\nu}\mathbf{i}^\mu\mathbf{i}_\nu \\ \mathbf{i}^\alpha \times \mathbf{i}^\beta &= e_{\alpha\beta\mu\nu}\mathbf{i}_\mu\mathbf{i}_\nu \end{aligned}\right\}. \tag{52.42}$$

There are similarly four different representations of a three-dimensional region, obtained by taking the scalar product of each of these four vector products with a unit vector. For example,

$$\mathbf{i}^\gamma \cdot \mathbf{i}^\alpha \times \mathbf{i}_\beta = e_{\alpha\beta\gamma\nu}\mathbf{i}^\nu. \tag{52.43}$$

Evidently still another scalar product with a unit vector is possible, leading to the quadruple scalar region. Since there is a multiplicity of results, as regards sign, these quantities are neither true vectors nor scalars, but pseudo-vectors and pseudo-scalars in the sense mentioned in §4.

The determinant $|A|$ of any four-by-four set of matrix elements $A_{\alpha\beta}$ is given by the equations:

$$e_{\alpha\beta\gamma\delta}\,|A| = e_{\mu\nu\sigma\rho}A_{\alpha\mu}A_{\beta\nu}A_{\gamma\sigma}A_{\delta\rho}. \tag{52.44}$$

If for A we substitute the Lorentz transformation matrix and use the fact that this matrix has a unit determinant, we can derive from Eq. 52.44 the identity:

$$e_{\alpha\beta\gamma\delta}L_{\gamma\sigma}L_{\delta\rho} = e_{\mu\nu\sigma\rho}L_{\alpha\mu}{}^{-1}L_{\beta\nu}{}^{-1}. \tag{52.45}$$

Now from any given antisymmetric matrix F, $F_{\mu\nu} = -F_{\nu\mu}$, we can construct a permuted form P, called the *dual* of F, such that

$$P^{\sigma\rho} = \tfrac{1}{2}e_{\mu\nu\sigma\rho}F_{\mu\nu}. \tag{52.46}$$

As the notation indicates, we can prove that the dual form of a contravariant tensor is a covariant tensor. Thus the dual of the transformed components of F is, by definition,

$$P^{\sigma\rho\prime} = \tfrac{1}{2}e_{\mu\nu\sigma\rho}F_{\mu\nu}{}'. \tag{52.47}$$

Using Eq. 52.39 (ii) for $F_{\mu\nu}{}'$, we see that

$$P^{\sigma\rho\prime} = \tfrac{1}{2}e_{\mu\nu\sigma\rho}L_{\mu\alpha}{}^{-1}F_{\alpha\beta}L_{\beta\nu}{}^{-1} = \tfrac{1}{2}e_{\mu\nu\sigma\rho}L_{\alpha\mu}{}^{-1}L_{\beta\nu}{}^{-1}F_{\alpha\beta},$$

where the symmetry of L^{-1} has been used at one step. From Eq. 52.45 this is

$$P^{\sigma\rho\prime} = \tfrac{1}{2}L_{\gamma\sigma}L_{\delta\rho}e_{\alpha\beta\gamma\delta}F_{\alpha\beta} = L_{\sigma\gamma}P^{\gamma\delta}L_{\delta\rho}. \tag{52.48}$$

From the symmetry of L this shows that P transforms according to Eq. 52.39(iv), the covariant rule.

Having defined vector products, Eq. 52.42, it is now possible to define the curl operation in four dimensions. Let **F** be any vector, and write

$$\begin{aligned}\Box\times\mathbf{F} &= (\partial/\partial x_\alpha)\mathbf{i}^\alpha\times F^\beta\mathbf{i}^\beta = (\partial F^\beta/\partial x_\alpha)\mathbf{i}^\alpha\times\mathbf{i}^\beta\\ &= e_{\alpha\beta\mu\nu}(\partial F^\beta/\partial x_\alpha)\mathbf{i}_\mu\mathbf{i}_\nu = \mathrm{F}_{\mu\nu}\mathbf{i}_\mu\mathbf{i}_\nu,\end{aligned} \tag{52.49}$$

where F is the permuted form, or dual of $\Box$**F**, thus:

$$\mathrm{F}_{\mu\nu} = e_{\alpha\beta\mu\nu}\partial F^\beta/\partial x_\alpha. \tag{52.50}$$

It will be appreciated that the analog of Gauss' theorem, transforming a volume integral into a surface integral in three-dimensional space, may be generalized to four-dimensional space-time and written in the form

$$\iiiint\Box\cdot\mathbf{F}dx_0dx_1dx_2dx_3 = \iiint\mathbf{F}\cdot d\boldsymbol{\sigma}, \tag{52.51}$$

where $d\boldsymbol{\sigma}$ is now interpreted as the vector three-dimensional region bounding the four-dimensional domain of the first integral. To visualize this more clearly, let the four-dimensional domain be represented by the history of all the events within a solid cube between times t_0 and t_1. The boundary of this history consists of (1) the instantaneous solid cube at time t_0, (2) the history of the six plane faces of the cube between the times t_0 and t_1, and (3) the instantaneous solid cube at time t_1. One cannot get "outside" of this domain without going through one part of this boundary, which therefore actually encloses the domain completely. Note specially the fact that the boundary is everywhere a three-dimensional region: such a region is often called a *hypersurface* in four dimensions. This is but one example of the many interesting theorems that can be formulated in four-dimensional analysis, but for our present purposes we do not need to go further into such matters.

Exercise V.52

52.1. Write out the nine dyads $\mathbf{i}^\alpha\mathbf{i}_\beta$ in the form of matrices and verify that their traces are given correctly by Eq. 52.11.

52.2. Prove that $E\cdot A = A\cdot E$, where A is any dyadic and E is the dyadic defined in Eq. 52.15.

52.3. Define the covariant components of the 4-vector interval given in Eq. 52.16 and find their values in terms of the contravariant components.

52.4. Derive Eq. 52.22 from Eq. 52.21 without using the dummy subscript notation.

52.5. Use Eqs. 52.23 and 52.21 to derive Eq. 52.26.

52.6. Verify Eq. 52.27.

52.7. Prove that any dummy subscript, *e.g.* the subscript α in Eq. 52.28, can be replaced by any other letter symbol not already occurring in the expression, *e.g.* $\alpha\rightarrow\mu$ in Eq. 52.28, without altering the meaning of the equation. This is the reason for the name "dummy."

52.8. Prove Eqs. 52.32 and 52.33 by taking scalar products of Eqs. 52.22 and 52.24 with an appropriate unit vector.

52.9. Derive Eq. 52.35.

52.10. Derive the transformation relations for mixed and covariant tensors given in Eq. 52.39 from Eq. 52.38.

52.11. Write down explicit matrix forms for the dyads $\mathbf{i}_\alpha\mathbf{i}_\beta$, $\mathbf{i}_\alpha\mathbf{i}^\beta$ and $\mathbf{i}^\alpha\mathbf{i}^\beta$ from the unit dyad $\mathbf{i}^\alpha\mathbf{i}_\beta$ and Eq. 52.2.

52.12. Find the covariant and mixed components of a dyadic in terms of its contravariant components.

52.13. Write down the four-dimensional divergence of a pure contravariant tensor.

52.14. Let the Lorentz transformation have the matrix form

$$L = \begin{pmatrix} \cosh\phi & \sinh\phi & 0 & 0 \\ \sinh\phi & \cosh\phi & 0 & 0 \\ 0 & 0 & 1 & 0 \\ 0 & 0 & 0 & 1 \end{pmatrix}.$$

Find L^{-1} in matrix form. Find the dyads $\mathbf{i}^{\alpha\prime}\mathbf{i}_\beta{}'$, $\mathbf{i}^{\alpha\prime}\mathbf{i}^{\beta\prime}$ and $\mathbf{i}_\alpha{}'\mathbf{i}^{\beta\prime}$ and $\mathbf{i}_\alpha{}'\mathbf{i}_\beta{}'$, using only the transformation rules of Eqs. 52.23 and 52.21.

52.15. Write down all the e-symbols in detail from Eq. 52.41, *e.g.* $e_{1320}=1$.

52.16. Write out in detail all the equations implied in Eq. 52.43, using Eq. 52.41, *e.g.* $\mathbf{i}^1\cdot\mathbf{i}^2\times\mathbf{i}_3=e_{2310}\mathbf{i}^0=-\mathbf{i}^0$.

52.17. Verify Eq. 52.44 and prove that

$$|A| = (1/4\,!)e_{\alpha\beta\gamma\delta}e_{\mu\nu\sigma\rho}A_{\alpha\mu}A_{\beta\nu}A_{\gamma\sigma}A_{\delta\rho}.$$

52.18. Derive Eq. 52.45 from Eq. 52.44.

52.19. Find the dual of a pure covariant antisymmetrical tensor and prove that it is a contravariant tensor under the proper Lorentz transformations.

52.20. Prove that the dual of a symmetrical tensor is identically zero, according to the definition Eq. 52.46.

52.21. Write out the covariant components of the dyadic $\Box\mathbf{F}$ as a matrix and find the dual of this matrix, using Eq. 52.46; verify Eq. 52.50.

V.53 Invariance of the Electromagnetic Field Laws. In this section we shall set up the laws of electromagnetism in Lorentz invariant form. In the last section we introduced a number of mathematical entities with different transformation properties. If we restrict ourselves to proper Lorentz transformations, *i.e.* either simple rotations in space or uniform relative velocities in various directions, we now have the following mathematical objects, each with its own characteristic behavior under these transformations: scalar, covariant vector, contravariant vector, covariant, contravariant and two kinds of mixed tensors, all in four dimensions of space-time. In order to represent a physical law, the terms of an equation must be homogeneous in their transformation properties in order to be invariant under Lorentz transformations: the equation must have every term a scalar, or every term a covariant vector, etc. An equation between a scalar on the one hand and a vector on the other could be true in one coordinate system, but would in general fail to be true in some other coordinate system.

If we wish to make a physical law invariant under a wider class of transformations—for instance, including the improper Lorentz transformations like reflections, inversions, or time reversals—then the equations must be even more carefully set up; *e.g.* every term must be not only scalar, but true scalar, as distinct from pseudo-scalar, or all pseudo-vectors of one type, etc. In our work we shall ignore these improper Lorentz transformations, but note in passing that they become important in nuclear theory.

In attempting to fit the electromagnetic field laws into this program, we must examine each field variable and decide on its status with respect to the Lorentz transformations. We can begin with the current density vector, which is usually interpreted as charge density multiplied by velocity. Since we already have a 4-vector corresponding to velocity, we can construct a 4-vector current density simply by postulating a scalar *proper charge density* ρ_0, and multiplying this into the 4-vector velocity. Thus we write following Eq. 50.20:

$$\mathbf{j} = \rho_0 \mathbf{V} = \rho_0(c, v_x, v_y, v_z)/(1-v^2/c^2)^{1/2} \tag{53.1}$$

for the 4-vector current density. If we now identify the three space-like components of this with the ordinary 3-vector current density, we must interpret the relative charge density as

$$\rho = \rho_0/(1-v^2/c^2)^{1/2}. \tag{53.2}$$

This implies that the total charge carried in an element of volume is an invariant: the density increases with velocity according to Eq. 53.2 while the volume decreases due to the Lorentz contraction in the direction of motion: $dx' = dx(1-v^2/c^2)^{1/2}$, so that

$$\rho dx'dy'dz' = \rho_0 dxdydz. \tag{53.3}$$

The electrostatic potential Φ and the 3-vector potential $\mathbf{A}$ make a natural foursome to identify with a 4-vector. To see just how this works out we can look first at the wave equations satisfied by the two potentials:

$$\nabla^2\Phi - \epsilon\mu\partial^2\Phi/\partial t^2 = -\rho/\epsilon, \tag{53.4}$$

$$\nabla^2\mathbf{A} - \epsilon\mu\partial^2\mathbf{A}/\partial t^2 = -\mu\mathbf{J}, \tag{53.5}$$

adapted from Eqs. 41.19 and 41.22 for empty space. If we multiply the first of these by the time-like unit vector $\mathbf{i}_0$, and the second equation by c, where of course $c = 1/\sqrt{\epsilon\mu}$, and add the two results, we find:

$$(\nabla^2 - \epsilon\mu\partial^2/\partial t^2)(\mathbf{i}_0\Phi + c\mathbf{A}) = -(\mathbf{i}_0 c\rho + \mathbf{J})/c\epsilon. \tag{53.6}$$

On the right we recognize the 4-vector current density as already set up in Eq. 53.1, while the differential operator on the left is exactly the D'Alembertian, of Eq. 52.20, $\Box\cdot\Box$. This means that Eq. 53.6 is an invariant relation provided we have a 4-vector potential,

$$\boldsymbol{\Phi} \equiv (\Phi,\ cA_x,\ cA_y,\ cA_z), \tag{53.7}$$

so that the propagation equations combine to form the single 4-vector Lorentz invariant relation:

$$\Box\cdot\Box\boldsymbol{\Phi} = -\mathbf{j}/c\epsilon. \tag{53.8}$$

Because $\mathbf{j}$ is derived in Eq. 53.1 from a contravariant displacement vector (see Eq. 52.16), we must interpret the components in Eq. 53.7 as giving the contravariant components of $\boldsymbol{\Phi}$. The covariant components are therefore

$$\boldsymbol{\Phi}_{\text{cov}} = (\Phi,\ -cA_x,\ -cA_y,\ -cA_z). \tag{53.9}$$

Consider the four-dimensional divergence of the 4-vector potential using Eq. 52.18. This involves the contravariant components in Eq. 53.7 and we have

$$\Box\cdot\boldsymbol{\Phi} = (1/c)\partial\Phi/\partial t + c\nabla\cdot\mathbf{A}. \tag{53.10}$$

This expression means that the gauge convention on the vector potential in Eq. 41.25, *viz.*: $\nabla\cdot\mathbf{A}=0$, is not an invariant equation in four-dimensional analysis. An invariant equation to replace Eq. 41.25 would be

$$\Box\cdot\boldsymbol{\Phi} = 0. \tag{53.11}$$

The gauge transformation in Eq. 41.23 must be supplemented, and we now note that if we set

$$\left.\begin{aligned} \Phi' &= \Phi - \partial f/\partial t \\ \mathbf{A}' &= \mathbf{A} + \nabla f \end{aligned}\right\}, \tag{53.12}$$

where f is an arbitrary function, then the expression for $\mathbf{E}$ remains the same in the new potentials as in the old: $\mathbf{E} = -\nabla\Phi' - \partial\mathbf{A}'/\partial t$. The invariance of Eq. 53.11 to the gauge transformation then requires only that f be a solution of the D'Alembert equation:

$$\Box\cdot\Box f = 0. \tag{53.13}$$

The field vectors $\mathbf{B}$ and $\mathbf{E}$ do not behave like space parts of two 4-vectors. This is apparent when we remember that uniform motion through a magnetic field creates an e.m.f.; but uniform motion is simply a Lorentz transformation, and if this can turn a magnetic field into an electric field, neither fields can be simple vectors. We must therefore seek a tensor representation, and since there are six components involved, at least, we may guess that an antisymmetric tensor will work.

Starting from the vectors already introduced—*i.e.* the potential 4-vector and the current density 4-vector, we can construct tensors by means of the gradient operations. Thus we have at once:

$$\square \times \mathbf{\Phi} = (\partial \Phi^{\beta}/\partial x_{\alpha}) e_{\alpha\beta\mu\nu} \mathbf{i}_{\mu} \mathbf{i}_{\nu} = F_{\mu\nu} \mathbf{i}_{\mu} \mathbf{i}_{\nu}, \tag{53.14}$$

an antisymmetric tensor. Putting in the covariant components of $\mathbf{\Phi}$ from Eq. 53.9 and making use of Maxwell's equations, Eq. 41.15 and Eq. 41.17, we find the following matrix of components for F:

$$F_{\mu\nu} = \begin{pmatrix} 0 & -cB_x & -cB_y & -cB_z \\ cB_x & 0 & E_z & -E_y \\ cB_y & -E_z & 0 & E_x \\ cB_z & E_y & -E_x & 0 \end{pmatrix}. \tag{53.15}$$

Therefore, if we can assert that the electric and magnetic vectors combine to form a contravariant tensor in accord with Eq. 53.15, then the two sets of Maxwell's equations, Eqs. 41.15 and 41.17: $\mathbf{B} = \nabla \times \mathbf{A}$ and $\mathbf{E} + \partial \mathbf{A}/\partial t = -\nabla \Phi$, are equivalent to the Lorentz invariant relation

$$F = \square \times \mathbf{\Phi}. \tag{53.16}$$

It is also easy to see that the identity:

$$\square \cdot \square \times \mathbf{\Phi} \equiv 0, \tag{53.17}$$

is equivalent to those of Maxwell's equations that similarly follow as identities from Eqs. 41.15 and 41.17, *viz.*: Eqs. 41.1 and 41.2.

We can now construct another antisymmetric tensor by forming the dual tensor to F, in accord with the definition Eq. 52.46. It is easy to show that this tensor has the following (covariant) components:

$$P^{\mu\nu} = \begin{pmatrix} 0 & E_x & E_y & E_z \\ -E_x & 0 & -cB_z & cB_y \\ -E_y & cB_z & 0 & -cB_x \\ -E_z & -cB_y & cB_x & 0 \end{pmatrix}. \tag{53.18}$$

Taking the divergence of the contravariant form of this tensor:

$$\square \cdot P = \mathbf{i}_{\beta}\, \partial P_{\alpha\beta}/\partial x_{\alpha}. \tag{53.19}$$

Writing the modified components from Eq. 53.18 and using Maxwell's equations from Eq. 41.3 and Eq. 41.4, we find that

$$\square \cdot P = \mathbf{j}/c\epsilon. \tag{53.20}$$

This is a Lorentz invariant equation because both sides are contravariant 4-vectors; and so we have found the invariant form of the other Maxwell equations, *viz.*: $\nabla \cdot \mathbf{E} = \rho/\epsilon$ and $\nabla \times \mathbf{B} = \mathbf{J}/c^2\epsilon + \dot{\mathbf{E}}/c^2$.

Summarizing, we introduce the 4-vector potential $\mathbf{\Phi}$ and accept the gauge convention $\square \cdot \mathbf{\Phi} = 0$, Eq. 53.11. This potential generates the field tensor $F = \square \times \mathbf{\Phi}$, Eq. 53.16, whose components are **E** and **B** as in Eq. 53.15. The dual tensor P is then connected with the 4-vector current density by $\square \cdot P = \mathbf{j}/c\epsilon$ in Eq. 53.19. All these relations are Lorentz invariant.

A number of hypotheses have been proposed here that had better be discussed more critically. In the first place, however, let us note that Maxwell's equations are experimentally invariant for transformations between different inertial systems—*i.e.* they are Lorentz invariant. This in itself can be taken as proof of the sufficiency at least, of the identifications we have made between the electromagnetic field variables and the corresponding mathematical objects. What we would like to have is a stronger statement as to the necessity of these identifications. It is of course never possible to assert the complete necessity of any physical hypothesis, but we can derive the greatest possible diversity of detailed consequences from a hypothesis and test them experimentally. The more tests a hypothesis can pass, the less likely it is that an alternative hypothesis could be found to replace it. Here we shall enumerate several detailed checks on the identifications that have been made.

The first assumption made was essentially that charge is a scalar under Lorentz transformations—Eq. 53.3. This assumption has been verified with very high precision in modern high-energy accelerating machines; even when an electron is given speeds approaching the velocity of light, its charge remains the same. All the other assumptions can be reduced essentially to the identification of **B** and **E** as the components of an antisymmetric tensor as listed in Eq. 53.15. We can test this by working out the detailed transformation rules for such a tensor, and finding how **E** and **B** should change due to relative motion.

From Eq. 52.39(ii) the transformation rule for the contravariant electromagnetic field tensor is given by:

$$F_{\mu\nu}' = L_{\mu\alpha}^{-1} F_{\alpha\beta} L_{\beta\nu}^{-1}. \tag{53.21}$$

We may use the L matrix from Ex. 52.14, in which we found

$$L_{\mu\nu}^{-1} = \begin{pmatrix} \cosh\phi & -\sinh\phi & 0 & 0 \\ -\sinh\phi & \cosh\phi & 0 & 0 \\ 0 & 0 & 1 & 0 \\ 0 & 0 & 0 & 1 \end{pmatrix}. \tag{53.22}$$

Equating the transformed matrix to the same form as Eq. 53.15, but with primed components, and working out the matrix products on the right of Eq. 53.21, we find the following relations:

$$\begin{aligned} B_1' &= B_1, & E_1' &= E_1, \\ B_2' &= \cosh\phi\cdot B_2+\sinh\phi\cdot E_3/c, & E_2' &= \cosh\phi\cdot E_2-c\sinh\phi\cdot B_3, \\ B_3' &= \cosh\phi\cdot B_3-\sinh\phi\cdot E_2/c, & E_3' &= \cosh\phi\cdot E_3+c\sinh\phi\cdot B_2. \end{aligned} \tag{53.23}$$

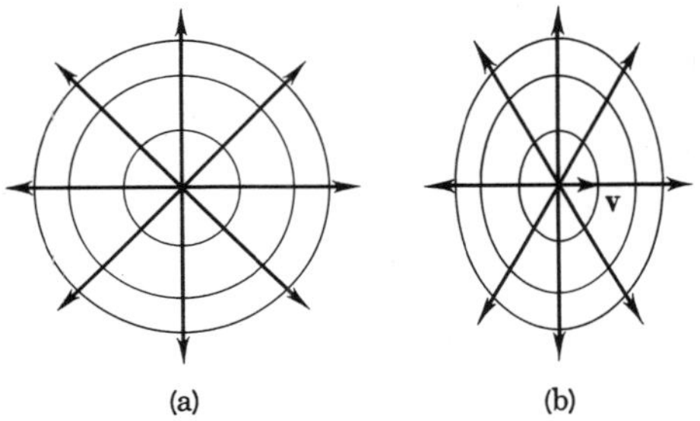

FIG. 53.1. The electric field surrounding (a) a stationary charge, and (b) a rapidly moving charge

This transformation can be expressed in terms of the relative velocity v of the two systems, where $\tanh\phi = v/c$ (see Eq. 49.28). We find then that the last two sets of equations become:

$$\left.\begin{aligned} B_2' &= (B_2+vE_3/c^2)(1-v^2/c^2)^{-1/2} \\ B_3' &= (B_3-vE_2/c^2)(1-v^2/c^2)^{-1/2} \end{aligned}\right\}, \tag{53.24}$$

$$\left.\begin{aligned} E_2' &= (E_2-vB_3)(1-v^2/c^2)^{-1/2} \\ E_3' &= (E_3+vB_2)(1-v^2/c^2)^{-1/2} \end{aligned}\right\}. \tag{53.25}$$

These relations reduce to the familiar relations for motion through a field when v^2/c^2 can be neglected compared with unity:

$$\mathbf{B}' \doteqdot \mathbf{B}-\mathbf{v}\times\mathbf{E}/c^2 \qquad \mathbf{E}' \doteqdot \mathbf{E}+\mathbf{v}\times\mathbf{B}. \tag{53.26}$$

The Lorentz contraction factor appearing in the exact relations has the effect of increasing the field components in directions perpendicular to the relative motion compared with the nonrelativistic approximation. These corrections become very large at the high speeds attained in high-energy accelerators, and calculations for the designs of these machines must take them fully into account. The mere fact that the machines operate according to specifications is abundant and decisive evidence for the accuracy of the relativistic formulas.

Exercise V.53

53.1. Show in detail that Eq. 53.9 gives the covariant components of the 4-vector potential.

53.2. Prove that the expression for **E** in terms of Φ and **A** is invariant for the gauge transformation, Eq. 53.12, provided the transformation is restricted by Eq. 53.13.

53.3. Derive Eq. 53.15 as outlined in the text.

53.4. Prove that Eq. 53.17 is an identity in four dimensions, and show that it is equivalent to Maxwell's equations, Eqs. 41.1 and 41.2.

53.5. Derive the covariant dual electromagnetic field tensor of Eq. 53.18 from the contravariant field tensor of Eq. 53.15.

53.6. Show that the antisymmetric part of the dyadic $\Box\boldsymbol{\Phi}$ is the dual of $\Box \times \boldsymbol{\Phi}$.

53.7. Show that the divergence of the antisymmetric part of $\Box\boldsymbol{\Phi}$ is the D'Alembertian of $\boldsymbol{\Phi}$ provided $\Box\cdot\boldsymbol{\Phi}=0$. Hence derive Eq. 53.20 from Eq. 53.11.

53.8. Derive Eq. 53.23 in detail as outlined in the text.

53.9. Note that the Lorentz transformation of Ex. 52.14 corresponds to motion in the positive x-direction. Run through the argument following Eq. 53.21 with v in the negative x-direction and find the corresponding forms of Eqs. 53.24 and 53.25. Remember that ct, x, y, z are contravariant components.

V.54 Einstein's Theory of Gravitation. In the restricted theory of relativity one attempts to express physical laws (laws of motion, electromagnetic field laws, etc.) in forms that remain invariant for Lorentz transformations between systems having relative uniform motion—the inertial systems. In his general theory of relativity Einstein was motivated to express the laws of motion in a form that remains invariant even when transformations are made between systems having relative accelerations. His work led him to an amazing theory of gravitation, and to understand why this was so, we shall first discuss the principle of equivalence between inertial and gravitational mass.

The principle of equivalence stems from the apparent fact that, on the basis of purely local laboratory measurements of a mechanical nature, an observer would be unable to distinguish between the effects of (a) a forced acceleration in one direction and (b) a gravitational field that would cause an equal acceleration in the opposite direction. Thus, an elevator accelerated vertically upward at 32.2 feet per sec^2 effectively doubles the weight of all objects in it; exactly as if the earth's gravitational field (downward) had been doubled. If the observer were not allowed to see anything outside the elevator he could not tell the difference. The exact equivalence of the two effects is due to the identity of inertial and gravitational mass.

The inertial mass m_i is measured by dynamical experiments and the equation:

$$\mathbf{F} = m_i\mathbf{a}, \tag{54.1}$$

where **F** is the force and **a** the acceleration. The gravitational mass m_g is measured by static experiments in a torsion balance:

$$W = Gm_g M_g / R^2, \tag{54.2}$$

where W is the weight, M_g the gravitational mass of the earth, R the radius of the earth, and G the universal gravitational constant (see Eq. 34.2). The weight acts to produce acceleration of magnitude g in a dynamic experiment:

$$W = m_i g \tag{54.3}$$

which is Eq. 54.1 for this case. Equating the two expressions for W we have a relation between the two kinds of mass, which can be written in the form

$$g = G(m_g/m_i)M_g/R^2. \tag{54.4}$$

The acceleration of free fall, g, thus depends on the ratio m_g/m_i and by comparing g for different bodies at the same place, we can find whether this ratio differs for different materials. In fact g is found to be independent of material. This is true even for heavy nuclei where the binding energy causes appreciable mass defect in accordance with the relation of Einstein, Eq. 51.10: both gravitational and inertial mass are reduced by the same amount.

From the classical point of view this circumstance is quite mysterious. Why should the inertia of a body always be exactly proportional to the gravitational "charge" of the body? Einstein approached this question as follows. In classical mechanics we have assumed that the natural undisturbed motion of a mass particle is uniform motion in a straight line, and force is required to change this state of motion; inertial mass is the reaction to this change, and equals the ratio of force to acceleration. Einstein questioned the validity of this assumption, and one can best appreciate his reasoning by asking what is meant by "undisturbed" motion? If "undisturbed" means "untouched by any other body," then free fall or satellite motion would be a truly undisturbed state of motion; to force a mass to move uniformly in a straight line near the surface of the earth requires that it be supported underneath and restrained by contact with a horizontal guide plane, and it is not logical to call this an undisturbed motion. The guide in this case must keep a force acting upward on the body exactly equal to its weight. In fact, this upward force actually is the body's weight—it is what we measure by means of scales and balances. The force supposed to be exerted downward by the earth's gravity is pure hypothesis invented to "explain" why the body does not naturally move in a straight line. Einstein then asks us to think of free fall and satellite motion as natural undisturbed motion, and weight as a force produced by contact with a support designed to prevent the natural motion. Interpreted in this way, weight is nothing but inertia—the force required to change the natural undisturbed motion; gravitational and inertial mass are not just equal to each other, they are identical.

For an observer in free fall or on a satellite in orbital motion round the earth, the sensation of weight would disappear—no supporting guide would be necessary. Neglecting the small gravitational field due to the satellite's own mass, the local gravitational field is zero in such a reference system. Transformation of coordinates, from a system at rest on the earth to a system accelerated with the satellite, has removed the gravitational field with respect to the latter system. From this one might be tempted to regard the gravitational field as in some sense fictitious, just as the centrifugal field of force produced by rotation is often spoken of as fictitious. However, we are already familiar with the fact that uniform relative motion is able to remove the magnetic field, and this has never been regarded as sufficient to demote the magnetic field to a fictitious status. The magnetic field is recognized as dependent on the relative motion between source and measuring system. Similarly we now recognize that gravitational fields depend on the relative acceleration between source and measuring system. Instead of saying that the gravitational field is just as fictitious as centrifugal and Coriolis forces, Einstein proposed to regard centrifugal and Coriolis forces as just as real as gravitational fields—in fact, that essentially they are gravitational in nature.

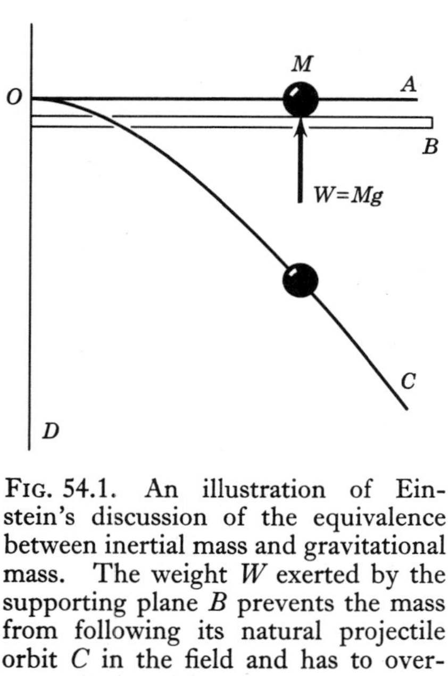

FIG. 54.1. An illustration of Einstein's discussion of the equivalence between inertial mass and gravitational mass. The weight W exerted by the supporting plane B prevents the mass from following its natural projectile orbit C in the field and has to overcome the inertial mass to do so

Rotational motion is relative in the same sense as rectilinear motion, and the rotation of the earth relative to the "fixed" galaxy of stars is made apparent by the response of Foucault's pendulum to the Coriolis forces caused by this rotation. The Coriolis forces observed on the earth are a direct gravitational result of the diurnal rotation of the stars relative to the coordinate system attached to the earth. Such a gravitational effect of relative motion is not included in Newton's theory of gravitation. On the classical picture an absolute standard of zero rotation was presumed to exist, and the Coriolis forces were indeed fictitious in the sense that they were corrections we had to make if we insisted on using a rotating reference system. In Einstein's picture, the distribution and motion of all the material in the universe determine the gravitational field at every point, the field in any reference system

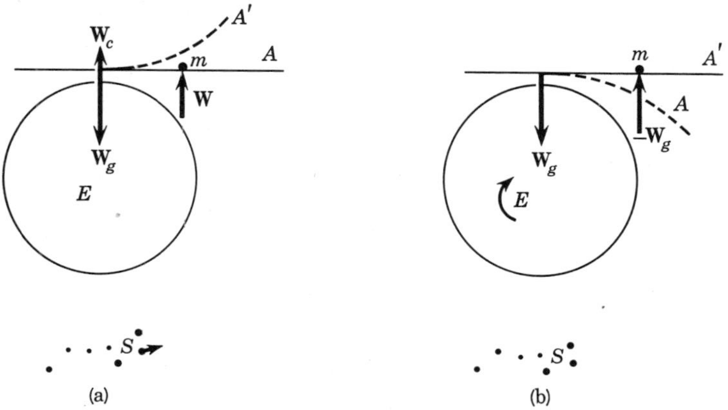

FIG. 54.2. The effect of rotation on the gravitational field on Einstein's theory compared with Newton's

(A) Coordinates with Earth at rest, stars revolving produce gravitational field $\mathbf{W}_c$ called centrifugal force. Support is needed to force the particle to follow a straight path OA in this system, $\mathbf{W}=\mathbf{W}_g-\mathbf{W}_c$.

(B) Coordinates with Earth rotating and stars at rest. No centrifugal field exists in this system, and support needed to force particle to follow straight path OA' is different in this system because the path is different, $\mathbf{W}_g$ alone is sufficient

depending on the relative acceleration of the system with respect to all the matter—*i.e.* all the sources of the field.

In order to give these ideas quantitative expression we need a mathematical scheme that treats the observed curvilinear accelerated and rotational motions of gravitationally attracting masses as "natural" in a unique sense, just as Euclidean geometry and Newtonian mechanics treated the straight line and uniform motion as uniquely natural in the absence of gravitational fields. We saw in Eq. 51.26 that in restricted relativity the law of motion was generalized to read: the natural motion of any particle is along a history of minimum absolute four-dimensional interval between given end events. In the generalized sense, this is still a "straight line," but the "space" is not the same as in Newtonian mechanics. The minimum absolute interval provides the criterion of uniqueness for the actual path. Einstein postulated that the principle of minimum absolute interval should characterize all undisturbed motion and that gravitational effects are due to a curvature of space-time: in other words, he retains the same law of motion, but modifies the continuum.

In three-dimensional geometry we are familiar, §8, with the concept of curvilinear coordinates. The concept of a curved two-dimensional region (surface) in three-dimensional Euclidean space is elementary. In differential geometry we learn how to set up the intrinsic properties of a curved surface without reference to the three-dimensional space in which it is embedded. For example, the radius of curvature of a surface can be expressed in terms of the difference between the area of a small circular element and $P^2/4\pi$ where P is its periphery. The concept of curved regions of higher dimensionality is a purely mathematical analogy. Thus in Euclidean three-dimensional geometry the ratio of volume to surface of a sphere is such that $A^{3/2}:V=6\sqrt{\pi}$. If experimental test showed that the area A and volume V were not related in this way, we would have to deny the applicability of Euclidean geometry.

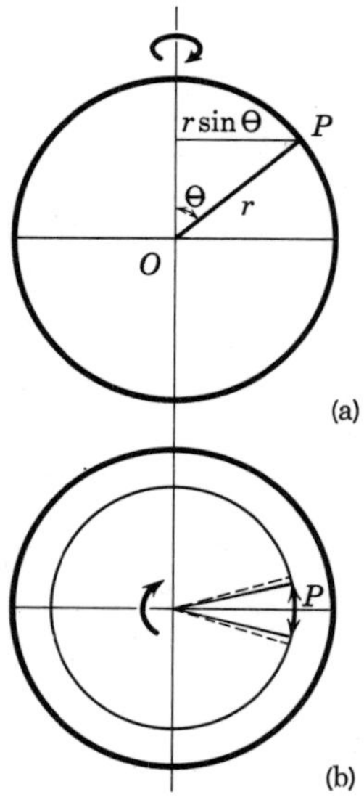

FIG. 54.3. Effect of rotation on the surface of a sphere. The measuring rod is placed on a line of latitude, suffers a Lorentz contraction due to speed $v=\omega r \sin\theta$; the circumference therefore seems longer than when the rod is not revolving with the sphere

Consider measurements made on the surface and volume of a sphere of radius r, by means of a rotating measuring system, having an angular velocity ω about an axis of the sphere. From the point of view of a nonrotating system, the length of a unit scale used in the rotating system to measure the periphery of the sphere around any line of constant latitude would suffer the Lorentz contraction in the ratio $1:[1-(\omega r/c)^2 \sin^2\theta]^{1/2}$. The total apparent length of a colatitude circle as measured by such a scale is then

$$2\pi r \sin\theta[1-(\omega r/c)^2 \sin^2\theta]^{-1/2}.$$

The radius of the sphere would be unaltered by the contraction because the scale does not change in directions perpendicular to the motion. The area of the sphere on the rotating system integrates to

$$A = 4\pi r^2[1+\tfrac{1}{2}(\omega r/c)^2+\cdots]. \tag{54.5}$$

The volume of the same sphere on the rotating system can be found by integrating this with respect to the radius:

$$V = (4/3)\pi r^3[1+(3/10)(\omega r/c)^2+\cdots]. \tag{54.6}$$

The relation of volume to area is now

$$A^{3/2}:V = 6\sqrt{\pi}[1+(1/5)(\omega r/c)^2+\cdots]. \tag{54.7}$$

For larger volumes the deviation from Euclidean becomes serious. We can only interpret this as a failure of Euclidean geometry for the rotating system. In fact, even the Lorentz transformation itself becomes questionable. If the rotating observer attempts to set up his system to distances where $\omega r > c$ he will find an impasse, at least if the Lorentz transformation is true in the presence of the rotation. We may interpret this difficulty as roughly analogous to the attempt to map the polar regions of the earth's surface on the projection of Mercator. It is even worse in fact: the "flat" theory of restricted relativity applied to the rotating systems of coordinates is unable to give a real set of coordinates to any events beyond the distance c/ω from the center.

We can speak of a curved three-dimensional space embedded in a space of more than three dimensions which is itself Euclidean. The rotating system of coordinates is a four-dimensional one, and to give this a curvature, we need a generalization of Euclidean space of more than four dimensions. A plane curve can be drawn in a flat two-dimensional space. A twisted curve requires a three-dimensional Euclidean space. The more complicated the curvature, the greater the excess dimensionality needed in the embedding Euclidean space. Einstein discovered that, in order to give space-time sufficient curvature, an embedding Euclidean space-time of ten dimensions is needed.

Speaking quite qualitatively, one may assert that by allowing all kinds of wild curvatures in space-time, one could produce paths of minimum interval as tortuous as he pleases. Nature does not behave in this way, and Einstein sought to express the law of gravitation as a limitation on the kinds of curvature existing in the physical world. This limitation must be expressed in invariant form—*i.e.* it must be a tensor equation in terms of the parameters describing the intrinsic curvature of the space-time manifold. He chose the simplest one that would not be trivial. Solutions of the equation can then be sought corresponding to different mass distributions.

In restricted relativity we have so far always used Cartesian type coordinates and written: $ds^2 = c^2dt^2-(dx^2+dy^2+dz^2)$. We could equally well use spherical polar coordinates in space and write instead:

$$ds^2 = c^2dt^2-dr^2-r^2d\theta^2-r^2\sin^2\theta d\phi^2. \tag{54.8}$$

This form of interval is of course still for flat space-time, in the sense that one can transform Eq. 54.8 back into the Euclidean-Cartesian form by means of a change in coordinate system without any volume distortion anywhere. There exists a coordinate system in Einstein's curved space-time in which the interval expression is similar with Eq. 54.8, *viz.*:

$$ds^2 = c^2\gamma dt^2 - dr^2/\gamma - r^2 d\theta^2 - r^2 \sin^2\theta d\phi^2, \tag{54.9}$$

where

$$\gamma = 1 - 2mG/rc^2,$$

and this is interpreted as the appropriate curved space-time interval in the neighborhood of a mass m (at the origin), producing a gravitational potential mG/r. This expression is then regarded as the quadratic form appropriate for the description of undisturbed planetary motion in the neighborhood of a central mass m. The law of motion is the same as in restricted relativity:

$$\delta \int ds = 0, \tag{54.10}$$

where the variation is between alternative paths or histories connecting two given end events. This law is exactly the same in form as the Hamilton variation principle, with a Lagrangian function that we can easily write down now that we have the quadratic form ds^2 in Eq. 54.9. Thus if we set

$$L = -m_0 c[c^2\gamma - (\partial r/\partial t)^2/\gamma - r^2(\dot{\theta}^2 + \sin^2\theta\dot{\phi}^2)]^{1/2}, \tag{54.11}$$

then Eq. 54.10 becomes

$$\delta \int L dt = 0. \tag{54.12}$$

This is the same as Eq. 24.2, and its solution is Eq. 24.15. Working this out in detail, it is discovered that the orbit of a planet or satellite is almost identical with that predicted by Newton's law of gravitation; however, a small precession of the major axis of the ellipse is predicted in addition to the effects expected from Newton's theory. This precession was observed in the orbit of the planet Mercury, whose path round the Sun is sufficiently eccentric for the effect to be measurable.

The path of a ray of light is given by equating the interval to zero, $ds = 0$, in Eq. 54.9. With $d\theta = d\phi = 0$, we have the radial velocity $dr/dt = c\gamma$, or

$$c_{\text{grav}} = c(1 - 2mG/rc^2). \tag{54.13}$$

The gravitational potential reduces the velocity of propagation of light, so that the field is equivalent to a refractive index greater than unity. Slight deviations are therefore expected in the apparent positions of stars seen near the limb of the Sun during total eclipse; this shift has been observed and measured exactly as predicted.

To make the law of motion in the gravitational field, Eq. 54.10, satisfy the principle of invariance, we must allow the coordinate transformations to leave ds invariant in Eq. 54.9. This expression was however only a particular case. In general, all possible solutions of Einstein's law of gravitation may lead to intervals having the general quadratic form

$$ds^2 = g_{\mu\nu} dx_\mu dx_\nu \tag{54.14}$$

where the coefficients $g_{\mu\nu}$ describe the details of the gravitational field. Einstein's law of gravitation restricts the functional form of these coefficients as functions of position and time. In order that this expression be invariant, it is necessary that the coefficients behave like a tensor set. This transformation rule causes the apparent gravitational field to depend directly on the relative motion, including acceleration and rotation of the observer with respect to the source of the field, which, as we have noted, is a novel feature of Einstein's theory of gravitation compared with the classical theory.

For a more complete discussion of Einstein's theory of gravitation it is necessary to understand Riemannian differential geometry, and it is not possible to go fully into this interesting branch of mathematics in the present course.

Exercise V.54

54.1. Check the derivations of Eqs. 54.5 and 54.6.

54.2. Find the relation between periphery and area of a circle for coordinates rotating with angular velocity ω about an axis through the center of the circle.

54.3. The standard meter is defined as equal exactly to 1,650,763.73 times the wavelength of the radiation emitted *in vacuo* by krypton atoms making the transition between levels $2p_{10}$ and $5d_5$. If the krypton source were mounted on an artificial satellite orbiting between 100 and 4000 miles above the surface of the Earth, what numerical effect would the radial motion of the source have in the wavelength measured against a standard meter at rest on the Earth? This problem involves the relativistic Doppler effect.

References

The following texts contain material for collateral reading with the contents of this chapter.

Relativity; The Special Theory, J. L. Synge, North Holland Co., 1956.

The Theory of Relativity, C. Møller, The Clarendon Press, Oxford, England, 1952.

Introduction to the Theory of Relativity, P. G. Bergmann, Prentice-Hall Inc., Englewood Cliffs, N.J., 1942.

The Mathematical Theory of Relativity, A. S. Eddington, Cambridge University Press, New York, 1930.

Relativity, The Special and General Theories, A. Einstein, Methuen, London, England, 1920.

Relativity, Thermodynamics and Cosmology, R. C. Tolman, The Clarendon Press, Oxford, England, 1938.

The Principles of Natural Knowledge, A. N. Whitehead, Cambridge University Press, New York, 1925.

The Classical Theory of Fields, L. Landau, Addison-Wesley Publishing Co., Inc., Reading, Mass., 1951.

Chapter VI

QUANTUM THEORY

Chapter VI

QUANTUM THEORY

VI.55 Wave-Particle Duality. In this course we first studied some selected topics in the classical continuum theory of matter (Chapter II), finding this theory generally adequate for the discussion of many phenomena on the gross scale as they are presented to our unaided senses. As soon as we begin to probe beneath the immediately obvious with the aid of the microscope and other scientific instruments, phenomena are discovered for the description of which it becomes necessary to include the molecular structure of matter. The molecular theory of matter (Chapter III) can be carried only so far without introducing the concepts of field theory, which is once again a continuum theory (Chapter IV). The bifurcation of the physical world into matter and field, the first being discrete and the second a continuum, was generally felt to be philosophically unsatisfactory and logically incomplete. Many attempts were made, following the spectacular successes of Maxwell's field theory of electromagnetic phenomena, to discover, within the mathematical formalism of the field theory, some form of singularity in the field that could be identified with atomic material particles. For example, at one time it was hoped that vortices in the ether could be identified with electrons. With the abandonment of the ether theory, these attempts were given up, but the idea of a unified mathematical model that will include both field and particles has never been given up. In fact, the search for such a mathematical model became more urgent with the discovery that the electromagnetic field was itself capable of exhibiting particle-like characteristics; in other words, the continuum theory was not a completely valid picture even of the field.

Historically, the first indications of the discrete characteristics of electromagnetic fields were seen in the fact that the energy distribution in black-body radiation in equilibrium with matter at a definite temperature could not be explained in terms of the continuum theory. Rayleigh and Jeans proved that no finite amount of energy in a continuum field could be in equilibrium with matter: the energy density required at high frequencies goes up indefinitely with frequency, and no theoretical limit exists to the frequency in a continuum. In actual fact, black-body radiation shows a frequency of maximum energy density, the frequency being proportional to the absolute temperature (Wien's law). Planck discovered in 1900 that the "ultraviolet catastrophe" was resolved if one assumed that matter can emit or absorb radiation of any specified frequency ν only in quanta of energy equal to $h\nu$, where h is now known as *Planck's constant* and has the value: $h = 6.6253 \times 10^{-27}$ erg sec. Planck's reasoning went somewhat as follows:

We consider an enclosure having perfectly reflecting walls, cubic in shape, with sides of length L. We investigate the possible standing waves of electromagnetic radiation in the enclosure—called the normal modes of the radiation field by analogy with the standing waves of an elastic membrane (*cf.* Chapter II, §10). We can build up electromagnetic radiation in the enclosure by means of a small piece of matter maintained at a constant temperature T, capable of emitting and absorbing radiation of all frequencies—it is black in color. When the radiation field and the piece of matter are in equilibrium at the temperature T, the radiation field is said to have the energy distribution of black-body radiation at this temperature. Planck assumed that the piece of matter can handle only discrete quanta of energy $h\nu$. Because of this discreteness, we can use statistical methods to determine the distribution in energy.

To find the normal modes of the radiation field in the enclosure we note that any plane wave of frequency ν may be represented by the function

$$\Phi(\mathbf{r}, t) = Ae^{-2\pi i\nu t}e^{2\pi i\mathbf{k}\cdot\mathbf{r}}, \tag{55.1}$$

where $\mathbf{k}$ is a vector normal to the wave front in the direction of propagation equal in magnitude to the reciprocal wavelength. The space factor in this expression may be expressed in the form of a product of sine functions, aside from an arbitrary phase angle, thus:

$$e^{2\pi i\mathbf{k}\cdot\mathbf{r}} \sim \sin(2\pi k_1 x)\cdot\sin(2\pi k_2 y)\cdot\sin(2\pi k_3 z). \tag{55.2}$$

This function will form standing waves by reflection at the walls of the enclosure only if each factor vanishes at the walls, *i.e.* if and only if

$$k_j = n_j/2L \quad (j = 1, 2, 3), \tag{55.3}$$

where n_j's are any three positive integers. Since the velocity of propagation is c, the same for all frequencies, this equation results in a restriction on the possible frequencies:

$$\nu\{n\} = (c/2L)(n_1{}^2 + n_2{}^2 + n_3{}^2)^{1/2} \tag{55.4}$$

for the frequency associated with the set of integers $\{n\}$.

The number of possible frequencies between two values ν and $\nu+d\nu$ is the number of integers that satisfy the condition

$$2L\nu/c \leqslant (n_1^2+n_2^2+n_3^2)^{1/2} \leqslant 2L(\nu+d\nu)/c. \qquad (55.5)$$

In visible radiation, the *wave numbers* (reciprocal wavelengths $=\nu/c$) are of the order 10^5 per cm, so the numbers n_j are very large. A change of unity is therefore infinitesimal, and we may estimate the number given by Eq. 55.5 as the volume of an octant spherical shell in n-space between two radii, $2L\nu/c$ and $2L(\nu+d\nu)/c$, which is

$$N(\nu)d\nu = 4\pi(L/c)^3\nu^2 d\nu. \qquad (55.6)$$

This is the number of stationary states, or normal modes, of the radiation field having frequencies between ν and $\nu+d\nu$.

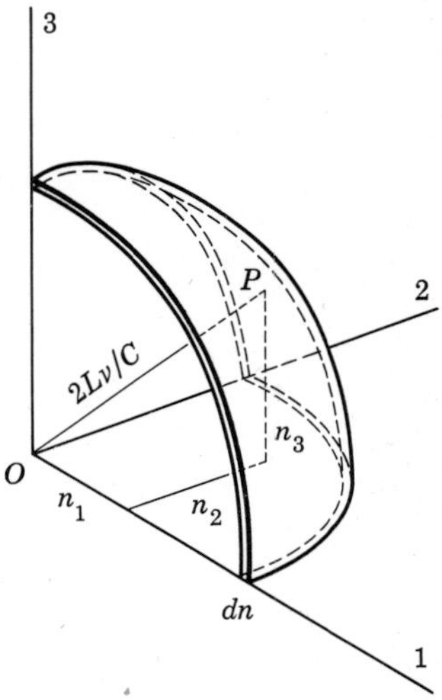

FIG. 55.1. Equation 55.3 illustrated

According to Planck's hypothesis, the piece of matter can provide only integral numbers of quanta $h\nu$, so that there can only be a quantity $rh\nu$ of energy at the frequency ν, where r is some integer. To find the distribution in energy we need to find the most probable value of r as a function of frequency. To do this we can imagine a large number of similar enclosures—*viz.* an ensemble of enclosures—and throw quanta of energy at random into these enclosures. Let the total number of enclosures be N, and their total energy be E, and apply the formulas of Eqs. 29.12 and 29.13.

Using Eq. 29.12 to obtain $e^{-\alpha}$, we can express Eq. 29.13 in the form:

$$E/N = \sum_j E_j e^{-\beta E_j} / \sum_j e^{-\beta E_j}, \qquad (55.7)$$

for the average in the ensemble for the energy per enclosure. Now we have for any given frequency ν, $E_j = jh\nu$, and therefore the two series in Eq. 55.7 can actually be summed easily:

$$\sum_j e^{-\beta E_j} = 1/(1-e^{-\beta h\nu}), \qquad (55.8)$$

and

$$\sum_j E_j e^{-\beta E_j} = -(\partial/\partial\beta)\sum_j e^{-\beta E_j} = -h\nu e^{-\beta h\nu}/(1-e^{-\beta h\nu})^2. \qquad (55.9)$$

Then Eq. 55.7 becomes:

$$E/N = h\nu/(e^{\beta h\nu}-1), \qquad (55.10)$$

for the average energy per enclosure associated with the frequency ν.

Combining this result with Eq. 55.6, we assert that, on the average in the ensemble, the energy present in one enclosure having frequency in the range ν to $\nu + d\nu$ is

$$E(\nu)d\nu = 4\pi(L/c)^3 h\nu^3(e^{\beta h\nu} - 1)^{-1}d\nu. \tag{55.11}$$

This is *Planck's formula for black-body radiation* and agrees perfectly with the observed character of radiation in equilibrium with matter at the temperature T given by Eq. 31.10, *viz.*: $\beta = 1/kT$.

In 1904, Planck's quantum hypothesis received a much more direct verification when Einstein showed that it explained the details of the photoelectric effect. Here the electrons ejected from metals by light waves always receive integral numbers of quanta. Even under a very feeble illumination where a single quantum would, on the continuum theory, be spread over a wide area, electrons are ejected always with whole quanta, never with a fraction of a quantum. There is nothing in the continuum theory of the electromagnetic field that can account for this sudden concentration of energy from the wave into the electron.

Still more convincing evidence was secured in 1923 by A. H. Compton's discovery of the shift in frequency of X-rays scattered by electrons. In this experiment, a quantum of X-radiation is absorbed by an electron whose initial energy is negligible compared with the X-ray quantum. The latter behaves like a particle of energy $h\nu$ and momentum $h\nu/c$:

$$E_{\text{photon}} = h\nu, \quad p_{\text{photon}} = h\nu/c = h/\lambda. \tag{55.12}$$

When behaving like a particle, the quantum of radiation is called a *photon*. Energy and momentum are conserved during the collision and the photon is scattered into a different direction after transferring some of its energy to the electron. Relativistic expressions for the energy and momentum of the electron have to be used in the calculation. In terms of Fig. 55.2 we have the following relations:
Conservation of energy demands that

$$h\nu + m_0c^2 = h\nu' + mc^2, \tag{55.13}$$

where m_0 is the rest mass of the electron and m is its mass in motion after the collision. Conservation of momentum in the x-direction requires:

$$h\nu/c = (h\nu'/c)\cos\beta + mv\cos\alpha, \tag{55.14}$$

and conservation of momentum in the y-direction requires:

$$0 = (h\nu'/c)\sin\beta - mv\sin\alpha. \tag{55.15}$$

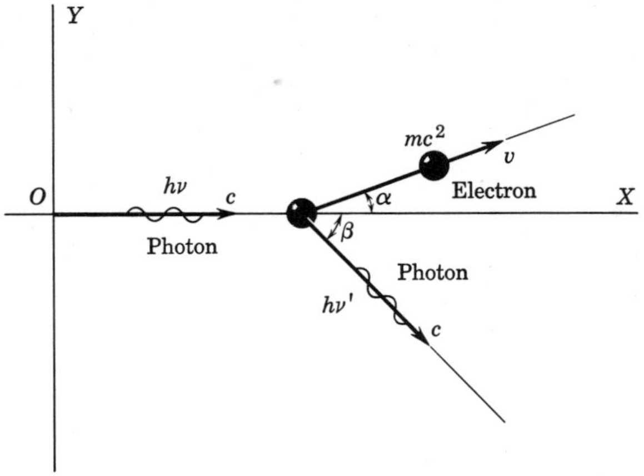

FIG. 55.2. The Compton effect on electron scattering of a photon

Eliminating α between the last two equations, we obtain

$$(h/c)^2(\nu^2 - 2\nu\nu' \cos\beta + \nu'^2) = m^2v^2, \qquad (55.16)$$

while from Eq. 55.13 we have

$$h^2(\nu-\nu')^2 + 2h(\nu-\nu')m_0c^2 + m_0{}^2c^4 = m^2c^4. \qquad (55.17)$$

Subtracting Eq. 55.16 from Eq. 55.17 and using Eq. 51.2 on the right, we eliminate v and find

$$m_0(\nu-\nu') - (h\nu\nu'/c^2)(1-\cos\beta) = 0. \qquad (55.18)$$

This may also be expressed in the form:

$$\lambda' - \lambda = (2h/m_0c)\sin^2(\tfrac{1}{2}\beta), \qquad (55.19)$$

giving the observed shift in wavelength as a function of the angle of scattering. The factor h/m_0c is called the *Compton wavelength*: the shift is equal to this Compton wavelength for scattering at right angles. Clearly the success of this theory of the Compton effect gives evidence supporting both the theory of relativity and the quantization of the electromagnetic field (see Eq. 51.24.)

To complete the collapse of classical ideas, de Broglie suggested in 1924 that material atomic particles might be the discrete aspects of a field, in analogy with the photon-wave duality of the electromagnetic field, and predicted that a matter-wave should be observable with a wavelength determined by the momentum of the atomic particle through the same formula that applied in the electromagnetic field, Eq. 55.12:

$$\lambda_{\text{atom}} = h/p_{\text{atom}}. \qquad (55.20)$$

This prediction was confirmed by the diffraction experiments of

Davisson and Germer in 1927, and G. P. Thomson in 1928. It is now clear that both "matter" and "field" have dual properties, continuum and discrete, and that the classical dichotomy of the physical world into continuum field and particulate matter is false. Nevertheless we are faced with perhaps an even more difficult problem—how to construct suitable mathematical entities that shall have both wave and particle characteristics needed to explain quantum phenomena. The search for a completely acceptable theory is still being pressed. Fantastic progress has in fact been made during the past thirty years, both in the discovery of new phenomena and in their theoretical interpretation; the wave-particle duality has been completely resolved.

Most of the discussion to be presented in this concluding chapter of our text will be based on frankly heuristic arguments. Some of the steps must also be presented without all the evidence, because they are based on recent developments that cannot be appropriately included in our course. It is felt, however, that this type of presentation is both more realistic and better calculated to stimulate study beyond the present text, which, after all is but an "introduction."

We shall attempt to rationalize quantum theory in terms of the following ideas. The interactions between matter and radiation, observed as the emission or absorption of quanta (photons), are regarded as elementary events of the type postulated in §49 in connection with the theory of relativity. The photons themselves are quasi-particles, abstractions invented for convenience in tracing the causal connections between emission and absorption events. In electromagnetic theory we have little difficulty in accepting the electromagnetic field as the "reality" that controls the situation. In fact, it is possible to interpret the intensity of the electromagnetic wave at any frequency, $I(\nu)$, as the probability of observing a photon of energy $h\nu$ in case an apparatus is set up to detect photons at the point where this intensity is defined. The electromagnetic field laws are interpreted as the probability calculus appropriate for describing the causal interrelations among this type of events.

Because electromagnetic waves were discovered and became familiar long before photons were discovered, and because familiar experience is still mostly concerned with the wave aspects of the electromagnetic field, it is easy to accept the photon as an abstraction, invented for convenience in the description of the interaction between matter and radiation. We do not doubt the "reality" of the photon just because it has properties that are not familiar in large-scale classical particles. The fact that the photon does not exist as a localizable entity, except at the moment of absorption or emission, does not detract from its respectability.

On the other hand, the fact that the particle character of atoms and electrons was discovered long before their wave-like characteristics may make it difficult for us to accept the fact that electrons and atoms, as particles, are only abstractions from experience, invented for the purpose of correlating certain events in the laboratory. Yet this is the clear implication of modern discoveries. The physically basic entities are events, scintillations, photographic traces, cloud-chamber tracks, etc., directly observed in experimental work. The proper theory of such events appears to be a field theory yielding a probability calculus describing the causal interrelations among the events. The fundamental particles, such as electrons, atoms, etc., are never observable directly; their properties have been invented for convenience in correlating events; if these properties are assumed to be completely analogous with those of massive particles, their usefulness disappears—they lead to contradictions. The basic reality is not describable in terms of such atomic particles, but must be a field controlling the probabilities of events associated with these particles. The discovery of the proper field laws controlling atomic events is the central problem of quantum theory.

Exercise VI.55

55.1. Review the discussion of §29, replacing the words "system or particle" by the phrase "normal modes of a radiation enclosure." By treating the energy states E_j of the normal mode of frequency ν as being integral multiples of quanta $jh\nu$, justify Planck's reasoning leading to Eq. 55.10. Discuss the assumption that the statistical parameter β in this formula is to be related to the absolute temperature of the black body by Eq. 31.10.

55.2. Find the frequency $\nu_{\max}$ at which the energy density $E(\nu)$ of Eq. 55.11 is a maximum. Discuss the result graphically.

55.3. Find the total energy density in black-body radiation, overall frequencies from zero to infinity, as a function of T. The result is *Stefan's law*.

55.4. Find the energy of the ejected (recoil) electron following Compton scattering, as a function of the frequency of the photon and the direction of ejection.

55.5. Classical results can be deduced from quantum theory formulas by letting Planck's constant h go to zero. Discuss the black-body radiation formula, Eq. 55.11, and the Compton effect, Eq. 55.19, from this point of view.

VI.56 Schrödinger's Wave Equation. The construction of a mathematical formalism that shall possess both continuous and discrete properties is not in principle a difficult problem. In fact, we have already seen just such a construction in the continuum theory of the vibrating membrane, §10. The motion is continuous, the displacement $u(x, y, t)$ is a continuous function of position and time; so far as the mechanical laws of motion are concerned any frequency of vibration is possible, but boundary conditions induce a discrete spectrum of normal modes. Quantization of a continuum by the imposition of boundary conditions is in fact a familiar classical situation. In 1926, Erwin Schrödinger discovered a mathematical formulation of Louis de Broglie's hypothesis of matter waves that made maximum use of this classical method of quantization. In this section we present Schrödinger's theory without attempting any historical perspective, emphasizing rather the postulational nature of the theory: the formalism has been accepted because it succeeds in producing many correct results that were formerly completely inexplicable.

Consider any mechanical problem, for example the motion of an electron in the electric field of an atomic nucleus. Set up the classical Hamiltonian for this mechanical system, Eq. 27.4, as a function of the momenta p_j and the coordinates q_j, $H(q_1, q_2, \cdots, p_1, p_2, \cdots)$. Wherever p_j appears in this function we replace it by the differential operator $-i\hbar\partial/\partial q_j$; $\hbar = h/2\pi$:

$$\mathbf{p} \to -i\hbar\nabla. \tag{56.1}$$

This has the effect of transforming the classical Hamiltonian function into a differential operator. Then we write down the classical equation between the Hamiltonian function and the total energy E:

$$H = T + V = E, \tag{56.2}$$

transform H into the operator, and replace E by the operator

$$E \to i\hbar\partial/\partial t, \tag{56.3}$$

thereby replacing Eq. 56.2 by the differential equation:

$$H(q, -i\hbar\nabla)\Psi = i\hbar\partial\Psi/\partial t, \tag{56.4}$$

where Ψ is some function of position and time to be determined by solving this differential equation. In fact, we shall think of Ψ as the amplitude of the de Broglie wave, and Eq. 56.4 as the *wave equation* for these waves. We shall first illustrate Schrödinger's prescription for setting up the wave equation in a few cases of special interest, before further discussing the physical meaning of the wave function Ψ.

Consider first the classical problem of the motion of a mass particle under no forces, but restricted to the interior of a cubical box. The potential function is zero inside the box, and the impenetrability of the walls may be represented by setting the potential discontinuously infinite there. The classical Hamiltonian in Cartesian coordinates here is $p^2/2m + V$, where V is the potential function. *Schrödinger's equation* is therefore

$$-(\hbar^2/2m)\nabla^2\Psi + V(x, y, z)\Psi = i\hbar\partial\Psi/\partial t, \tag{56.5}$$

where the potential function is zero at all points inside the cube, but becomes discontinuously infinite at the walls. The only solution with V infinite is obviously $\Psi = 0$. Therefore we can look for solutions of:

$$-(\hbar^2/2m)\nabla^2\Psi = i\hbar\partial\Psi/\partial t \tag{56.6}$$

that satisfy the boundary condition $\Psi = 0$ on the walls. We shall look only for solutions that are separable in the time and the space variables:

$$\Psi(x, y, z, t) = u(x, y, z)\Phi(t). \tag{56.7}$$

Substituting this form into Eq. 56.6 we find that the variables can be separated:

$$\frac{-(\hbar^2/2m)\nabla^2 u(x, y, z)}{u(x, y, z)} = \frac{i\hbar\partial\Phi(t)/\partial t}{\Phi(t)} = \hbar\omega \tag{56.8}$$

with ω a separation constant. The solution of the time equation for Φ is then

$$\Phi(t) = e^{-i\omega t}. \tag{56.9}$$

Thus ω is the angular frequency of the de Broglie wave. Using the operator for total energy in Eq. 56.3 on this function yields the relation:

$$E\Phi = i\hbar(\partial/\partial t)e^{-i\omega t} = \hbar\omega e^{-i\omega t} = \hbar\omega\Phi. \tag{56.10}$$

Thus the total energy is equivalent to $\hbar\omega = h\nu$, and this is the de Broglie hypothesis, relating the frequency of the wave with the energy of the particle.

We can now write the space-part of Eq. 56.8 in the form:

$$-(\hbar^2/2m)\nabla^2 u = Eu. \tag{56.11}$$

This equation can also be solved by separating the variables x, y, and z. The result can easily be shown to have the following form:

$$u = {\sin \atop \cos}(k_1 x)\ {\sin \atop \cos}(k_2 y)\ {\sin \atop \cos}(k_3 z), \tag{56.12}$$

where the notation means that either the sine or the cosine function may be chosen for each factor, and where

$$k_j^2 = 2mE_j/\hbar^2\ (j = 1, 2, 3;\quad E_1 + E_2 + E_3 = E), \tag{56.13}$$

the E's being separation constants. We are to accept only such solutions as vanish on the walls of the box. Let the box have its center at the origin, and sides of length L. Then the function u must vanish wherever

$$x = \pm\tfrac{1}{2}L,\quad y = \pm\tfrac{1}{2}L,\quad z = \pm\tfrac{1}{2}L. \tag{56.14}$$

This can be ensured if we restrict the energies in such a way that in Eq. 56.13

$$k_j = n_j\pi/L,\quad (j = 1, 2, 3), \tag{56.15}$$

where n_j are integers. Wherever n_j is an even integer we must use the sine function in that factor, and wherever n_j is an odd integer we must use the cosine factor. For example, we could have

$$u = \sin(2\pi x/L)\cos(\pi y/L)\cos(3\pi z/L),$$

which certainly vanishes at $x = \pm\frac{1}{2}L$, $y = \pm\frac{1}{2}L$, $z = \pm\frac{1}{2}L$. The condition on the energy implied in Eqs. 56.15 and 56.13 is:

$$E = (n_1^2 + n_2^2 + n_3^2)h^2/8mL^2, \tag{56.16}$$

and this is the quantization rule for the energy of the matter waves that accompany the particle in the box. The lowest possible energy is that given by taking each integer equal to unity. Note that if we took any of the integers zero, an even value, we would be obliged to use the sine function, and this would vanish completely—so there would be no matter-wave at all. The lowest energy possible for the matter-wave is then:

$$E_{\min} = 3h^2/8mL^2. \tag{56.17}$$

This is called the *zero-point* energy of a particle in the box. It is one of the most strikingly novel features of quantum theory compared with classical mechanics, which of course would allow the particle to have absolutely no energy at all, as the minimum value.

Combining the functions u in Eq. 56.12 with the time function Φ in Eq. 56.9, as required by Eq. 56.7, we see that the solutions are all standing wave solutions with fixed nodes. They are the so-called steady states of matter waves in the box, completely analogous with the standing waves one can have of electromagnetic radiation in a resonance chamber.

Next let us consider the quantum mechanical theory of the linear harmonic oscillator. Here the potential energy is proportional to the square of the distance from the center of equilibrium, and the classical Hamiltonian is $H = p_1^2/2m + \frac{1}{2}Kx^2$, where p_1 is the x-momentum. Schrödinger's equation is then

$$-(\hbar^2/2m)\partial^2\Psi/\partial x^2 + \tfrac{1}{2}Kx^2\Psi = i\hbar\partial\Psi/\partial t. \tag{56.18}$$

Again we look only for solutions that separate in the two variables:

$$\Psi(x, t) = u(x)\Phi(t), \tag{56.19}$$

and find again that

$$\Phi = e^{-iEt/\hbar}. \tag{56.20}$$

The equation for $u(x)$ is then:

$$-(\hbar^2/2m)d^2u/dx^2 + \tfrac{1}{2}Kx^2u = Eu. \tag{56.21}$$

This equation can be solved by expansion of u as a power series in x.

If we impose only the boundary condition that u shall vanish at great distances, $u \to 0$ as $x \to \infty$, it is found that no solution exists unless E is carefully chosen. In fact we have to make

$$E_n = (n+\tfrac{1}{2})h\nu, \qquad \nu = (1/2\pi)(K/m)^{1/2}, \tag{56.22}$$

where n is some non-negative integer. The quantity ν is the frequency of the classical oscillator, and we can see from Eq. 56.20 that the frequency of the de Broglie wave, E/h, is not equal to the classical frequency of the oscillator, but $(n+\tfrac{1}{2})$ times this. Note again the existence of a finite, non-zero value for the lowest possible state of the quantum mechanical oscillator:

$$E_{\text{min}} = \tfrac{1}{2}h\nu, \tag{56.23}$$

called the *zero-point* energy of the oscillator. Further details of the wave functions for the oscillator will be found in texts on quantum mechanics.

Finally, consider the motion of an electron, electric charge $-e$, moving in the central field of force due to a fixed nucleus of charge Ze at the origin. After separating out the time variable in the same way as in the last two examples, Schrödinger's equation yields

$$-(\hbar^2/2m)\nabla^2 u(r, \theta, \phi) - (Ze^2/r)u = Eu, \tag{56.24}$$

where E is the total energy of the electron. To solve this we write the Laplacian operator in spherical polar coordinates and separate the variables r, θ, and ϕ. See Ex. 8.11(c). The details of this procedure can be found in the texts referred to at the end of this chapter. One separation constant λ has to be introduced on separating the radial function from the angular function. Another separation constant m is introduced on separating the two angles. The function of the angle ϕ turns out to be $e^{im\phi}$ and to keep this a single-valued function of position we have to make the constant m an integer. (This constant is not to be confused with the mass of the electron, which is represented by the same symbol.) The equation for the function of θ then turns out to have well-behaved solutions that remain finite and single-valued for all values of the angle θ only if the separation constant λ has the special values $\lambda = l(l+1)$ where l is a non-negative integer not less than the magnitude of the integer m. The solutions are *Legendre polynomials*. Finally, using these allowed values of λ the radial equation can be solved, the solution vanishing at infinity, only if the total energy E has a certain spectrum of values:

$$E_n = -mZ^2e^4/2\hbar^2n^2, \tag{56.25}$$

where m here is the electronic mass, n an integer not less than $l+1$.

The total wave function is a product of the various factors, and they all depend on the integers n, l, and m that may be selected. The lowest energy is given by Eq. 56.25 when $n=1$. There is only one wave function with this energy, namely that given by choosing $l=0$ and $m=0$:

$n=1$, $l=0$, $m=0$

$$u_{100} = (\pi)^{-1/2}a^{-3/2}e^{-r/a}, \tag{56.26}$$

where $a=\hbar^2/me^2$ is the radius of the first "Bohr orbit." The next energy level is given by choosing $n=2$, and for this choice there are two choices of l, namely 0 and 1. When $l=1$ we may choose m to be either $+1$, 0, or -1; while for $l=0$ we have only $m=0$ possible. There are then four different wave functions for this energy level:

$n=2$, $l=0$, $m=0$

$$u_{200} = \tfrac{1}{4}(2\pi)^{-1/2}a^{-3/2}(2-r/a)e^{-r/2a}, \tag{56.27}$$

$n=2$, $l=1$, $m=0$

$$u_{210} = \tfrac{1}{4}(2\pi)^{-1/2}a^{-5/2}re^{-r/2a}\cos\theta, \tag{56.28}$$

$n=2$, $l=1$, $m=\pm 1$

$$\begin{aligned} u_{21,+1} &= \tfrac{1}{4}(2\pi)^{-1/2}a^{-5/2}re^{-r/2a}\sin\theta\cos\phi, \\ u_{21,-1} &= \tfrac{1}{4}(2\pi)^{-1/2}a^{-5/2}re^{-r/2a}\sin\theta\sin\phi. \end{aligned} \tag{56.29}$$

The number of wave functions that have the same energy is called the *degeneracy* of that energy value. We must picture the H atom as possessing the steady states represented by these wave functions, which are called the *eigenfunctions* of the system—they are analogous to the normal modes of simple classical systems. If left undisturbed, the atom can remain in any one of its eigen states indefinitely, but may, either spontaneously or under the influence of an external disturbance, jump from one state to another. In jumping down from a state of energy E_a to an energy E_b, the difference in energy is emitted in the form of electromagnetic energy of frequency given by

$$\nu_{ab} = (E_a - E_b)/h = (2\pi^2 me^4/h^3)(1/n_b{}^2 - 1/n_a{}^2). \tag{56.30}$$

The various possible integers n_a and n_b give the series of spectral frequencies observed in the emission spectrum of hydrogen.

A complete theory of Zeeman and Stark effects has also been worked out in terms of this model. The presence of disturbing fields, magnetic and electrostatic, adds terms to the Hamiltonian for the problem, and this modifies the differential equation to be satisfied by the eigen functions, resulting both in changes of the eigenfunctions and of the eigenvalues of the energy. The most significant changes involve the removal of the degeneracy to a considerable degree: whereas in the absence of the disturbing field, all the eigenfunctions with various l- and m-values for a given value of n had the same energy given by Eq. 56.25; with the disturbing field, many of these eigenfunctions have different energies, and the spectral lines split into several components. The observed splitting of the lines due to magnetic or electric fields is well explained by the theory.

As far as we have taken it here, the Schrödinger theory gives no explanation of why the electron can remain in any one steady state without radiating its energy continuously as demanded by the classical theory of electromagnetism. This aspect of the problem is however connected with the quantum theory of fields; Schrödinger's quantization is concerned only with the matter-wave accompanying the particle and, as such, at least succeeds in deriving the correct energy levels of the electron. It is necessary to quantize the radiation field, and develop the theory of photons, to understand the interaction between matter and radiation. This would take us beyond the scope of the present course.

Exercise VI.56

56.1. Find the zero-point energy in ergs of a particle of mass one gram in a cubic box one centimeter on a side.

56.2. Find the zero-point energy in electron volts of an electron in a box of side 1.00×10^{-7} cm. Find the temperature T that would give an equivalent thermal energy kT.

56.3. Solve the problem for a particle in a box whose impenetrable walls are at the positions $x=0, L$; $y=0, L$; $z=0, L$.

56.4. Set up the Schrödinger equation for a three-dimensional oscillator with a potential function $\frac{1}{2}(K_1x^2+K_2y^2+K_3z^2)$. Find the energy eigenvalues and the zero-point energy for this system.

56.5. Show that Schrödinger's equation for a rigid body constrained to rotate about a fixed axis may be written in the form

$$-(\hbar^2/2I)\partial^2\Psi/\partial\phi^2 = i\hbar\partial\Psi/\partial t,$$

where I is the moment of inertia, and ϕ is the angle of rotation about the axis. Assuming that Ψ is a single-valued function on the circle, so $\Psi(\phi)=\Psi(\phi+2\pi)$, find the steady-state eigenfunctions and the eigenvalues of the energy.

VI.57 The Meaning of Schrödinger's Wave Function. The intensity of the matter-wave at any point is interpreted as giving the probability per unit volume, $P(\mathbf{r}, t)$, of finding a material particle at that point $\mathbf{r}$ and time t. Thus, if the wave function $\Psi(\mathbf{r}, t)$ is a complex quantity, the intensity of the wave is the square modulus of the function:

$$P(\mathbf{r}, t) = \bar{\Psi}(\mathbf{r}, t)\Psi(\mathbf{r}, t). \tag{57.1}$$

If the wave function represents a steady state, it has the form implied in Eqs. 56.7 and 56.9, and the time factor cancels from the square modulus so the probability density is independent of time:

$$P(\mathbf{r}, t) = |u(\mathbf{r})|^2. \tag{57.2}$$

If the experimental arrangement for detecting the particle is such that we are certain we shall find the particle somewhere within a certain region R of space, we normalize the wave function so that the integrated probability over the whole region is unity:

$$\int_R \bar{\Psi}(\mathbf{r}, t)\Psi(\mathbf{r}, t)d\tau = 1. \tag{57.3}$$

Such a normalization is always possible for steady-state wave functions because the time factor cancels, and because an arbitrary numerical factor present in all solutions of the Schrödinger equation can be adjusted to satisfy Eq. 57.3.

We can, however, prove that the same normalization is possible for the general case where Ψ is not a steady-state wave function. To do this we have to prove that the integrated probability is independent of time in any case. Now we have

$$(\partial/\partial t)\iiint P(\mathbf{r}, t)d\tau = \iiint[\bar{\Psi}(\partial\Psi/\partial t)+\Psi(\partial\bar{\Psi}/\partial t)]d\tau. \tag{57.4}$$

Apply Schrödinger's equation, Eq. 56.5, to $(\partial\Psi/\partial t)$ in this expression, and the conjugate of Eq. 56.5 to $(\partial\bar{\Psi}/\partial t)$. The right-hand side then becomes

$$(i\hbar/2m)\iiint[\bar{\Psi}\nabla^2\Psi - \Psi\nabla^2\bar{\Psi}]d\tau = (i\hbar/2m)\iiint\nabla\cdot(\bar{\Psi}\nabla\Psi - \Psi\nabla\bar{\Psi})d\tau. \tag{57.5}$$

From Gauss' theorem this becomes a surface integral over the boundary of the region R, so

$$(\partial/\partial t)\iiint P(\mathbf{r}, t)d\tau = (i\hbar/2m)\iint(\bar{\Psi}\nabla\Psi - \Psi\nabla\bar{\Psi})\cdot d\boldsymbol{\sigma}. \tag{57.6}$$

When the region is bounded by impenetrable walls, so that the presence of the particle is ensured, the wave function vanishes on the boundary and the surface integral is necessarily zero. This proves that the integrated probability is independent of time, and the wave function can therefore be permanently normalized to unity.

The interpretation of Ψ as a probability amplitude implies that we can carry out, at least in thought, identical measurements on a large number of copies of the given system, *i.e.* on an ensemble in the sense of §28. If there are N systems in the ensemble and a given result turns up n times among these N systems, the probability of that result in the ensemble is n/N. The average value of any variable α in the ensemble can be written down in terms of this probability. If the possible values of α are $\alpha_j (j = 1, 2, \cdots)$ and the number of times the value α_k occurs in the ensemble is n_k, then the average value of α in the ensemble is defined as

$$\begin{aligned} \langle\alpha\rangle &= \sum_k n_k \alpha_k \div \sum_k n_k \\ &= \sum_k (n_k/N)\alpha_k = \sum_k P_k \alpha_k. \end{aligned} \tag{57.7}$$

Thus, suppose that the configuration of the system, *e.g.* the position $\mathbf{r}$ of the particle in its enclosure, is the variable to be measured; the mean value of $\mathbf{r}$ in the ensemble must be

$$\langle\mathbf{r}\rangle = \iiint P(\mathbf{r}, t)\mathbf{r}\,d\tau = \iiint \overline{\Psi}(\mathbf{r}, t)\mathbf{r}\Psi(\mathbf{r}, t)\,d\tau. \tag{57.8}$$

Because the variable is continuous rather than discrete, the sum that is appropriate in Eq. 57.7 has been written as an integral in Eq. 57.8.

When the variable to be observed is represented by an operator rather than an ordinary function, the order of the factors in the integrand becomes significant—a circumstance that did not occur in classical probability theory. We must here proceed by trying all alternatives to find which one works. Thus if we seek the average momentum, represented by the operator in Eq. 56.1, we could try either of the two alternatives:

$$\langle\mathbf{p}\rangle = \left\{ \begin{matrix} -i\hbar\iiint \nabla(\overline{\Psi}\Psi)\,d\tau \\ \text{or} \\ -i\hbar\iiint \overline{\Psi}\nabla\Psi\,d\tau \end{matrix} \right\}.$$

It turns out that the first alternative gives an imaginary value to the mean momentum in cases where it obviously should be real. The second alternative is accepted and, in general for any observable represented by an operator α, the mean value in an ensemble is defined in quantum mechanics by the relation

$$\langle\alpha\rangle \equiv \iiint \overline{\Psi}\alpha\Psi\,d\tau, \tag{57.9}$$

where the wave function represents the state of every copy of the system when the observations are carried out.

Schrödinger's equation written out for stationary states in operator form:

$$H\Psi = E\Psi, \tag{57.10}$$

is a typical eigenvalue equation. The operator H acting on Ψ results in a constant multiplied by Ψ. The equation is true only for a particular set of functions like Ψ, called the *eigenfunctions* of the operator H, and the corresponding values of E are called its *eigenvalues*. They may form a discrete set, and we may write Eq. 57.10 in the form:

$$H\Psi_n = E_n\Psi_n, \tag{57.11}$$

where n is a numerical index of a particular eigenfunction. We have seen examples of eigenfunctions of H in the last section, namely those for the harmonic oscillator, the hydrogen atom, etc.

If the system is prepared in a state represented by an eigenfunction Ψ_n of H, the mean value of its energy is the eigenvalue E_n; thus

$$\langle H\rangle_n \equiv \int\int\int\overline{\Psi}_n H\Psi_n d\tau = E_n\int\overline{\Psi}_n\Psi_n d\tau = E_n, \tag{57.12}$$

because of the normalization of the wave function, Eq. 57.3.

The average deviation from the mean, of the results of measuring any observable, can be expressed in terms of the quantum mechanical probability. Thus, if $\langle\alpha\rangle$ is the mean value of the observable α in some state Ψ, the mean square deviation would be given by the mean value of $(\alpha - \langle\alpha\rangle)^2$. By definition, this is

$$\int\overline{\Psi}(\alpha - \langle\alpha\rangle)^2\Psi d\tau = \int\overline{\Psi}\alpha^2\Psi d\tau - 2\langle\alpha\rangle\int\overline{\Psi}\alpha\Psi d\tau + \langle\alpha\rangle^2\int\overline{\Psi}\Psi d\tau.$$

Using Eq. 57.9, we then have

$$\langle(\alpha - \langle\alpha\rangle)^2\rangle = \int\overline{\Psi}\alpha^2\Psi d\tau - (\int\overline{\Psi}\alpha\Psi d\tau)^2. \tag{57.13}$$

Let us try this on systems in the eigenstate Ψ_n of the Hamiltonian H, and ask for the mean square deviation of energy measurements:

$$\langle(H - \langle H\rangle)^2\rangle = \int\overline{\Psi}_n H^2\Psi_n d\tau - (\int\overline{\Psi}_n H\Psi_n d\tau)^2.$$

Remembering that $H^2\Psi_n = HE_n\Psi_n = E_nH\Psi_n = E_n^2\Psi_n$, we have at once

$$\langle(H - \langle H\rangle)^2\rangle = E_n^2\int\overline{\Psi}_n\Psi_n d\tau - (E_n\int\overline{\Psi}_n\Psi_n d\tau)^2 = 0. \tag{57.14}$$

We conclude quite generally that the eigenfunctions of any observable represent states in which the system certainly has the corresponding eigenvalue of the observable, and that measurements in an ensemble will always yield a perfectly precise result for that particular observable.

Exercise VI.57

57.1. Show that e^{ikx} is an eigenfunction of the momentum operator corresponding to an eigenvalue $\hbar k$ for the momentum in the x-direction.

57.2. Use e^{ikx} as the momentum eigenfunction in the rejected trial expression for the mean value of the momentum preceding Eq. 57.9 and show that it leads to an unreasonable value for $<\mathbf{p}>$.

57.3. Find the mean momentum in the x-direction for a particle in an energy eigenstate for a box, as given by Eqs. 56.12 and 56.15.

57.4. Find the mean square momentum in the above example.

57.5. What is the mean square deviation of the momentum in the last two exercises?

57.6. Find the numerical factor required to normalize to unity the energy eigenfunctions displayed in Eqs. 56.12, 56.15.

VI.58 The Principle of Superposition. It can be shown that the set of normalized eigenfunctions of any operator of interest in quantum mechanics can be used as unit functions or unit vectors in function space in the sense of §3. We shall prove first that the eigenfunctions of the Hamiltonian operator satisfy the orthogonality condition analogous to Eq. 3.11: two eigenfunctions corresponding to different eigenvalues are mutually orthogonal.

Let the eigenfunctions be Ψ_n and Ψ_m. Write down the eigenvalue equation for Ψ_n and the complex conjugate equation for Ψ_m:

$$H\Psi_n = E_n\Psi_n, \quad \bar{H}\bar{\Psi}_m = E_m\bar{\Psi}_m, \tag{58.1}$$

remembering that the eigenvalues are real quantities. Multiply the first of these equations by $\bar{\Psi}_m$ and the second equation by Ψ_n, and subtract the two equations; then integrate them over the domain of the functions:

$$\int(\bar{\Psi}_m H\Psi_n - \Psi_n\bar{H}\bar{\Psi}_m)d\tau = (E_n - E_m)\int\Psi_n\bar{\Psi}_m d\tau. \tag{58.2}$$

We now prove that the left side of this equation integrates to zero. Thus

$$H = -(\hbar^2/2m)\nabla^2 + V, \tag{58.3}$$

where V is an unspecified potential function of position only. Then the left-hand side of Eq. 58.2 becomes:

$$-(\hbar^2/2m)\int(\bar{\Psi}_m\nabla^2\Psi_n - \Psi_n\nabla^2\bar{\Psi}_m)d\tau$$
$$= -(\hbar^2/2m)\int\nabla\cdot(\bar{\Psi}_m\nabla\Psi_n - \Psi_n\nabla\bar{\Psi}_m)d\tau.$$

Using Gauss' theorem, this reduces to:

$$-(\hbar^2/2m)\int(\bar{\Psi}_m\nabla\Psi_n - \Psi_n\nabla\bar{\Psi}_m)\cdot d\boldsymbol{\sigma} = 0. \tag{58.4}$$

This vanishes for the same reason as given in the discussion following Eq. 57.6.

When the eigenvalues are unequal, $E_n \neq E_m$, Eq. 58.2 leads to the theorem to be proved:

$$\int\Psi_n\bar{\Psi}_m d\tau = \delta_{nm}. \tag{58.5}$$

The Kronecker delta implies also that, when $n = m$, the eigenfunctions are normalized to unity.

Although we have proved this only for the Hamiltonian operator, it is clear from Eq. 58.2 that any operator that satisfies the identity:

$$\int(\bar{\Psi}_m\alpha\Psi_n - \Psi_n\bar{\alpha}\bar{\Psi}_m)d\tau = 0, \tag{58.6}$$

will have just the same property. Any such operator α is called an *Hermitian operator*. We can prove that the eigenvalues of any Hermitian operator are necessarily real numbers, thus: let Ψ_n be the eigenfunctions of α corresponding respectively to eigenvalues α_n. Then we have:

$$\int \bar{\Psi}_m \alpha \Psi_n d\tau = \alpha_n \int \bar{\Psi}_m \Psi_n d\tau = \alpha_n \delta_{nm}. \tag{58.7}$$

Taking the complex conjugate of this equation and using Eq. 58.6 on the left side, we see at once that $\alpha_n = \bar{\alpha}_n$, so that α_n is real. This result has been accepted as a basic guiding principle in seeking operators to represent physical observables when we know these to be real quantities: they are to be represented by Hermitian operators.

It is worth checking that the operator $-i\hbar\nabla$ for the momentum is a Hermitian operator. If it is, we must, according to Eq. 58.6, have

$$-i\hbar \int \bar{\Psi}_m \nabla \Psi_n d\tau = i\hbar \int \Psi_n \nabla \bar{\Psi}_m d\tau, \tag{58.8}$$

and therefore

$$\int (\bar{\Psi}_m \nabla \Psi_n + \Psi_n \nabla \bar{\Psi}_m) d\tau = 0$$

or

$$\int \nabla (\Psi_m \bar{\Psi}_n) d\tau = 0. \tag{58.9}$$

Even if we did not know the eigenfunctions, we could proceed to apply a special form of Gauss' theorem to this integral, transforming it to the surface integral $\int \Psi_m \bar{\Psi}_n d\boldsymbol{\sigma}$. Then if the functions vanish over the boundary of the region, the integral vanishes as required.

The essential content of the superposition principle makes use of the fact that the eigenfunctions of a Hermitian operator form a complete orthonormal set of unit vectors in function space and asserts that: Let a system be prepared in some state represented by any arbitrary well-behaved function Ψ; then we may express this state function as a linear combination of the eigenfunctions Ψ_n of any specified Hermitian operator corresponding to some particular physical observable:

$$\Psi = \sum_n a_n \Psi_n, \tag{58.10}$$

where the coefficients a_n $(n=1, 2, \cdots)$ are constants. Let α be the operator; then the expectation or mean value $\langle \alpha \rangle$ of the observable represented by α when the system is in the state Ψ is by definition Eq. 57.9:

$$\langle \alpha \rangle = \int (\sum_n \bar{a}_n \bar{\Psi}_n) \alpha (\sum_m a_m \Psi_m) d\tau. \tag{58.11}$$

Bringing the sums and constant factors through the operator and through the integral, we have

$$\begin{aligned} \langle \alpha \rangle &= \sum_{n,m}\sum \bar{a}_n a_m \int \bar{\Psi}_n \alpha \Psi_m d\tau = \sum_{n,m}\sum \bar{a}_n a_m \alpha_m \delta_{nm} \\ &= \sum_n \bar{a}_n a_n \alpha_n = \sum_n |a_n|^2 \alpha_n, \end{aligned} \tag{58.12}$$

where α_n are the eigenvalues of α. Comparing this result with Eq. 57.7 we see that

$$|a_n|^2 = P_n \tag{58.13}$$

is the probability that the system be found in the state Ψ_n with the eigenvalue α_n in the ensemble.

We say, in summary, that if the system is prepared in any arbitrary state, we may make measurements of any observable—*e.g.* the energy—and our result must be one of the possible eigenvalues of that observable; the probability that the result be some particular eigenvalue is the square modulus of the amplitude component of the state function expressed in terms of the eigenfunctions of the observable in question.

From Eq. 58.10 we have, by multiplying into $\overline{\Psi}_m$ and integrating:

$$\int \Psi \overline{\Psi}_m d\tau = \sum_n a_n \int \Psi_n \overline{\Psi}_m d\tau = a_m,$$

and therefore we can compute the probability of finding the mth eigenvalue from the formula

$$|a_m|^2 = \int \Psi \overline{\Psi}_m d\tau \cdot \int \overline{\Psi} \Psi_m d\tau. \tag{58.14}$$

Exercise VI.58

58.1. Consider a particle free to move only on the x-axis in the range $-\frac{1}{2}L \leqslant x \leqslant \frac{1}{2}L$, and let it be in an energy eigen state

$$\Psi = (2/L)^{1/2} \sin(2\pi x/L). \tag{58.15}$$

Compare this with Eq. 56.12, and show that it is normalized to unity in the stated domain. Note that the integral domain is linear and the volume integrations in equations like 58.5 etc., are replaced by line integrals. Show that the energy eigenvalue for this state is

$$E = h^2/2mL^2. \tag{58.16}$$

Next consider the momentum eigenfunction:

$$\Psi_m = (1/L)^{1/2} e^{2\pi i x/L}, \tag{58.17}$$

and prove this is normalized to unity in the same range. Show that its momentum eigenvalue is $+h/L$. Use Eq. 58.14 to determine the probability of finding the momentum $+h/L$ when the particle is in the energy state of Eq. 58.15, where its momentum can be either $+h/L$ or $-h/L$ according to Eq. 58.16.

58.2. Use the momentum eigenfunction of Eq. 58.17 in Eq. 58.6 to show that the momentum operator is Hermitian.

VI.59 The Uncertainty Principle. In classical mechanics the state of a particle, or system of particles, is specified by its phase, in the sense of §23. We must know the momentum components and position coordinates at any time for the whole system. If we have this information with full precision at any instant, the entire history, past and future, of the system is given by the Lagrange equations of motion.

In quantum mechanics we have only a wave function to specify the state of a system. In this section we shall see to what extent we can use the quantum mechanical state to specify the classical phase of a particle. We can set up a particle in an eigenstate of the momentum, Eq. 58.17, where the momentum is known precisely, *e.g.* $p_x = h/L$. The mean position and mean square deviation in position are then given by Eqs. 57.9 and 57.13, respectively, thus:

$$\langle x \rangle = (1/L) \int_{-L/2}^{L/2} e^{-2\pi i x/L} x e^{2\pi i x/L} dx = 0, \tag{59.1}$$

$$\langle (x - \langle x \rangle)^2 \rangle = (1/L)(x^3/3) \Big|_{-L/2}^{L/2} = L^2/12. \tag{59.2}$$

For an indefinitely large enclosure, the precision of the momentum is the same, $\Delta p_x = 0$; but the mean deviation in the position increases without limit, $\Delta x \to \infty$, as $L \to \infty$. In general, if we repeat precise measurements of both momentum and position of a particle prepared in any given quantum mechanical state of the momentum, there will be zero deviation in the momentum readings, but wide deviations in the coordinate readings.

We next try to improve the situation regarding the location of the

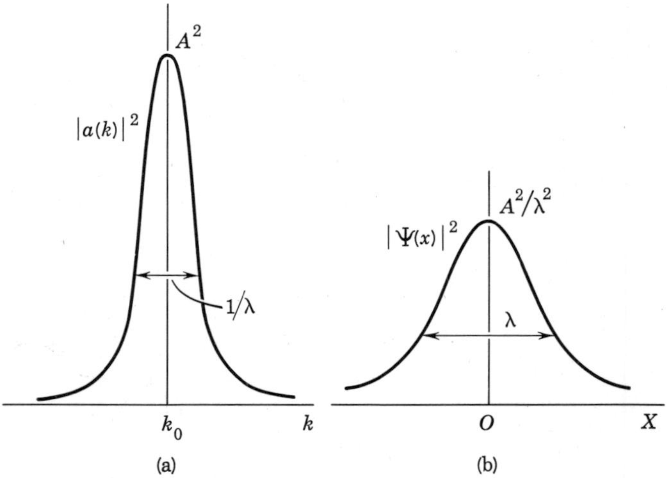

FIG. 59.1. The Gaussian wave packet. (A) represents the distribution in wave number, Eq. 59.5, (B) represents the distribution in space, Eq. 59.8. The product of the two half-widths is unity, a constant, which leads to the Heisenberg uncertainty relation

particle in space by relaxing the precision of our specification of its momentum. To do this we can set up the particle in a state Ψ that is a superposition of momentum eigen states in which $p_x = \hbar k$, and k is different in different eigen states:

$$\Psi = (1/\sqrt{2\pi})\sum_k a_k e^{ikx}. \tag{59.3}$$

Let the enclosure be so large, and the successive eigenvalues so close together, that k has almost a continuous spectrum; then we can write an integral superposition in place of Eq. 59.3:

$$\Psi(x) = (1/\sqrt{2\pi})\int_{-\infty}^{\infty} a(k)e^{ikx}dk. \tag{59.4}$$

Here, as in Eq. 58.13, $|a(k)|^2$ is the probability of finding the momentum value $\hbar k$. If we choose

$$a(k) = Ae^{-\lambda^2(k-k_0)^2/2}, \tag{59.5}$$

we have a Gaussian distribution in momentum about a mean value k_0. The width, or root-mean-square deviation, of the distribution is $1/\lambda$, so we have

$$\Delta p_x = \hbar\Delta k = \hbar/\lambda. \tag{59.6}$$

Inserting Eq. 59.5 in Eq. 59.4 and integrating over k, we have

$$\Psi(x) = \int_{-\infty}^{\infty} (A/\sqrt{2\pi})e^{-\lambda^2(k-k_0)^2/2}e^{ikx}dk = (A/\lambda)e^{ik_0x}e^{-x^2/2\lambda^2}. \tag{59.7}$$

The probability, per unit extent, of locating the particle at x is then

$$|\Psi(x)|^2 = (A/\lambda)^2e^{-x^2/\lambda^2}. \tag{59.8}$$

The quantity A in Eq. 59.5 is an arbitrary normalizing factor; to make the integrated probability equal to unity

$$\int_{-\infty}^{\infty} |\Psi(x)|^2dx = 1,$$

we have to choose

$$A^2 = \lambda/\sqrt{\pi}. \tag{59.9}$$

The factor $(1/\sqrt{2\pi})$ in Eq. 59.3 is a normalization factor on the momentum eigenfunction in the infinite domain.

The significant thing about this result is that the given superposition state $\Psi(x)$ gives a Gaussian distribution both in the momentum, through Eq. 59.5, and in the position, through Eq. 59.8. Moreover, the mean deviations in the two measurements are related. From Eq. 59.8 the width of the Gaussian distribution in the position is λ:

$$\Delta x = \lambda. \tag{59.10}$$

Comparing Eqs. 59.10 and 59.6, we have the relation

$$\Delta x \Delta p_x = \hbar, \tag{59.11}$$

which is known as *Heisenberg's uncertainty relation.*

We may emphasize again the interpretation of this relation. An ensemble is set up containing a large number of copies of the particle, all in the same state, defined quantum mechanically by the function $\Psi(x)$ of Eq. 59.3. Measurement of the momentum of the particles results in a Gaussian distribution with a mean deviation $\Delta p_x \sim \hbar/\lambda$. An identical ensemble set up in the same state and used for measuring the position, results also in a Gaussian distribution, but with a mean deviation $\Delta x \sim \lambda$. If the ensemble is set up to give precise reproducible results for the momentum, the results for position will be completely random, and vice versa. It is impossible to find a quantum mechanical state which will yield greater simultaneous precision in both momentum and position than is prescribed by Heisenberg's uncertainty relation. Indeed, if we include all the other sources of error due to experimental techniques, the uncertainty principle becomes an inequality setting an upper limit to the precision:

$$\Delta x \Delta p_x \geqslant \hbar. \tag{59.12}$$

A similar relation holds between energy and time measurements at any given position. Let us consider the eigenfunctions of the energy operator $i\hbar\partial/\partial t$:

$$\Psi(x, t) = (1/\sqrt{2\pi})e^{-iEt/\hbar}u(x), \tag{59.13}$$

and let us form a superposition of these eigenfunctions to represent a state of the particle:

$$\Psi(x, t) = (1/\sqrt{2\pi})\sum_E a(E)e^{-iEt/\hbar}u(x). \tag{59.14}$$

Again we shall imagine the enclosure very large, so that the spectrum is very closely spaced; then we can rewrite the sum in Eq. 59.14 as an integral, with E a continuous variable. We shall now choose the coefficients $a(E)$ so that the probability of finding any value of the energy has a Gaussian distribution:

$$|a(E)|^2 = A^2 e^{-(E-E_0)^2/G^2}. \tag{59.15}$$

The width of the distribution is G, and this measures the mean deviation in the results of energy measurements on the particle in the state, Eq. 59.14:

$$\begin{aligned}\Psi(x, t) &= (A/\sqrt{2\pi})u(x)\int_0^\infty e^{-(E-E_0)^2/2G^2}e^{-iEt/\hbar}dE \\ &= AGu(x)e^{-t^2G^2/2\hbar^2}e^{-iE_0t/\hbar}.\end{aligned} \tag{59.16}$$

The probability density is then a function of the time:

$$|\Psi(x, t)|^2 = P(x, t) = A^2G^2|u(x)|^2e^{-t^2G^2/\hbar^2}. \qquad (59.17)$$

The distribution in time is again Gaussian, and measurements of the time have a mean deviation equal to $\hbar/G$:

$$\Delta t = \hbar/G. \qquad (59.18)$$

From Eq. 59.15 the deviation in energy measurements for the same state is

$$\Delta E = G, \qquad (59.19)$$

so that the two deviations are related:

$$\Delta E \Delta t \geqslant \hbar. \qquad (59.20)$$

Again we interpret this in terms of an ensemble set up with every particle in the state represented by the function in Eq. 59.14: The measurements of the energy at a certain point and of the time of arrival at the same point suffer correlated uncertainties. It is impossible to find a quantum mechanical state in which both energy and time can be measured simultaneously with zero dispersion in a series of measurements performed on identical states.

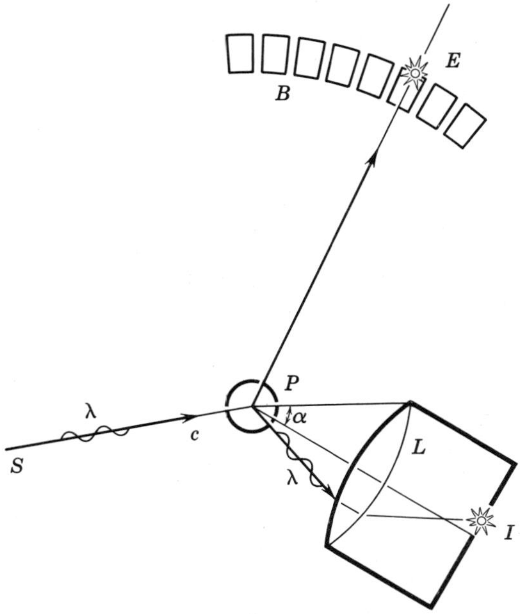

FIG. 59.2. Heisenberg's thought experiment to measure position and momentum of an electron simultaneously. Repetition yields scatter of the points I and E, related to each other by the uncertainty principle

We illustrate the meaning of Eq. 59.12 in terms of a "thought experiment" originally visualized by Heisenberg. In Fig. 59.2 the circle at P represents a device that injects electrons one at a time and with very low speed into the field of view of an optical microscope L. The region at P is illuminated from a source of light S. According to geometrical optics the image of any point within the region P would be at some point such as I. However, from the wave nature of light we know that the image of such a point is actually a diffraction pattern in the neighborhood of I and that the minimum diameter of the central disk of the pattern is $\lambda/\sin \alpha$, where λ is the wavelength of the light and α is the half angle of the lens subtended at P when the object and the image are at the same distance from the lens.

An electron at P scatters a photon into the lens and produces a single scintillation in the receiver. With sufficiently feeble illumination these scintillations are individually observed, and each such scintillation is taken as a reading of the position of the image of the electron that caused the scattering. Continued illumination allows repeated observation, and the scintillations gradually build up the diffraction pattern. The mean deviation of these successive readings is the width of the diffraction ring, viz. $\lambda/\sin \alpha$:

$$\Delta x = \lambda/\sin \alpha. \tag{59.21}$$

The Compton effect occurs when an electron scatters a photon into the lens, and we may use this in an attempt to compute the momentum of the electron after it has scattered the photon. However, to do this, we can see from §55 that we must know the direction of scattering. We do know the direction of the incident photon, but the scattered photon is known only to enter the lens, and this leaves its direction and hence its momentum in the direction of x uncertain, by

$$\Delta p_x = (\hbar/\lambda) \sin \alpha. \tag{59.22}$$

The uncertainty in the momentum of the electron is exactly the same as this, and hence we can combine Eqs. 59.21 and 59.22 to confirm Eq. 59.12. Now we shall include a battery of electron counters placed strategically at E to catch and observe the ejected electrons (see Fig. 59.2), recording their positions and times of arrival at some distance from P. If there is no field or obstruction between P and E, we can compute the momentum of the electrons during their transit from P to E. With this apparatus we can assign simultaneous values to the position and momentum of an electron suffering Compton scattering in the region P. The uncertainty relation of Eq. 59.12, to which these values are subject, asserts that on repeated observations the scatter or deviations of the position and momentum readings are related and that it is impossible to reduce one without increasing the other. For example, we could reduce the scatter in the readings of position by using a short wave illumination and a wide angle lens, but, from Eq. 59.22, this would inevitably increase the scatter in the momentum readings.

It is also important to realize that even though we have here a device for recording simultaneously the position and momentum of an electron, the information is not of the type needed to specify the classical phase of the electron: the information concerns only the state of the electron just before the observation; immediately after the observation the electron may no longer exist as an individual entity because, in the scintillating screen used to detect it, the electron may have entered an atom and become part of its structure. At best, the electron will be strongly deflected by the scintillator, and another complicated piece of apparatus will be needed to find its new momentum after the observation. Thus the best we can do as regards finding the classical phase of the electron is to make a running commentary on what it was a moment ago; the means of observation that we are obliged to use, themselves disturb the electron so violently that we have only very incomplete information about the classical phase subsequent to any observation.

Exercise VI.59

59.1. In deriving the mean position in Eq. 59.1, we used the momentum eigenfunction of Eq. 58.17. This eigenfunction cannot be restricted to a finite domain. If the particle is restricted to a box of length L, as assumed in Eq. 59.1, we should use a superposition of two momentum eigenfunctions:

$$\Psi' = (1/\sqrt{2L})(e^{-2\pi i x/L} - e^{2\pi i x/L}).$$

Verify that Ψ' vanishes at $x = \frac{1}{2}L$ and $x = -\frac{1}{2}L$. Use this function to compute $\langle x \rangle$, Δx, and Δp_x.

59.2. Carry through the integrations in Eqs. 59.7 and 59.16.

59.3. Give the detailed deduction of Eq. 59.22.

59.4. The injection equipment at P, in Fig. 59.2, has an opening 0.1 mm in diameter. Find the uncertainty in transverse momentum of an electron as it arrives at P. If $\lambda = 10^{-5}$ cm and $\alpha = 45°$, find the uncertainty in momentum of the electron after it has scattered a photon.

59.5. The position of a pointer in a measuring instrument is recorded to within a precision of 0.1 mm. The mass of the pointer is 1 gram. Find the uncertainty in the momentum of the pointer.

VI.60 Relativistic Quantum Mechanics. There are many effects observed in atomic spectra, having a relativistic origin, that cannot be understood in terms of Schrödinger's wave equation. We may mention particularly the effects ascribed to the existence of electron spin. In attempting to include these effects in quantum mechanics, we may note that Schrödinger's equation is not invariant for Lorentz transformations, and by the principle of invariance is not acceptable as a general physical law. This fact is clear from the lack of symmetry of the equation with respect to space and time: for Lorentz invariance the variable ict ought to occur in the same way as x, y, or z. In Schrödinger's equation for the particle, Eq. 56.5, time appears in a first-order differential operator, while the position coordinates appear in the second-order Laplacian operator. For Lorentz invariance it is necessary for both time and position coordinates to appear homogeneously either in first- or second-order differential operators.

Fortunately the operational prescription used to derive Schrödinger's equation from the classical Hamiltonian can be expressed in a Lorentz invariant way. If we write down the 4-vector momentum $\mathbf{p}^*$ from Eq. 51.1 and the 4-vector gradient operator $\Box$ from Eq. 52.17, the operational prescriptions of Eqs. 56.1 and 56.3 become combined in one 4-vector (and therefore Lorentz invariant) equation:

$$\mathbf{p}^* \rightarrow i\hbar \Box \tag{60.1}$$

In Schrödinger's wave equation, the wave function has been treated as a scalar. However $|\Psi|^2$ is defined as the probability per unit volume in space and, in relativity, volume in space is not a scalar. If we wish the probability in any given region to be a scalar in four dimensions, then the probability density $|\Psi|^2$ must be a scalar divided by a three-dimensional volume of space. But (see Eq. 52.43) a scalar is obtained when we multiply a three-dimensional volume by a time component; therefore a scalar divided by a three-dimensional volume must be a time-component of a 4-vector. We therefore need to identify $|\Psi|^2$ with the time-component of a 4-vector in order to give the wave equation its proper relativistic invariance. This implies two things: first, that Ψ itself must be a new kind of entity, with a square-root transformation rule compared with the transformation of the time-component of a vector; and secondly, there must be three other wave functions whose square moduli are associated with $|\Psi|^2$ to form the four components of a 4-vector. To build a Lorentz invariant wave equation it is therefore necessary to set up four wave equations, one for each wave function, such that the set as a whole transforms in a well-defined fashion. The four wave functions form a mathematical object called a *spinor*; the spinor transformation rule is roughly a square-root transformation rule compared with the Lorentz transformation; where the Lorentz transformation rotates a system through some angle, the spinor transformation associated with it rotates it through one half the angle. We do not have space in this text to go into the very interesting details of spinor analysis. The name "spinor" is related with the fact that this reasoning led Dirac to the discovery of a first-order differential equation for the electron wave functions that included the phenomena ascribed to electron spin. This equation also led to the prediction of the existence of positive electrons (positrons) and implied the possibility of creation and anihilation of electron-positron pairs with the absorption or emission of gamma-ray photons.

As a solution to the problem of inventing a wave equation that shall be Lorentz invariant, Dirac's equation is not unique. It is easy to write down a relativistic second-order differential wave equation. From Eq. 51.1 we can write down the four-dimensional scalar equation:

$$\mathbf{p}^*\cdot\mathbf{p}^* = m_0^2c^2, \tag{60.2}$$

and using Eq. 60.1 in this we have the single scalar wave equation that is invariant for Lorentz transformations:

$$-\hbar^2\Box^2\Psi = m_0^2c^2\Psi. \tag{60.3}$$

This is called the *Klein-Gordon equation.* For Lorentz invariance, the wave function must be a single four-dimensional scalar function, and so cannot be identified with a probability density per unit volume. It is not suitable for the electron wave equation, therefore. It could, however, be interpreted as a probability scalar per unit volume per unit time, which would be a scalar in four-dimensional analysis. Thus if we interpret the equation as pertaining to the occurrence of some repetitive events rather than the appearance (once for all) of some particle, the equation may have a good physical meaning. In fact, it has been used by Yukawa to describe the field of force between nuclei—the so-called scalar meson field.

A time-independent solution of Eq. 60.3 may have the form:

$$\Psi = e^{-x/\lambda},$$

where

$$\lambda = \hbar/m_0c. \tag{60.4}$$

This wave function becomes very small for distances $x > \lambda$; it is known that nuclear force fields become negligible at distances greater than about 10^{-13} cm. If we set $\lambda = 10^{-13}$ in Eq. 60.4 and use the known values of Planck's constant and the velocity of light, we find m_0 must be about one-tenth the mass of a hydrogen nucleus. This circumstance led Yukawa to predict that the nuclear forces were accompanied by a new particle, the meson, having a mass about 200 times that of an electron. These particles would be analogous to the photons that accompany electromagnetic fields. The events whose recurrent probability is given by the scalar $|\Psi|^2$ are pictured here as the continual emission and absorption of the mesons by the interacting nuclei.

Experimental discovery of the meson has confirmed Yukawa's

picture, but at the same time many more particles of different masses have been discovered, and a major problem has arisen to discover enough different kinds of wave equations to describe all their individual properties.

Exercise VI.60

60.1. Prove that any Hermitian 2×2 matrix $\mathbf{s}$ can be expressed in the form

$$\mathbf{s} = \sum_{\alpha} s_{\alpha}\sigma_{\alpha} \ (\alpha = 0, 1, 2, 3), \tag{60.5}$$

where s_{α} are real numbers and σ_{α} are the matrices

$$\left.\begin{aligned} \sigma_1 &= \begin{pmatrix} 0 & 1 \\ 1 & 0 \end{pmatrix} \qquad \sigma_2 = \begin{pmatrix} 0 & -i \\ i & 0 \end{pmatrix} \\ \sigma_3 &= \begin{pmatrix} 1 & 0 \\ 0 & -1 \end{pmatrix} \qquad \sigma_0 = \begin{pmatrix} 1 & 0 \\ 0 & 1 \end{pmatrix} \end{aligned}\right\}. \tag{60.6}$$

60.2. Prove that the determinant of the matrix $\mathbf{s}$ is

$$s_0^{\,2} - s_1^{\,2} - s_2^{\,2} - s_3^{\,2} = \mathrm{Det}\cdot\mathbf{s}. \tag{60.7}$$

60.3. A linear transformation of the matrix $\mathbf{s}$ is effected by means of a unimodular matrix U:

$$\mathbf{s}' = \mathbf{U}\cdot\mathbf{s}\cdot\tilde{\mathbf{U}}. \tag{60.8}$$

If

$$\mathbf{U} = \begin{pmatrix} \cos\beta/2 & \sin\beta/2 \\ -\sin\beta/2 & \cos\beta/2 \end{pmatrix}$$

show that

$$\left.\begin{aligned} s_1' &= s_1\cos\beta + s_3\sin\beta \\ s_3' &= s_3\cos\beta - s_1\cos\beta \end{aligned}\right\}. \tag{60.9}$$

60.4. Prove that under the transformation of the last exercise:

$$s_0^{\,2} - s_1^{\,2} - s_2^{\,2} - s_3^{\,2} = s_0'^{2} - s_1'^{2} - s_2'^{2} - s_3'^{2}. \tag{60.10}$$

60.5. Find the transformation relations for s_{α}' when

$$\mathbf{U} = \begin{pmatrix} e^{\alpha/2} & 0 \\ 0 & e^{-\alpha/2} \end{pmatrix}. \tag{60.11}$$

Compare the result with the Lorentz transformation.

60.6. Let Φ be a two-component vector, with complex components, that transforms according to the rule

$$\Phi' = \mathbf{U}\Phi. \tag{60.12}$$

Show that the dyadic product of two such vectors, $\Phi\cdot\Psi$, transforms like the 2×2 matrix $\mathbf{s}$ in Eq. 60.8.

60.7. The vector Φ of the last exercise is called a two-component spinor and the dyadic product $\Phi\Psi$ has four components. Show that these four components undergo the Lorentz transformation like a 4-vector when Φ and Ψ undergo the transformation of Eq. 60.12.

VI.61 Philosophical Implications. The philosophical implications of the successes of relativity and quantum theories have been revolutionary. A vast literature has emerged dealing with this revolution in human thought, and, in bringing the present course to a close, we can do no more than alert the student to the existence of this lively discussion and recommend that he take part in it. Every philosophical problem connected with the interpretation of science can be reduced in the last analysis to the question of the meaning of words, in particular the meaning of the word "meaning" itself. In physics we are able to give an operational meaning to almost everything we measure. These operational meanings prescribe certain laboratory procedures appropriate for the measurement of length, mass, time, force, electric charge, etc.; they are usually learned in high school or college courses of laboratory instruction and have become a tradition that is seldom subject to philosophical discussion. Philosophical discussion naturally arises only when no traditionally accepted definition exists, or when a traditionally accepted definition is proven to imply relationships with other concepts which are false. We shall discuss very briefly only two such questions as illustrative of these problems: (*a*) Is the quantum mechanical wave function "real"? and (*b*) Does the uncertainty principle imply a breakdown in "causality"?

The two words whose definitions are vague enough here to leave room for argument are "real" and "causality." There are so many different shades of meaning to the word "real" in everyday use that the question of the reality of any purely scientific entity is almost impossible to answer. Dictionary definitions are always circular. For example: "real—being an actual thing, with objective existence." Then: "actual—existing in fact, real." And again: "exist—to have actual being." Evidently this leads nowhere, and we may try again from "objective": "Objective—pertaining to that which can be known, free from personal feelings." But what is "that which can be known"? We see next: "knowledge—acquaintance with facts, truths . . . , perception of a body of truths or facts." But "perception = cognition" and "cognition" is the "act or process of knowing." So we are back to where we started from. Finally let us follow "thing." There are at least a dozen different meanings in any good dictionary, none of which help. Instead we may try "thing-in-itself" from Kantian philosophy, defined as "reality as it is apart from experience," so we are back again to reality without knowing what it is.

Language has not been constructed by any logical process. We learn to associate words with corresponding experiences and we "know" what they mean simply by this association. No dictionary definition is either necessary or possible. The function of the dictionary is to point out close relationships between the meanings of various word clusters, and somewhere in this field of words and phrases the reader is expected to find some that are already closely coupled with his own experience. The meaning of "reality" is especially elusive because it is learned so early in our experience; it is perhaps one of the most primitive concepts in terms of which we tend to define all others. About the best one can say is that we tend to associate "reality" with all experiences that repeat themselves independently of our desires or fancy. Whether any particular scientific discovery feels "real" to us depends completely upon our intellectual familiarity with it. The atom was a fiction to the Greeks; it is real to the modern chemist, although no scientist has actually seen one.

It is clear that, for our question concerning the reality of the quantum mechanical wave function to have an answer, we must set up a stricter definition of "reality" than our discussion has achieved up to this point. Margenau has suggested that "physical reality" may be defined as those aspects of "reality," in the ordinary vague sense, that are found by scientific investigation to be subject to laws and regularities such as the conservation principle. In this narrower sense, the quantum mechanical wave function describes a physical reality; in fact, it is better qualified to be so classed than are the fundamental particles, whose behavior is apparently subject to chance, and whose existence is now known, from positron-electron anihilation, to be ephemeral. If we do not at first feel that the wave function is real, this is merely because we have not yet become sufficiently familiar in our experience with its properties. The question of whether the wave function is real or not is perhaps not a question about the wave function at all, but about the word "real". We know well enough what the wave function is, but we are not entirely agreed upon what "real" means.

As our other example of a philosophical problem, we ask about the implications of the uncertainty principle for causality. The notion of causality is associated in our thinking with such concepts as determinism and free will. The fact that quantum mechanics reduces all physical laws to statements about the probability function has been felt by many people to imply that physical laws have become acausal, with chance the only controlling factor. To examine this question, we must discuss the meaning of "causality." It is even more amusing to scan a dictionary for the meaning of "cause": "cause = that which produces an effect," and "effect = that which is produced by some cause." In everyday understanding the notion of cause implies an uncertainty in a situation. We say that *A* causes *B* IF *B* always follows *A*, with an emphasis on the IF. In the absence of a possible alternative to *A*, the notion of cause is void. There must be the possibility of giving meaning to the statement: "IF A does not occur, then B will not occur either." From this point of view, determinism is the complete negation of causality, because there are no alternatives—it is impossible for A not to occur, and so we cannot test whether B would fail in that case.

Classical mechanics was completely deterministic; from the present state of the system we can deduce both the entire future and the entire past. It would be as reasonable to say that the present causes the past as to claim that the present causes the future. There is no room for true causality in classical mechanics. Chance is essential to causality, because only by admitting chance is it possible to give any meaning to alternative situations yielding different results. We conclude, therefore, that the uncertainty principle has strengthened the causal nature of physics rather than the reverse.

In conclusion, it seems fitting to point out that the philosophical implications of modern physics arise chiefly for the reason that the precision of thought required for the understanding of modern physics makes us dissatisfied with the hazy concepts that have, traditionally, been accepted as the objects of philosophical speculation.

Exercise VI.61

61.1. Write up an argument for a conclusion contrary to that drawn in the text concerning the reality of quantum mechanical matter waves. Is there a distinction between matter-waves and the wave function?

61.2. Discuss whether a complete picture of the universe can be constructed if the picture itself is part of the universe? Can you see any relation between this question and the uncertainty principle?

61.3. In what sense do you consider the question of the "reality of free will" a meaningful question? Does the uncertainty principle necessarily have any connection with this question?

References

Quantum Mechanics, L. I. Schiff, McGraw-Hill Book Co., Inc., New York, 1955.

Quantum Theory, D. Bohm, Prentice-Hill, Inc., Englewood Cliffs, N.J., 1951.

The Fundamental Principles of Quantum Mechanics, E. C. Kemble, McGraw-Hill Book Co., Inc., New York, 1937.

Philosophy of Science, P. Frank, Prentice-Hall, Inc., Englewood Cliffs, N.J., 1957.

Concepts and Methods of Theoretical Physics, R. B. Lindsay, D. Van Nostrand Co., Inc., Princeton, N.J., 1951.

Nature of Physical Reality, H. Margenau, McGraw-Hill Book Co., Inc., New York, 1950.

Logic of Modern Physics, P. W. Bridgman, The Macmillan Co., New York, 1927.

Readings in the Philosophy of Science, H. Feigl and H. Brodbeck, Appleton-Century-Crofts, Inc., New York, 1953.

Causality and Chance in Modern Physics, D. Bohm, D. Van Nostrand Co., Inc., Princeton, N. J., 1957.

INDEX